Design, Analysis and Control of Cable-Suspended Parallel Robots and Its Applications

Bin Zi · Sen Qian

Design, Analysis and Control of Cable-Suspended Parallel Robots and Its Applications

Springer

Bin Zi
Hefei University of Technology
Hefei
China

Sen Qian
Hefei University of Technology
Hefei
China

ISBN 978-981-10-1752-0 ISBN 978-981-10-1753-7 (eBook)
DOI 10.1007/978-981-10-1753-7

Library of Congress Control Number: 2016963640

Printed on acid-free paper

This Springer imprint is published by Springer Nature
The registered company is Springer Nature Singapore Pte Ltd.
The registered company address is: 152 Beach Road, #21-01/04 Gateway East, Singapore 189721, Singapore

Foreword

Robotics is an application technology integrated with multiple disciplines, including mechanics, electronics, information, control, etc. It is the intercrossing and expansion of mechanism, artificial intelligence and computer technology that have nurtured the development of robotics. Cable-suspended parallel robots have been developed as a novel type of parallel manipulator in the last decades, which are driven by flexible cables instead of rigid links. Due to the high stiffness, precision and load capacity inherited from parallel robots and the low inertia and large workspace inherited from cable-driven mechanisms, cable-suspended parallel robots have become more and more widely utilized in many engineering fields, such as the feed cable-suspended structure for large telescope, cargo hoisting, rehabilitation, bionics, wind tunnel facilities, etc.

This book is a summary and extension of the work accomplished by the authors in the field of cable-suspended parallel robots. The latest representative studies are referred. The main characteristic of the book is its comprehensive contents. As the organization principles of this book are from easy to hard, audiences can easily access the keys to understanding the theories of this book. Several important aspects on cable-suspended parallel robots are involved, in terms of mechanism innovation, dynamic modelling, integrated design, key performance indicator and error analysis, obstacle avoidance path planning, coordinated control, etc. These researching activities are of extensive applicability for analysis on cable-suspended parallel robots with various configurations. The other characteristic of the book is combining theory and application. Based on the theories and methods on multi-body dynamics, mechanism and control engineering, cable-suspended parallel robots are analyzed systematically on kinematics, dynamics, prototyping and control, combined with the typical engineering cases, including the feed cable-suspended structure for large telescope, the hybrid-driven-based cable-suspended parallel robot and the cooperative cable parallel manipulators for multiple mobile cranes.

The contents of this book can be served as a valuable tool and guide to researchers and engineers with backgrounds of design and manufacturing of mechanical and electronic equipments, especially cable-suspended parallel devices. However, there are still a lot of problems that are urgently need to be solved in the development and application of cable-suspended parallel robots. I expect the promoting effect of this book on the research and application of cable-suspended parallel robots and more achievements from the authors.

Xi'an, China
November 2016

Baoyan Duan

Preface

This book focuses on the innovative design, mechanics, control, development and applications of cable-suspended parallel robots, which is a summary and an extension of the work accomplished by the authors in the field of cable-suspended parallel robots. Some typical mechanical architectures of cable-suspended parallel robots in practical applications are provided and analyzed in this book, including the feed cable-suspended structure for super antenna, the hybrid-driven based cable-suspended parallel robot and the cooperative cable parallel manipulators for multiple mobile cranes. On the basis of the typical applications, the fundamental mechanics of cable-suspended parallel robots is addressed, including the kinematics, dynamics and trajectory tracking control of the feed cable-suspended structure for super antenna. In order to improve the performance of the traditional cable-suspended parallel robots, a novel hybrid-driven based cable-suspended parallel robot is proposed with integrated mechanism design method. Comparative study on error and performance indices of hybrid-driven based and traditional cable-suspended parallel robots is conducted. The book not only includes the basic issues but also references novel conceptions and approaches, i.e. obstacle avoidance planning, coordinate control and force distribution of the cooperative cable parallel manipulators for multiple mobile cranes. Numerical analyses and experiments of related case studies are conducted to help the audience master the main ideas of the book in both theory and practice. The aim of the book is to serve as a textbook in graduate-level courses and a reference to the practicing engineer in the field of cable-suspended parallel robots.

The main features of the book are as follows:

- *Introduces the development, classification, operating principle and design method of cable-suspended parallel robots;*
- *Focuses on the mechanics, control and applications of cable-suspended parallel robots, and provides a large number of mechanical architectures of cable-suspended parallel robots available to be used in practical applications.*
- *Kinematics, dynamics and trajectory tracking control of the feed cable-suspended structure for super antenna.*

- *Integrated mechanism design and control, error and performance indices analysis of the hybrid-driven based cable-suspended parallel robot.*
- *Development, dynamics, obstacle avoidance planning and coordinated control of the cooperative cable parallel manipulators for multiple mobile cranes.*

This book would not have been possible without the help and involvement of many people. In particular, the authors would like to thank Prof. Baoyan Duan, Academician of Chinese Academy of Engineering, who is the supervisor of Prof. Bin Zi in Xidian University. The authors also would like to thank Prof. Zhencai Zhu and Prof. Jingli Du for the comments and valuable suggestions for improving the quality of the book, postgraduate students Jianbin Cao, Bin Zhou, Xia Wu, Jun Lin, Huihui Sun, Xue Han, Bingyao Wang, Qiao Chen and Yuan Li for their research on cable-suspended parallel robots. The authors also gratefully acknowledge the continuous financial support of the National Natural Science Foundation of China (NSFC), especially, under the grant 50905179, 51275515, and 51575150.

Hefei, China
November 2016

Bin Zi
Sen Qian

Contents

Chapter 1
Introduction

1.1 The History and Development of Cable-Suspended Parallel Robots

Cable-suspended parallel robots (CPR) are parallel manipulators in which the end-effectors are driven by cables instead of rigid links. The challenges faced in the modern engineering application area of the CPRs are similar to those encountered in Stewart parallel manipulators. Indeed, the CPRs can be required not only for operations with greater flexibility, but also for large reachable workspace and high payload [1–5]. An important characteristic of cable-driven manipulators is that cables can only actuate unilaterally through tension and not compression (positive cable force). An n degrees-of-freedom (DOF) system that is actuated by m cables can be classified as being incompletely restrained if $m < n + 1$, completely restrained when $m = n + 1$, or redundantly restrained for $m > n + 1$ [6].

For fully constrained CPRs, the pose of the end-effector only depends on kinematic equations. For instance, Zi et al. [7] presented and analyzed a CPR with hybrid-driven planar five-bar mechanism. In [8], a sensorless position control method was proposed for a fully constrained CPR. For underconstrained CPRs, the robot tends to approach where the gravitational potential energy is minimal. In [9], the stiffness of a three cable-suspended parallel robot was evaluated throughout the workspace considering the cable mass. In [10] and [11], the kinematics and stability of cooperative CPRs for multiple quadrotors were derived. Zi et al. [12] addressed the cooperative localization and automatic leveling of multiple mobile cranes, which can be treated as an underconstrained four cable parallel robot. Tang and Yao [13] addressed dimensional design on a five-hundred meter aperture spherical radio telescope driven by six cables.

Research on cable-suspended parallel manipulators originates from America in 1984. Landsberger in MIT designed a cable-controlled parallel manipulator for underwater operation. The control linkages are cables, spooled by hydraulic motors,

B. Zi and S. Qian, *Design, Analysis and Control of Cable-Suspended Parallel Robots and Its Applications*, DOI 10.1007/978-981-10-1753-7_1

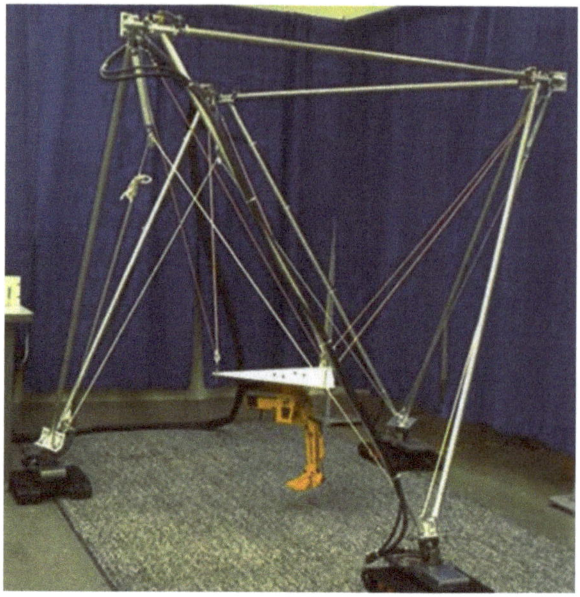

Fig. 1.1 NIST RoboCrane model

which operate in a parallel manner to determine the position and attitude of the platform. The kinematics, stiffness and workspace are analyzed preliminarily.

In 1989, National Institute of Standards and Technology in America started the NIST RoboCrane project [14–16]. The RoboCrane design utilizes the basic idea of the Stewart Platform parallel link manipulator. The unique feature of the NIST approach is the use of cables as the parallel links and the use of winches as the actuators [17–19]. The prototype is developed based on a variety of wire driven parallel robot system, which is successfully used in processing machinery, port cargo handling, bridge construction, welding and other areas, as shown in Fig. 1.1.

In the late 1980s, the SkyCam videotape system is invented by August Design company, as shown in Fig. 1.2, which is widely used for live broadcast [20]. The camera in SkyCam is suspended by 4 cables, the speed of which can be up to 44.8 km/h. It is possible to realize high-speed photography with SkyCam in large scale.

In 2001, cable array robots is proposed by Gorman from Pennsylvania State University [21], which are a class of robotic mechanisms which utilize multiple actuated cables to manipulate objects. The defining characteristics and the important issues relating to the class of robots is discussed. Experimental results are provided for the system, showing that the cable array robot is capable of tracking a trajectory within several centimeters, which is reasonable performance for many industrial applications.

In 2002, Abdullah B from University of Delaware proposed a cable suspended robot [22]. The kinematic and dynamic models, workspace, trajectory planning and feedback controllers for this robot is described. The results are demonstrated

Fig. 1.2 Cable parallel camera robot Skycam

through simulation and experiments on a six degree-of-freedom cable suspended robot. Oh and et al. designed the prototype and the feedback linearization controller [23].

Since the late 1990s, Germany, Japan, China and other countries also begin to pay close attention to the research on cable-suspended robots. In 1998, Maier from University of Rostock proposed a parallel robot driven by 3 cables [24]. A generalization of the inverse kinematic problem is introduced for a kinematically underconstrained cable suspension manipulator that supports a payload platform by three cables with computer-controlled winches. A mathematical model for tendon-driven Stewart platform manipulators is developed by Verhoeven from Duisburg [25]. Conditions for the workspace are given, including tension limits, stiffness and singularities. It is proved that there are six classes of manipulators that exert the same wrenches all over the workspace.

FALCON project is implemented in Japan with cable-driven parallel robots as a super-velocity robot, as depicted in Fig. 1.3, which is driven by 7 cables and 1 rigid link [26]. It achieves peak accelerations of up to 43G and maximum velocities of

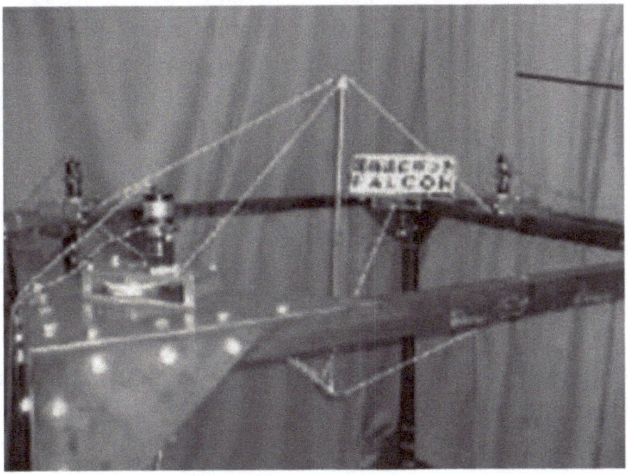

Fig. 1.3 FALCON prototype

Fig. 1.4 Large spherical radio telescope

13 m/s, even if considerably small DC motors (60 W) are used. Internal force control is employed among wires to reduce the vibration by utilizing nonlinear elasticity from the wire mechanism. Experimental results on point to point control using the prototype system are presented under linear PD feedback, to illustrate the effectiveness of the system [27].

In 1999, Duan et al. from Xidian University presented a new design of the line feed structure for a large spherical radio telescope [28] (see Fig. 1.4). Integrated mechanical, electronic, optic and automatic control technologies are employed to make considerable improvement upon the Arecibo aspherical radio telescope in Puerto Rico. Nonlinear dynamic analysis of the suspended cable system is carried out with some sensible results that could be useful to the real engineering of the large spherical radio telescope [2, 28].

Cable-suspended parallel manipulators are categorized as a type of parallel manipulator systems. Research in the topic of the CPRs is highly motivated and has a very strong interest as the modern engineering demand for such manipulators is growing rapidly. Cables are flexible members that can support very large tensile loads per unit weight, and cables can also be wound on spools. Therefore, the CPRs have potential advantages in terms of light weight, large reachable workspace, fast motion and heavy payload capability, and high reconfigurability [29–33]. Given their attractive properties, the CPRs have widely used in many engineering fields, such as large spherical radio telescopes [34], high speed robot [26], camera system [35], robot rehabilitation [36], large-scale manipulation [37], haptic device [38], and material handling [39] (see Fig. 1.5). In the last decades, research on cable-suspended parallel robots focus on the configuration innovation and performance improvement.

In recent years, many novel cable-driven parallel robots are proposed on the basis of systematic approach of type synthesis and modularization. Structure synthesis can generate different types of mechanisms with specified number of links and degrees of freedom, and it is a systematic and efficient method for the design of mechanisms. Reconfigurable robots have received much attention in the past decades in response to the fast increasing manufacturing demands. The topology of

Fig. 1.5 Material handling with cooperative multiple cranes

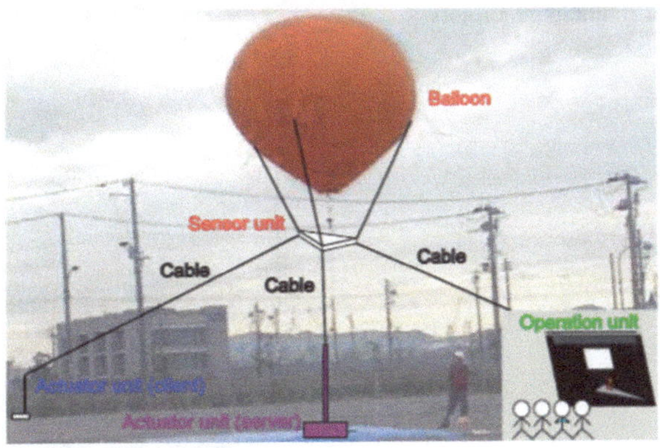

Fig. 1.6 The balloon-cable type robot

reconfigurable robots can be changed to improve productivity and save costs for flexibility, easy maintenance and rapid deployment.

Takemura designed the search and rescue work and information collection and other single and three cable-balloons parallel mechanism for earthquake disaster (see Fig. 1.6). The experiments are conducted to assess the attitude stability of the sensor unit and the movable scope computation is effective to determine the installation location for the actuator units [40, 41].

Heyden and Woernle [42] proposed a kinematically undetermined cable suspension manipulator named CableV (see Fig. 1.7), moving a payload platform in space by several cables with computer-controlled winches, The nonlinear quasi-static feedback of actual state variables is obtained based on the differential-algebraic equations. A control method combines the flatness-based feedforward controller is designed.

Fig. 1.7 Cable parallel robot CableV

Fig. 1.8 Towing equipment of multiple quadrotors

Kumar et al. address the kinematics of cooperative transport of payloads suspended by multiple aerial robots with cables, as shown in Fig. 1.8 [43–46]. In order to obtain all possible solutions, an efficient analytic algorithm based on dialytic elimination is presented. In addition, a numerical procedure is developed to determine the set of allowable tensions. An approach for stability analysis is developed, and the stability of all equilibrium configurations is analyzed.

Castelli et al. propose cable system for improving mobility (CaSIMo), which has been developed with the aim of improving the mobility of end-users in an urban environment as regards traversing architectural barriers like canals or roads, when classical solutions such as bridges cannot be adopted, as shown in Fig. 1.9. In particular, modeling for kinetostatic and dynamic analyses are proposed and discussed. The proposed system does not require structures of large dimensions, and its environmental impact can thus be greatly reduced with regard to other analogous devices [47].

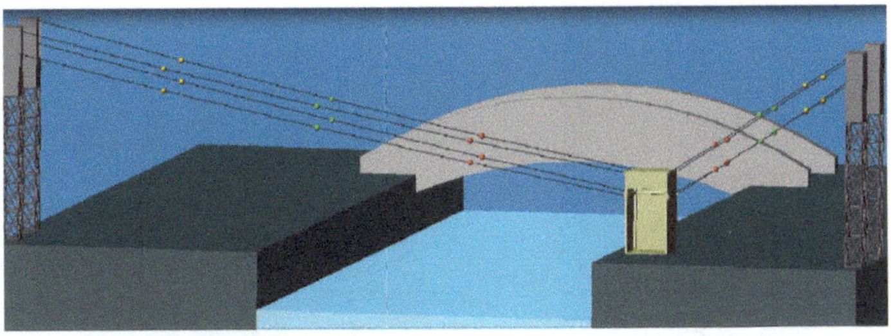

Fig. 1.9 A novel cable parallel bridge crane

1.2 The Application of Cable-Suspended Parallel Robots

1.2.1 Engineering

Merlet et al. presents the preliminary development of a full scale, portable, modular, fully autonomous parallel wire crane that is intended to be used for rescue operations [48]. This victim localization device can lift at least 100kg load through 2 to 6 cables even in a reduced configuration and may play an important role in rescue operations, as shown in Fig. 1.10.

The forward kinematics problem of a four-cable-driven under-constrained parallel mechanism are studied. the sinking winch mechanism is taken as prototype [49]. A new traversal algorithm is proposed to solve the tensions in cables and the pose of platform, when only the lengths of cables are obtained.

Fig. 1.10 Modular parallel wire robot for rescue operation

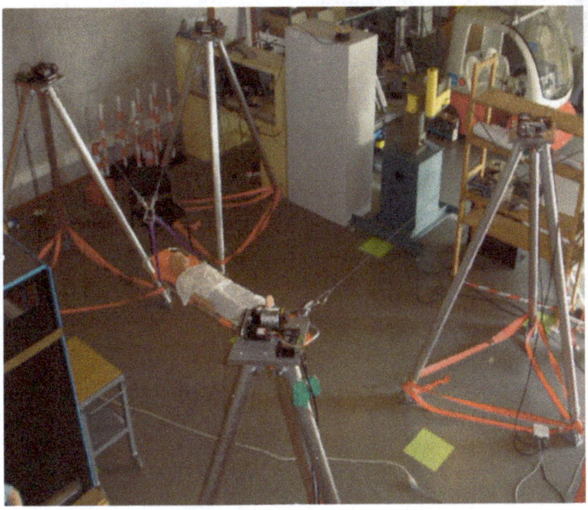

Fig. 1.11 Hybrid-driven-based cable parallel mechanism

The CPRs are required not only for operations with lower inertia and high payload, but also for output with greater flexibility. Zi et al. present and analyze a cable parallel manipulator with and without hybrid-driven planar five-bar mechanism (HDPM). Comparative study of kinematics and dynamics of the CPRs with and without HDPM are studied. Drive torques and drive powers are given to compare the load carrying capacity of the two manipulators, and workspace, stiffness performance, singularity analysis are also carried out [7, 50].

Barnett and Gosselin [51] introduce a large-scale 3D printer (see Fig. 1.11), which uses a six-degree-of-freedom cable-suspended robot for positioning, with polyurethane foam as the object material and shaving foam as the scaffolding. The cable-driven 3D foam printer are easily reconfigurable, lighter and transportable, compared to the traditionally used positioning systems. The accuracy, robustness and control are also investigated.

In recent years, many researchers paid great attention to the use of cable-driven mechanism in wind tunnel test due to its fewer interference on the streamline flow, as shown in Fig. 1.12. Some successful achievements have been made in the suspension active pour soufflerie (SACSO) project about the cable-driven parallel suspension system in low-speed wind tunnels with 8 cables [52]. Another application is WDPSS-8 developed by Zheng in wind tunnel test [53].

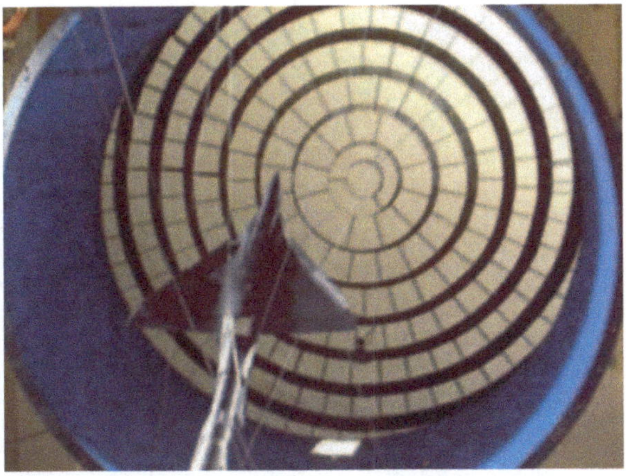

Fig. 1.12 SACSO prototype

1.2.2 Bionics

Mustafa and et al. from Singapore technology research institute intelligent presents the self-calibration of a novel biologically-inspired cable-driven robotic arm [54–56] (see Fig. 1.13). A self-calibration model is formulated based on the differential change in the cable end-point distances. A computationally efficient algorithm using iterative least-squares is employed to identify the errors in the geometric model parameters.

Capua et al. presents a novel concept of a mobile cable-suspended robot called SpiderBot, consisting of four cable-mechanisms and a central platform [57]. Each cable-mechanism includes a cable, a dispensing and rolling mechanism, and a gripper at the end of the cable. SpiderBot's motion is based on adjusting the cable lengths with cable-mechanism, which includes a cable, a dispensing and rolling mechanism, and a gripper. The inverse kinematics and motion planning are conducted, as shown in Fig. 1.14.

A generalized model for multilink cable-driven serial manipulators with an arbitrary number of links that allow for arbitrary cable routing is presented by

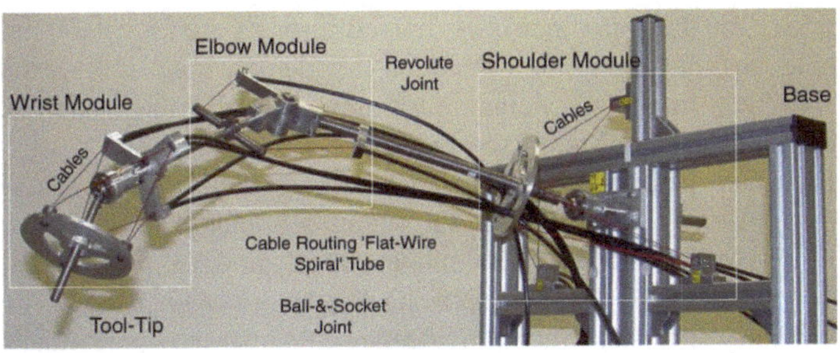

Fig. 1.13 Biologically-inspired cable-driven robotic arm

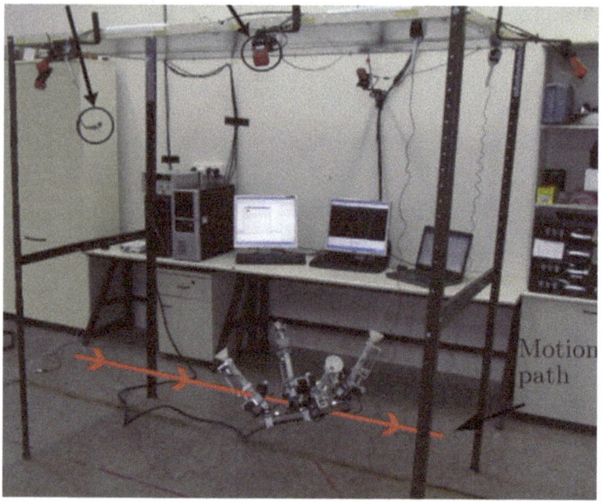

Fig. 1.14 SpiderBot prototype

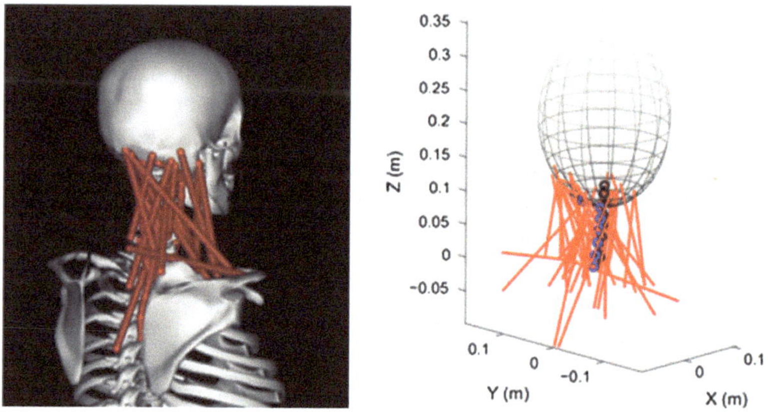

Fig. 1.15 8-link 8-spherical 76-cable neck model

Lau et al. [58]. The Jacobian matrix is derived through the kinematic analysis and two example manipulators are simulated to demonstrate the versatility and advantages of the formulation, including a simple 2-link 6-cable system and a more complex 8-link 76-cable system. (see Fig. 1.15).

A cable-driven parallel platform with a pneumatic muscle active support (CPPPMS) is presented by Zhao et al. for imitating human necks, where cables and pneumatic muscle are adopted to imitates neck muscles and spinal muscles, respectively [59]. A sliding mode controller has been designed based on the stiffness and dynamics analysis, which provides feasible operation performances for the CPPPMS.

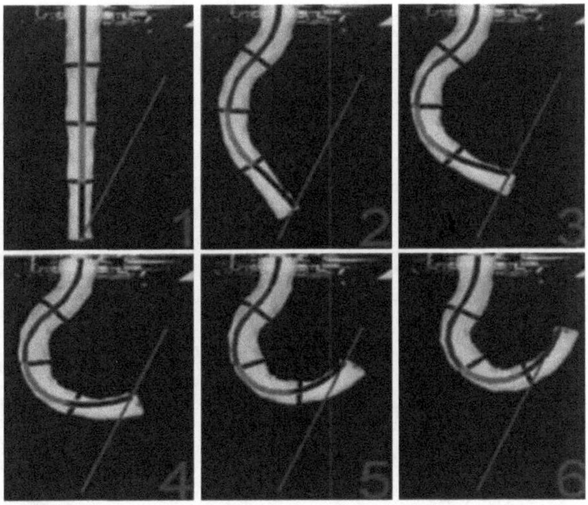

Fig. 1.16 Hyper redundant elephant trunk robot

Continuum robot is a new class of biologically inspired robot with good compliance and high dexterity, that can exhibit a large number of kinematic degree of freedoms, but not all of these degree of freedoms are directly actuated. It can not only grasp object with different end effectors, but also grasp object with the robot body itself. Different from discrete robot such as industrial robot, continuum robot is inspired from flexible animal organs like elephant trunk and octopus arms and characterized by flexible structures instead of any joints and rigid links. Due to this advantage, continuum robot can change the shape according the environment, which has excellent bending ability, good adaptability, especially in unstructured and highly congested environments.

Walker et al. present a novel hyper redundant elephant trunk robot (see Fig. 1.16), which features a segmented 'backbone' with a total of 32 degrees of freedom [60, 61]. Actuation is provided via a series of tendons routed through the structure. The previously designed kinematic model is applied to describe the kinematics of the robot and some examples of intelligent manipulation using the robot is provided.

Simaan et al. report the analysis and modeling results obtained while developing a tele-robotic system for minimally invasive surgery of the throat [62]. The snake-like unit is one of the main components, which uses a novel multi-flexible backbone design, as shown in Fig. 1.17.

Choi et al. [63] introduces a modified endoscope structure that employs a spring as its backbone. The compression and steering motions can be obtained by pulling the three cables around the spring backbone. Modeling and control of the spring backbone micro endoscope are carried out.

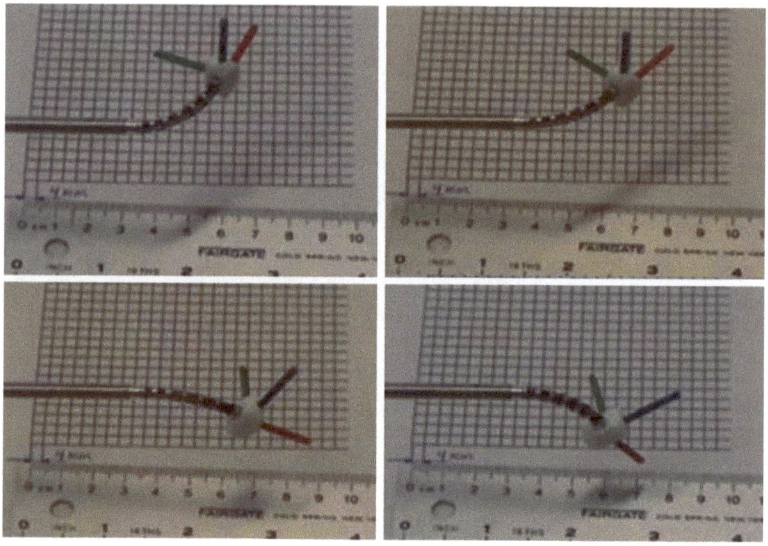

Fig. 1.17 Tele-robotic system for minimally invasive surgery

1.2.3 Rehabilitation

Due to the flexibility of cable-suspended parallel robot and the elasticity of cable itself, cable-suspended parallel robots are more and more widely used in rehabilitation. However, because of the complexity of human body structure and motion, the control and arrangement of the cables have become key problems during the application of cable-suspended parallel robots for rehabilitation of different parts of human body.

Mao et al. designed cable-suspended arm exoskeleton (CAREX) to achieve desired forces on the hand, i.e., both pull and push, in any direction as required in neural training, which is attached to the limb segments of a five degree-of-freedom anthropomorphic arm instrumented with joint sensors [36, 64–66]. Kinematic model considering the glenohumeral joint center rotation is established and a new continuous cable tension planner is designed for smooth tension trajectories during natural training of human subjects (see Fig. 1.18).

Zanotto et al. [67] present the application of the string-operated planar haptic interface for arm rehabilitation (Sophia-3), which is a planar cable-driven device with a tilting working plane (see Fig. 1.19). The kinematics and control algorithms and design of the new device are presented. The experiment conducted on healthy subjects confirms the feasibility of the semi-adaptive design.

Dynamic modeling and mechanics analysis of a novel pneumatic muscle driven parallel mechanism for imitating human pelvis (PMPMHP) is investigated by Zhao and Zi [68], which has some advantages for both parallel structure and pneumatic muscle actuator. Genetic algorithm is adopted for optimizing the parameters of the

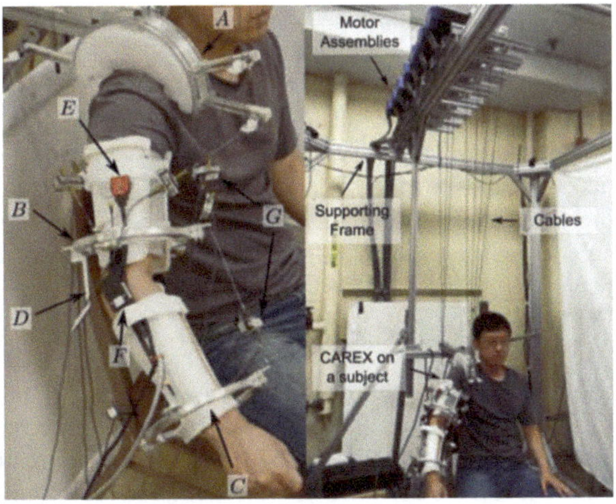

Fig. 1.18 Overview of the CAREX on a healthy subject

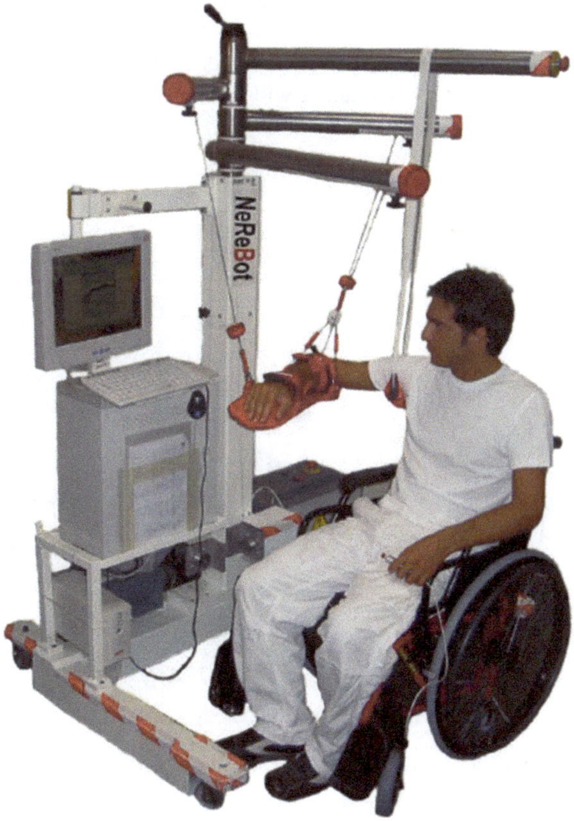

Fig. 1.19 Sophia-3

mechanism. The kinematic performance is investigated including workspace, singularity and stiffness analysis.

Chen et al. [69] develop a cable-driven wrist robotic rehabilitator (CDWRR) for motion training or assistance to subjects with motor disabilities. The CDWRR utilizes the wrist skeletal joints and arm segments as the supporting structure and takes advantage of cable-driven parallel design to build the system, which brings the properties of flexibility, low-cost, and low-weight. Experiments on a healthy subject are carried out to demonstrate the performance of the controller and the feasibility of the CDWRR on wrist motion training or assistance, as shown in Fig. 1.20.

Surdilovic and Bernhardt present a novel robotic prototype for advanced gait rehabilitation, integrating sophisticated robotic technology with control algorithms [70, 71]. The cable driven structure provides the requirements of controlling the posture in 6-DOFs. During the rehabilitation, the patient can be loosely harnessed with the minimum wire tension required to monitor the motion, as shown in Fig. 1.21.

Yamaura et al. [72] developed a flexible wire driven robot for finger rehabilitation for patients suffering from paralysis or contracture. The combination of the cable-driven mechanism and closed four-link mechanism overcomes the limitations of the arm structure and a wider range of joint motion can be obtained, as shown in Fig. 1.22.

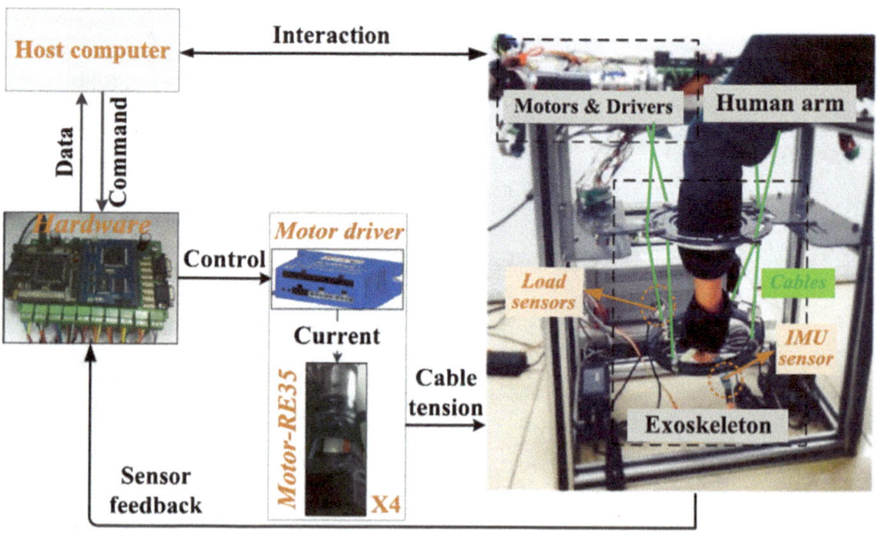

Fig. 1.20 Cable-driven wrist robotic rehabilitator

Fig. 1.21 STRING-Man system

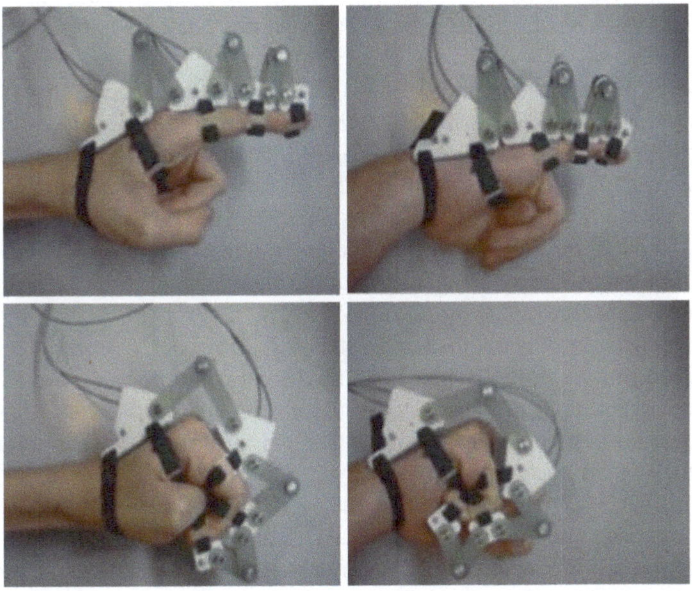

Fig. 1.22 Operation of little/index finger

1.3 Scope and Organization of This Book

This book focuses on the innovative design, mechanics, control, development and applications of cable-suspended parallel robots, which is a summary and an extension of the work accomplished by the authors in the field of cable-suspended parallel robots. Some typical mechanical architectures of cable-suspended parallel robots in practical applications are provided and analyzed in this book, including the feed cable-suspended structure for super antenna, the hybrid-driven based cable-suspended parallel robot and the cooperative cable parallel manipulators for multiple mobile cranes. On the basis of the typical applications, the fundamental mechanics of cable-suspended parallel robots are addressed, including the kinematics, dynamics and trajectory tracking control of the feed cable-suspended structure for super antenna. In order to improve the performance of the traditional cable-suspended parallel robots, a novel hybrid-driven based cable-suspended parallel robot is proposed with integrated mechanism design method. Comparative study on error and performance indices of hybrid-driven based and traditional cable-suspended parallel robots is conducted. The book not only includes the basic issues, but also references novel conceptions and approaches, i.e. obstacle avoidance planning, coordinate control, and force distribution of the cooperative cable parallel manipulators for multiple mobile cranes. Numerical analyses and experiments of related case studies are conducted to help the audience master the main ideas of the book in both theory and practice. The aim of the book is to serve as a textbook in graduate-level courses and a reference to the practicing engineer in the field of cable-suspended parallel robots. To reach this goal, the book is organized as follows:

In Chap. 2, design and experiment of the feed cable-suspended structure for the next generation large spherical radio telescope are presented. The optimization design of the telescope structure is performed and the connection between stable cables and twisting stiffness is verified. The inverse kinematic and dynamic model of CPR with non-negligible cable mass is carry out. Furthermore, the inverse dynamic formulation of the mechanical system including servomechanism dynamics is established by using Lagrange's Equations.

In Chap. 3, the fuzzy plus proportional-integral control (FPPIC) method, which can enhance the control performance for steady state errors, is utilized to control the wind-induced vibration of the trajectory tracking of the feed. Moreover, the simulation and experiment are respectively deployed on the 5-m and 50-m scaled model to test the different dynamic modeling present in former chapter and justify the efficiency of purposed control strategy.

In Chap. 4, the dynamic model of the HDCPR is developed on the basis of Lagrange method. Then, an adaptive iterative learning control strategy is designed for the high-precision trajectory tracking. Furthermore, the stability of the controller is proved by means of Lyapunov function. In order to improve the dynamic performance of the HDCPR system, a methodology of simultaneous optimal design of mechanism and control for the HDCPR is presented. The dynamic modeling of the HDCPR is performed based on Newton-Euler method, and the workspace of the manipulator is also analyzed.

In Chap. 5, the design model of the HDCPR is described. Then, mechanics analysis and error modelling of the HDCPR is described based on the differentiation of kinematic equations. Sensitivity analysis of the HDCPR is provided. Illustrative simulation studies highlight its performances. Finally, conclusions and future studies are summarized.

In Chap. 6, to illustrate the advantage of the performance indices, firstly, hybrid-driven based cable-suspended parallel manipulators (HDCPR) is presented, then Kinematics and dynamics of the cable parallel manipulators with and without

HDPM are studied based on Lagrange method separately. Finally, some performance indices, such as conditioning index, stiffness index, and payload index are investigated and compared based on above analysis.

In Chap. 7, the design, dynamic modelling, and workspace of the CPRMC are presented. The design model of the CPRMC is presented. Then, kinematics and dynamics of the CPRMC are performed. Workspace analysis of the CPRMC is provided and illustrative simulation studies highlight its performances. Finally, some concluding remarks are summarized.

In Chap. 8, design model and cooperative localization scheme of the CPRMC are presented, and the improved localization algorithm based on multilateration method is designed. The global path planning of the CPRMC is performed based on the artificial potential field method with the grid method, and the sensor technology is applied to the cooperative obstacle avoidance, and combined with Matlab and LabVIEW, a co-simulation platform is built for cooperative obstacle avoidance analysis. A four-point collaborative leveling method is adopted for automatic leveling control of the platform for the CPRMC.

In Chap. 9, the complete kinematics and dynamics of the cooperative cable parallel manipulators for multiple mobile cranes is studied, which is the foundation for the further research of this new promising and significant engineering equipment. The kinematics of the cooperative three mobile cranes with point-mass is established with elimination method, including both forward and inverse kinematics. Based on Lagrange equation and the kinematics, the complete dynamic model of the CPRMC is derived. A robust iterative learning controller (ILC) of the CPRMC is designed on the basis of the linearization of the dynamics. Numerical simulation is conducted in order to investigate the dynamics and the designed robust ILC of the CPRMC.

References

1. Carricato M, Merlet JP (2013) Stability analysis of underconstrained cable-driven parallel robots. IEEE Trans Rob 29(1):288–296
2. Zi B, Duan BY, Du JL et al (2008) Dynamic modeling and active control of a cable-suspended parallel robot. Mechatronics 18(1):1–12
3. Castelli G, Ottaviano E, Rea P (2014) A cartesian cable-suspended robot for improving end-users' mobility in an urban environment. Robot Comput Integr Manuf 30(3):335–343
4. Lahouar S, Ottaviano E, Zeghoul S et al (2009) Collision free path-planning for cable-driven parallel robots. Robot Auton Syst 57(11):1083–1093
5. Korayem MH, Tourajizadeh H, Zehfroosh A et al (2014) Optimal path planning of a cable-suspended robot with moving boundary using optimal feedback linearization approach. Nonlinear Dyn 78(2):1515–1543
6. Ming A, Higuchi T (1994) Study on multiple degree-of-freedom positioning mechanism using wires. I: Concept, design and control. Int J Jpn Soc Precis Eng 28(2):131–138
7. Zi B, Cao J, Zhu H et al (2014) Comparative study of cable parallel manipulators with and without hybrid-driven planar five-bar mechanism. Appl Math Model 38(24):5994–6017

8. Kino H, Yahiro T, Taniguchi S et al (2009) Sensorless position control using feedforward internal force for completely restrained parallel-wire-driven systems. IEEE Trans Rob 25 (2):467–474
9. Arsenault M (2013) Workspace and stiffness analysis of a three-degree-of-freedom spatial cable-suspended parallel mechanism while considering cable mass. Mech Mach Theory 66:1–13
10. Jiang Q, Kumar V (2013) The inverse kinematics of cooperative transport with multiple aerial robots. IEEE Trans Rob 29(1):136–145
11. Michael N, Kim S, Fink J, et al (2009) Kinematics and statics of cooperative multi-robot aerial manipulation with cables. ASME 2009 international design engineering technical conferences and computers and information in engineering conference. American Society of Mechanical Engineers, pp 83–91
12. Zi B, Lin J, Qian S (2015) Localization, obstacle avoidance planning and control of a cooperative cable parallel robot for multiple mobile cranes. Robot Comput Integr Manuf 34:105–123
13. Tang X, Yao R (2011) Dimensional design on the six-cable driven parallel manipulator of FAST. J Mech Des 133(11):111012
14. Albus J, Bostelman R, Dagalakis N (1992) The NIST robocrane. J Robot Syst 10(5)
15. Bostelman R, Albus J, Dagalakis N, et al (1994) Applications of the NIST RoboCrane. Proceedings of the 5th international symposium on robotics and manufacturing, pp 14–18
16. Lytle AM, Saidi KS, Bostelman RV et al (2004) Adapting a teleoperated device for autonomous control using three-dimensional positioning sensors: experiences with the NIST RoboCrane. Autom Constr 13(1):101–118
17. Rippey WG, Falco J A (1997) The NIST automated arc welding testbed. NIST Special Publication SP, pp 203–212
18. Amatucci E, Bostelman R, Dagalakis N, et al (1997) Summary of modeling and simulation for NIST RoboCrane applications
19. Williams RL (2005) Novel cable-suspended RoboCrane support. Ind Robot Int J 32(4):326–333
20. Tanaka M, Seguchi Y, Shimada S (1988) Kineto-statics of skycam-type wire transport system. Proceedings of USA-Japan symposium on flexible automation, Crossing bridges: advances in flexible automation and robotics
21. German JJ, Jablokow KW, Cannon DJ (2001) The cable array robot: theory and experiment. IEEE international conference on robotics and automation, 2001. Proceedings 2001 ICRA. IEEE, vol 3. pp 2804–2810
22. Alp AB, Agrawal SK (2002) Cable suspended robots: design, planning and control. IEEE international conference on robotics and automation, 2002. Proceedings ICRA'02. IEEE, vol 4. pp 4275–4280
23. Oh S, Agrawal SK (2005) A reference governor-based controller for a cable robot under input constraints. IEEE Trans Control Syst Technol 13(4):639–645
24. Maier T, Woernle C (1998) Inverse kinematics for an underconstrained cable suspension manipulator. Advances in robot kinematics: analysis and control. Springer, Netherlands, pp 97–104
25. Verhoeven R, Hiller M, Tadokoro S (1998) Workspace, stiffness, singularities and classification of tendon-driven stewart platforms. Advances in robot kinematics: analysis and control. Springer, Netherlands, pp 105–114
26. Kawamura S, Choe W, Tanaka S, et al (1995) Development of an ultrahigh speed robot FALCON using wire drive system. 1995 IEEE international conference on robotics and automation, 1995. Proceedings. IEEE, vol 1. pp 215–220
27. Kawamura S, Kino H, Won C (2000) High-speed manipulation by using parallel wire-driven robots. Robotica 18(1):13–21
28. Duan BY (1999) A new design project of the line feed structure for large spherical radio telescope and its nonlinear dynamic analysis. Mechatronics 9(1):53–64

29. Gosselin C (2014) Cable-driven parallel mechanisms: state of the art and perspectives. Mech Eng Rev 1(1):DSM0004–DSM0004
30. Zi B, Zhu Z, Du JL (2011) Analysis and control of the cable-supporting system including actuator dynamics. Control Eng Pract 19(5):491–501
31. Otis MJD, Perreault S, Nguyen-Dang TL et al (2009) Determination and management of cable interferences between two 6-DOF foot platforms in a cable-driven locomotion interface. IEEE Trans Syst Man Cybern Part A Syst Hum 39(3):528–544
32. Lim WB, Yang G, Yeo SH et al (2011) A generic force-closure analysis algorithm for cable-driven parallel manipulators. Mech Mach Theory 46(9):1265–1275
33. Hassan M, Khajepour A (2011) Analysis of bounded cable tensions in cable-actuated parallel manipulators. IEEE Trans Rob 27(5):891–900
34. Duan BY, Qiu YY, Zhang FS et al (2009) On design and experiment of the feed cable-suspended structure for super antenna. Mechatronics 19(4):503–509
35. Cone LL (1985) Skycam-an aerial robotic camera system. Byte 10(10):12
36. Mao Y, Agrawal SK (2012) Design of a cable-driven arm exoskeleton (CAREX) for neural rehabilitation. IEEE Trans Rob 28(4):922–931
37. Alikhani A, Behzadipour S, Alasty A et al (2011) Design of a large-scale cable-driven robot with translational motion. Robot Comput Integr Manuf 27(2):357–366
38. Sato M (2002) Development of string-based force display: SPIDAR. 8th international conference on virtual systems and multimedia
39. Pott A, Mütherich H, Kraus W et al (2013) IPAnema: a family of cable-driven parallel robots for industrial applications. Cable-driven parallel robots. Springer, Berlin, pp 119–134
40. Takemura F, Enomoto M, Tanaka T et al (2005) Development of the balloon-cable driven robot for information collection from sky and proposal of the search strategy at a major disaster, IEEE/ASME international conference on advanced intelligent mechatronics. IEEE, pp 658–663
41. Takemura F, Maeda K, Tadokoro S (2006) Attitude stability of a cable driven balloon robot. 2006 IEEE/RSJ international conference on intelligent robots and systems. IEEE, pp 3504–3509
42. Heyden T, Woernle C (2006) Dynamics and flatness-based control of a kinematically undetermined cable suspension manipulator. Multibody SysDyn 16(2):155–177
43. Michael N, Fink J, Kumar V (2011) Cooperative manipulation and transportation with aerial robots. Auton Robots 30(1):73–86
44. Jiang Q, Kumar V (2012) Determination and stability analysis of equilibrium configurations of objects suspended from multiple aerial robots. J Mech Robot 4(2):021005
45. Jiang Q, Kumar V (2013) The kinematics of 3-D cable-towing systems. 21st century kinematics. Springer, London, pp 161–188
46. Lindsey Q, Mellinger D, Kumar V (2012) Construction with quadrotor teams. Auton Robots 33(3):323–336
47. Castelli G, Ottaviano E (2014) A cartesian cable-suspended robot for aiding mobility. Computational kinematics. Springer, Netherlands, pp 369–376
48. Merlet J, Daney D (2010) A portable, modular parallel wire crane for rescue operations. 2010 IEEE international conference on robotics and automation (ICRA). IEEE, pp 2834–2839
49. Zhu ZC, Chen PCY, Shao XG et al (2014) Forward kinematics analysis and experimental validation of a four-cable-driven under-constrained parallel mechanism. Proc Inst Mech Eng Part C J Mech Eng Sci 228(7):1235–1247
50. Zi B, Ding H, Wu X et al (2014) Error modeling and sensitivity analysis of a hybrid-driven based cable parallel manipulator. Precis Eng 38(1):197–211
51. Barnett E, Gosselin C (2015) Large-scale 3D printing with a cable-suspended robot. Add Manuf 7:27–44
52. Farcy D, Llibre M, Carton P et al (2007) SACSO: wire-driven parallel set-up for dynamic tests in wind tunnel–review of principles and advantages for identification of aerodynamic models for flight mechanics. 8th ONERA-DLR Aerospace Symposium, Göttingen

53. Zheng Y (2009) Force-measuring experiment for the scale model of WDPSS in low-speed wind tunnel. J Huaqiao Univ (Natural Science) 2(2)
54. Yang G, Lin W, Kurbanhusen MS, et al (2005) Kinematic design of a 7-DOF cable-driven humanoid arm: a solution-in-nature approach. IEEE/ASME international conference on advanced intelligent mechatronics (AIM), Monterey, CA, July 2005, pp 24–28
55. Mustafa SK, Yang G, Yeo SH, et al (2007) Self-calibration of a biologically-inspired cable-driven robotic arm. 2007 IEEE/ASME international conference on advanced intelligent mechatronics. IEEE, pp 1–6
56. Mustafa SK, Yang G, Yeo SH, et al (2008) Kinematic calibration of a 7-DOF self-calibrated modular cable-driven robotic arm. IEEE international conference on robotics and automation, 2008. ICRA 2008. IEEE, pp 1288–1293
57. Capua A, Shapiro A, Shoval S (2014) SpiderBot: a cable-suspended walking robot. Mech Mach Theory 82:56–70
58. Lau D, Oetomo D, Halgamuge SK (2013) Generalized modeling of multilink cable-driven manipulators with arbitrary routing using the cable-routing matrix. IEEE Trans Rob 29 (5):1102–1113
59. Zhao X, Zi B, Qian L (2015) Design, analysis, and control of a cable-driven parallel platform with a 481 pneumatic muscle active support. Robotica 1–22
60. Hannan MW, Walker ID (2000) Analysis and initial experiments for a novel elephant's trunk robot. 2000 IEEE/RSJ international conference on intelligent robots and systems, 2000. (IROS 2000). Proceedings. IEEE, vol 1. pp 330–337
61. Hannan MW, Walker ID (2003) Kinematics and the implementation of an elephant's trunk manipulator and other continuum style robots. J Robotic Syst 20(2):45–63
62. Simaan N (2005) Snake-like units using flexible backbones and actuation redundancy for enhanced miniaturization. Proceedings of the 2005 IEEE international conference on robotics and automation. IEEE, pp 3012–3017
63. Choi DG, Yi BJ, Kim WK (2007) Design of a spring backbone micro endoscope. 2007 IEEE/RSJ international conference on intelligent robots and systems. IEEE, pp 1815–1821
64. Agrawal SK, Dubey VN, Gangloff JJ et al (2009) Design and optimization of a cable driven upper arm exoskeleton. J Med Devices 3(3):031004
65. Mao Y, Jin X, Agrawal SK (2014) Real-time estimation of glenohumeral joint rotation center with cable-driven arm exoskeleton (CAREX)—a cable-based arm exoskeleton. J Mech Robot 6(1):014502
66. Mao Y, Jin X, Dutta GG et al (2015) Human movement training with a cable driven arm exoskeleton (carex). IEEE Trans Neural Syst Rehabil Eng 23(1):84–92
67. Zanotto D, Rosati G, Minto S et al (2014) Sophia-3: A semiadaptive cable-driven rehabilitation device with a tilting working plane. IEEE Trans Rob 30(4):974–979
68. Zhao X, Zi B (2013) Design and analysis of a pneumatic muscle driven parallel mechanism for imitating human pelvis. Proc Inst Mech Eng Part C J Mech Eng Sci 0954406213489410
69. Chen W, Cui X, Zhang J et al (2015) A cable-driven wrist robotic rehabilitator using a novel torque-field controller for human motion training. Rev Sci Instrum 86(6):065109
70. Surdilovic D, Bernhardt R. STRING-MAN: a new wire robot for gait rehabilitation. 2004 IEEE international conference on robotics and automation, 2004. Proceedings. ICRA'04. IEEE, vol 2. pp 2031–2036
71. Surdilovic D, Zhang J, Bernhardt R (2007) STRING-MAN: wire-robot technology for safe, flexible and human-friendly gait rehabilitation. 2007 IEEE 10th international conference on rehabilitation robotics. IEEE, pp 446–453
72. Yamaura H, Matsushita K, Kato R, et al (2009) Development of hand rehabilitation system using wire-driven link mechanism for paralysis patients. 2009 IEEE international conference on robotics and biomimetics (ROBIO). IEEE, pp 209–214

Chapter 2
Kinematics and Dynamics of the Feed Cable-Suspended Structure for Super Antenna

2.1 Preamble

So far the largest antenna is the Arecibo spherical radio telescope with the diameter of 305 m, which was built in the 1970s and located in Puerto Rico, USA. As for Arecibo-type telescope, there exist three problems: high cost, narrow frequency band and spherical error. Due to the above problems, one plan proposed by Chinese astronomers and engineers is to build a set of large spherical reflectors in the extensively existing karst landform in southwest China. A new-generation large radio telescope (LT) nowadays is referred to as the square kilometer array (SKA). Many different technological solutions have been brought forward and studied by institutes participating in the SKA [1, 2]; and the papers [3–6] propose a multi-disciplinary design project and made nonlinear analysis about it with many reasonable results. Figure 2.1 shows a number of outstanding innovations of the design project: (1) the unique karst landform in Guizhou province, southwest China; (2) the main active reflector enabling to overcome both shortcomings of narrow bandwidth due to the line feed, and spherical error due to spherical reflector; and (3) the light-weight feed cabin driven by cables plus stable Stewart platform, by which, the tracing accuracy of feed mounted on the movable plate of Stewart platform can be up to millimeters. As a multi-science platform, LT will provide treasures to astronomers, as well as bring prosperity to other research, for instance, space weather study, deep space exploration, and national security.

The millimeter positioning precision of the feed fixed at the moving platform of Stewart platform can be realized by varying the six legs' lengths. Although this project can make it easier to implement the LT engineering, it raises a much higher requirement for the control engineering. This is due to the use of the large span cables to drive the cabin structure weighing 20–30 t. For such a large lag and heavy inertia control system, it is difficult to guarantee the requirement for the tracking accuracy (less than 4 mm).

B. Zi and S. Qian, *Design, Analysis and Control of Cable-Suspended Parallel Robots and Its Applications*, DOI 10.1007/978-981-10-1753-7_2

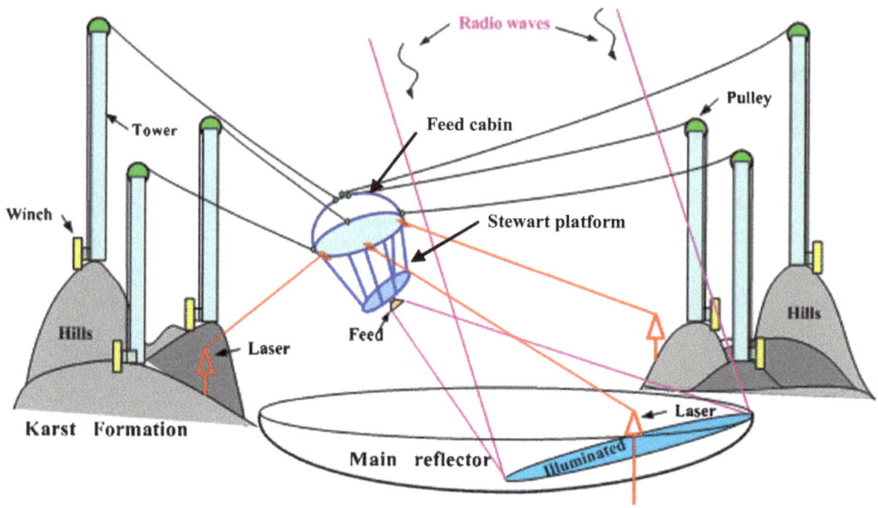

Fig. 2.1 Schematic diagram of a new design project of LT

The cooperative variation of lengths of six long cables pulls the feed cabin to track some radio source with six-DOFs. Similar to a parallel robot, the cable-cabin flexible structure of LT can be viewed as a cable-supporting parallel robot (CPR). Conventional robots with serial or parallel structures are impractical for some applications since the workspace requirements are higher than what the conventional robots can provide. CPR uses cables instead of links to manipulate objects. The cables are so light that actuators of CPR have only to drive their loads. Furthermore, CPR gives a wide range of motion, because drums of the mechanism can wind long cables. For the above reasons, cable-driven mechanisms have received attention and have been recently studied since 1980s [7–11].

As you know, dynamics is a huge field of study devoted to studying the forces required to cause motion. In order to accelerate the CPR from rest, glide at a constant end-effector velocity, and finally decelerate to a stop, a complex set of torque functions must be applied by the joint actuators [12]. One method of controlling the CPR to follow a desired path involves calculating these actuator torque functions by using the dynamic equations of motion of the CPR. In addition, as demonstrated in [13], servomechanism dynamics constitute an important component of the complete robotic dynamics. Therefore, the dynamics of the servomotors and its gears must be modeled for control design. However, the literature on the control of the CPR system including the actuator dynamics is sparse. In the following, we develop dynamic equations of the CPR system including actuator dynamics, which may be used to control and simulate the motion of the CPR.

This chapter presents design and experiment of the feed cable-suspended structure for the next generation large spherical radio telescope. To begin with, optimization design of the structure is performed to meet the requirement that the orientation of the feed cable-suspended structure need range from 40° to 60°. And aiming at the drawback of weak twisting stiffness of the feed cable-suspended system, a method to add stable cables is proposed to enhance the twisting stiffness of the structure. The experiments that developed combining with the inner FAST model of 5-m demonstrate the effectiveness and feasibility of the proposed method. And since the dynamic modeling of CPR (Cable-suspended parallel robot) is the precondition for its motion control. One objectives of this book is to find the precise kinematic and dynamic model of FAST and other similar CPRs (Cable-suspended parallel robot). Hence, according to the 5-m scaled model, the inverse dynamic formulation of CPR with non-negligible cable mass is established by using inverse kinematics analysis and Lagrangian dynamic formulation. Moreover, in this chapter, according to the 50-m scaled model of FAST, the inverse dynamic formulation of the mechanical system including servomechanism dynamics is established by using Lagrange's Equations. In Sect. 2.2, the optimization design of the telescope structure is performed and the connection between stable cables and twisting stiffness is verified. In Sect. 2.3, the inverse kinematic and dynamic model of CPR with non-negligible cable mass is carry out. Furthermore, the inverse dynamic formulation of the mechanical system including servomechanism dynamics is established in Sect. 2.4 by using Lagrange's Equations.

2.2 On Design and Experiment of the Feed Cable-Suspended

2.2.1 Mechanics Equations of the Feed Cable-Suspended Structure

Since the cables, in Fig. 2.1, are several hundred meters in length, and the movement of the cabin is pretty slow, the feed cabin can be viewed as static state on the points of trajectory. The cable-suspended system (CSS) suspends an object by six wires and restrains all motion degrees of freedom for the object using the wires and gravitational force when the cabin moves within the workspace. This type of mechanism is called "incompletely restrained wire-suspended mechanism," [14] and is able to control three-dimensional position and orientation of object by changing only length of wires. Experimental studies on the inner LT 5-m and outer LT 50-m model demonstrate the conclusion as shown in Figs. 2.2 and 2.3. For example, the results of LT 50-m model show that the angular tracking error and the position error can be less than 2° and 25 mm when the cabin reaches a pose angle of ±40°, respectively.

Fig. 2.2 The model of feed cable-suspended structure for LT 5-m

Fig. 2.3 The model of feed cable-suspended structure for LT 50-m

We have already developed mechanics equations of the feed cable-suspended structure [15], and briefly outline here.

$$\mathbf{A}(X)\mathbf{H} = \mathbf{B}(X) \tag{2.1}$$

where $\mathbf{A} \in R^{6\times6}$ is Jacobin matrix. $\mathbf{H} = [H_1\, H_2\, H_3\, H_4\, H_5\, H_6]^{\mathrm{T}}$ is the vector describing the horizontal components of the active cable tensions. $\mathbf{B} \in R^{6\times1}$ is the load vector acting on the structure. For more details on the specification of the feed cable-suspended structure, refer to [15].

2.2.2 *Optimization of the Feed Cable-Suspended Structure*

In order to achieve a design scheme to meet the requirement of the orientation of 60°, it is necessary to analyze the original one, as shown in Fig. 2.1. In this scheme, six cable towers are designed as the same height, one group of the cables is connected to the bottom circle of the feed cabin through the top of the three towers 120° apart; the other is connected to a small circle on the top of the feed cabin to eliminate the singularity [15]. In the following discussion, the cables connected to the bottom circle of the cabin are called lower cable, and the tower, through which the lower cable connected is called lower cable tower. As the same way upper cable and upper cable tower are also defined (see in Fig. 2.4). The cross-sectional shape and structure of cables is spiral strands, as shown in Fig. 2.5.

As the orientation of the feed cabin increased to 60° from 40°, the original scheme is not feasible from research results. So we have to find a new way from the original design. Two improved schemes can lead to this purpose from the results of computer simulation. One is to lower the lower cable towers and heighten the upper cable towers; the other is to change the connection way of the feed cable-suspended structure. So it is necessary to obtain the optimal design scheme to achieve the orientation of 60° by virtue of optimization strategy.

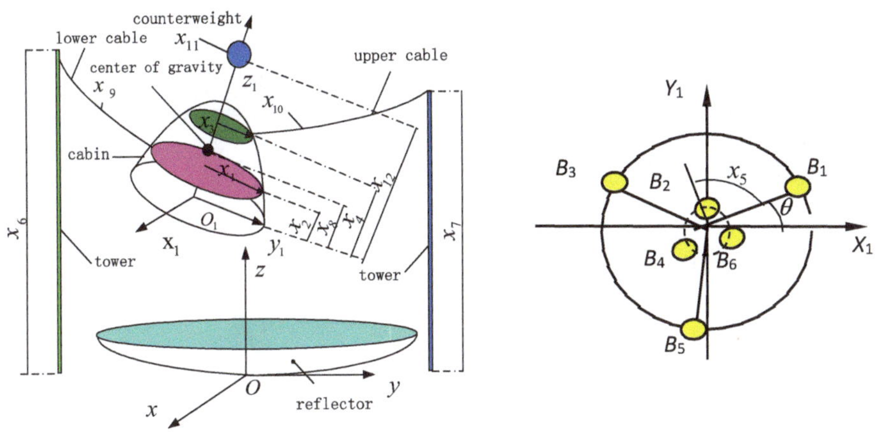

Fig. 2.4 Sketch map of design variables

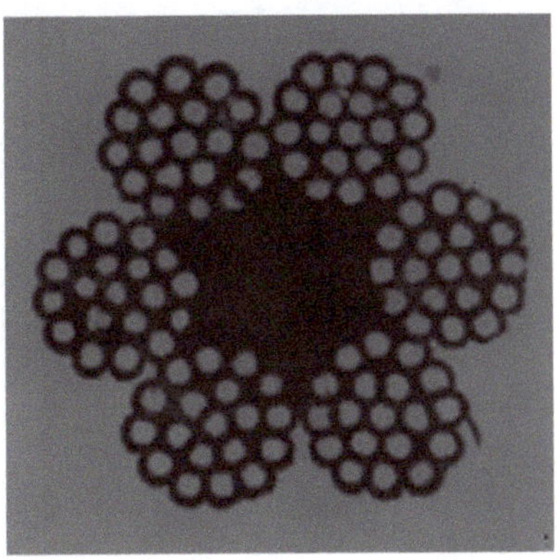

Fig. 2.5 Sketch map of the structure of cable

A. **Optimization model of the feed cable-suspended structure**

In order that the optimized structure can meet the requirement for the orientation of 60°, little cable tension difference, and all kinds of the constraint conditions, the following work needed to be performed.

Firstly, we should determine definitely that parameters can be chosen as design variables. As shown in the Fig. 2.4, design variables include the radius (x_1) and the height (x_2) of bottom circle where lower cables are connected to the cabin, the radius (x_3) and the height (x_4) of top circle where upper cables are connected to the cabin, angle (x_5) between the adjacent lower cable and upper cable,the height of lower and upper cable towers (x_6 and x_7), the weight center of cabin (x_8), and the cross sectional area of the lower and upper cables (x_9 and x_{10}). The meaning of the counterweight (x_{11} and x_{12}) is to be given in detail afterward.

Secondly, the constraint condition considering Eq. (2.1) in the whole workspace is written as

$$\mathbf{A}(X)\mathbf{H} = \mathbf{B}(X) \tag{2.2}$$

The problem of pseudo drag of the cable has to be avoided. As shown in Fig. 2.6, it shows the condition that cables are not pseudo drag, which requires the vertical component V of cable tension F being not less than zero; and Fig. 2.6b describes the condition that cables are pseudo drag, where the vertical component V is negative.

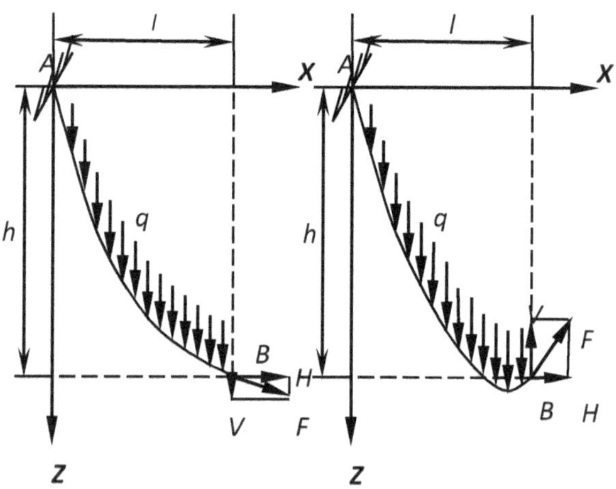

(a) Not pseudo drag (b) Pseudo drag

Fig. 2.6 Sketch map of state of cable

However, cables are of the unique property—carrying loads in tension but not in compression [16]. Therefore, this needs to meet the requirement for the lower bound of F corresponding to H and V, denoted as $\underline{H}$ and $\underline{V}$. So the constraint that cables are not pseudo drag is written as follows

$$g_1(X) = -H + \underline{H} \leq 0 \tag{2.3}$$

$$g_2(X) = -V + \underline{V} \leq 0 \tag{2.4}$$

In addition, cables should meet the requirement for material strength, written as,

$$g_{3e}(X) = \sigma_e - [\sigma] \leq 0, \quad (e = 1, 2, \ldots 6) \tag{2.5}$$

Of course, all the variables have their own lower and upper bounds, i.e.,

$$\underline{x_i} \leq x_i \leq \overline{x_i}, \quad (i = 1, 2, \ldots, n) \tag{2.6}$$

where n represents the total number of design variables.

Finally, the optimal objective is to make lower and upper cable tension difference as small as possible, which is beneficial to the control and economizing energy. Therefore, objective function can be formulated as

$$f(X) = \frac{2}{3}(F^{(l)}_{\max} - F^{(l)}_{\min}) + \frac{1}{3}(F^{(u)}_{\max} - F^{(u)}_{\min}) \tag{2.7}$$

where superscript l represents lower cable, u—upper cable, and F—cable tension that equals to the vector sum of horizontal tension H and vertical tension V. The reason why neither one coefficient nor the other is equal to $1/2$ is that the lower cables play an important role in the orientation of the cabin. As a result, optimization problem can be mathematically stated as

$$\begin{array}{ll} \text{(P1)} & \\ \text{Find} & \\ & X = (x_1, x_2, \ldots, x_n)^{\mathrm{T}} \\ \text{Min} & \\ & f(X) = \frac{2}{3}(F^{(l)}_{\max} - F^{(l)}_{\min}) + \frac{1}{3}(F^{(u)}_{\max} - F^{(u)}_{\min}) \\ \text{s.t.} & \\ & A(X)H = B(X) \\ & g_1(X) = -H + \underline{H} \leq 0 \\ & g_2(X) = -V + \underline{V} \leq 0 \\ & g_{3e}(X) = \sigma_e - [\sigma] \leq 0, \quad (e = 1, 2 \ldots 6) \\ & \underline{x} \leq x_i \leq \overline{x}, \quad (i = 1, 2 \ldots n) \end{array} \tag{2.8}$$

It is noted that problem **P1** is concerned over whole space and it is difficult for cabin to reach the 60° orientation. So the process in detail is that the horizontal circle, whose orientation is 60°, is divided into 360 points equally. Each point is a corresponding structure. If an optimum design can be found for each point, there will be an optimum design, $X^* = \left(x_1^*, x_2^*, \ldots, x_n^*\right)^{\mathrm{T}}$, for the whole working space.

B. **Solution of the optimal problem**

Problem **P1** is a nonlinear programming problem, and it is difficult to calculate its derivative. In recent years, textbooks are published and papers have appeared on genetic algorithm (GA) and its applications [17, 18]. Therefore, the optimal problem is solved with GA here.

The initial value, lower and upper bounds of variables and optimal results are shown in Table 2.1. The population size is 100; the number of generations is 600; the crossover rate is 0.6, and the mutation rate is 0.07. For the cable may be special made, the 10 design variables are continuous, and the size of the combinatorial space is mainly considered for the engineering. The time needed for one single evaluation is about 400 ms, and the first population is generated randomly. The initial values are the original design values before optimization.

The weight of the cabin is 300 kN; the material of cables is steel wire, whose allowable stress is 1850 N/mm^2, and cables should avoid pseudo drag. Hence, $\underline{H}$ is equal to 200 N, and $\underline{V}$ is equal to 10 kN.

Figure 2.7 illustrates the variation of cable tension F and cable length L. The maximum, minimum cable tension and cable tension difference of lower cables (A_1, A_2, A_3) are 391, 20 and 371 kN, respectively, corresponding values for those of upper cables (A_4, A_5, A_6) are 358, 165 and 193 kN.

It can also be known that the variation of cable tension is smooth in the whole space, without spine and mutation. Therefore, it is surely driven by servo system.

Figure 2.8 shows the combination of cabin and six cables in the end. Compared with what it was, it extends truss from the top and bottom of the cabin. Obviously, it becomes easily to realize the orientation requirement of the cabin.

Table 2.1 Initial values and optimal results of the variables

Variables	Initial value	Optimal result	Lower bound	Upper bound	Unit
*x*1	5.0	5.32	5.0	10.0	m
*x*2	0.0	−1.5	−1.5	0.0	m
*x*3	0.5	0.71	0.5	3.0	m
*x*4	4.5	6.89	3.0	10.0	m
*x*5	6.0	26.8	0.0	60.0	deg
*x*6	280	298.5	235	300	m
*x*7	280	260.8	235	288	m
*x*8	1.0	2.65	0.0	2.7	m
*x*9	5.0	5.3	2.0	10.0	cm^2
*x*10	5.0	3.2	1.0	10.0	cm^2

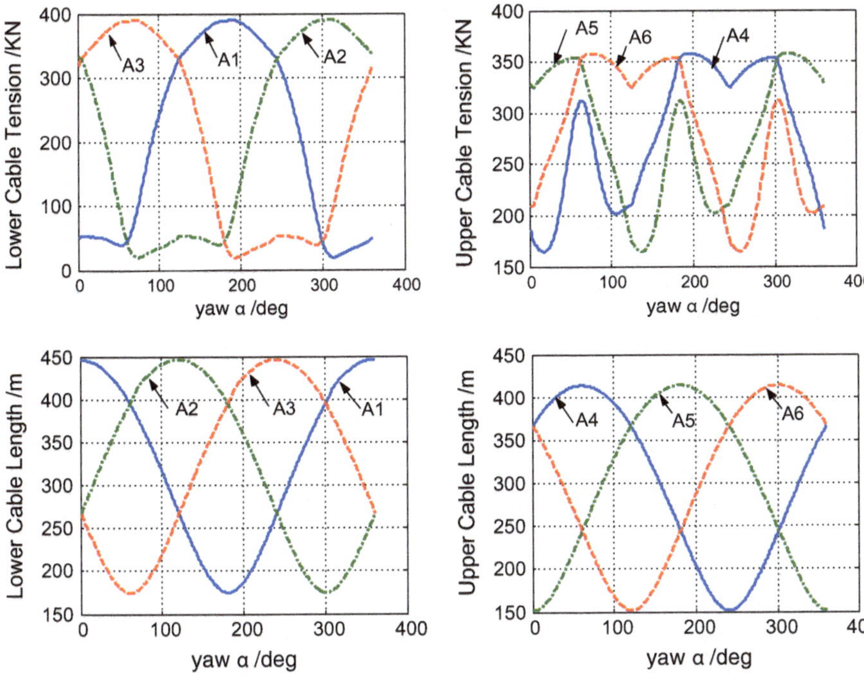

Fig. 2.7 Variation of cable tension and length with γ

Fig. 2.8 Connection between cables and cabin after optimization

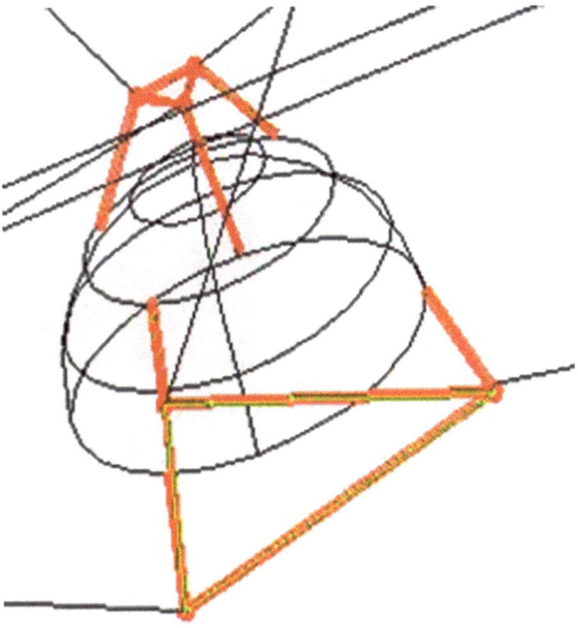

C. **Optimization of counterweight and further reduction of cable tension difference**

Based on the preceding analysis, the orientation angle of cabin can be increased up to 60° from 40° corresponding to the change of the structure of cabin (Fig. 2.4). Note that the tension difference of the cable is still large. The upper cable tension difference is 193 kN, and the lower is 371 kN.

In engineering, the clogs with 165 and 20 kN can be added to the upper and lower cable towers separately (Fig. 2.9) to reduce energy expense. As for the current results obtained, the required maximum power of servo motor decreased from 358 and 391 kN to 193 and 371 kN, respectively, which is our pursuance.

To reduce the tension difference of the cable further and make cable tension as smooth as possible, a counterweight is added on the cabin, which is proposed by Professor R.D. Nan in National Astronomical Observatories, Chinese Academy of Sciences. As shown in Fig. 2.4, variable x_{11} is the mass of counterweight and x_{12} is the distance from the counterweight to the bottom of the cabin. The objective is to minimize the tension difference by optimizing x_{11} and x_{12} in the whole workspace. This will be beneficial to the increment of orientation angle of the cabin. The process of calculation in detail is as follows. When γ is equal to 60°, x_{11} and x_{12} are taken as design variables, while other parameters keep the optimal results unchanged, as given in Table 2.1. The variables of x_{11} and x_{12} are optimized. Then, γ is given as 50°, 40°, 30°, 20°, 10° and 0°; x_{11} is constant; x_{12} is varied, and the optimal objective is to minimize the tension difference of the cable.

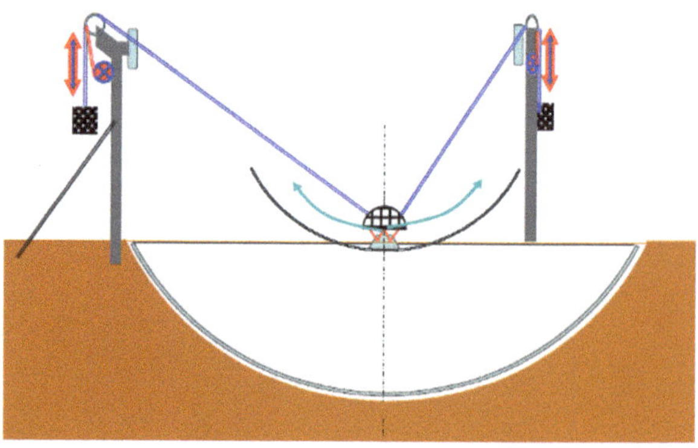

Fig. 2.9 Clogs are added on cable towers

So problem **P1** is changed into **P2**.

$$
\begin{aligned}
&\text{(P2)} \\
&\text{Find} \quad x_{12} \\
&\text{Min} \quad f(\mathbf{X}) = \tfrac{2}{3}\left(F_{\max}^{(l)} - F_{\min}^{(l)}\right) + \tfrac{1}{3}\left(F_{\max}^{(u)} - F_{\min}^{(u)}\right) \\
&\text{s.t.} \quad A(X)H = B(X) \\
&\qquad g_1(\mathbf{X}) = -H + \underline{H} \le 0 \\
&\qquad g_2(\mathbf{X}) = -V + \underline{V} \le 0 \\
&\qquad g_{3e}(\mathbf{X}) = \sigma_e - [\sigma] \le 0, \quad (e = 1, 2 \ldots 6) \\
&\qquad \underline{x} \le x_i \le \overline{x}, \quad (i = 1, 2 \ldots n)
\end{aligned}
\tag{2.9}
$$

Not that $\underline{x_{12}}$ is 0, $\overline{x_{12}}$ is equal to x_{12}^{*}, which is the optimal results when $\gamma = 60°$. The results can be obtained by virtue of solving **P2** with GA, as shown in Fig. 2.10. Figure 2.10a shows the variation of distance from the counterweight to the top of cabin. We are told that when $x_{11} = 29.7$ kN and orientation varies from 0° to 60°, the distance varies from 0 to 5.03 m.

Figure 2.10b illustrates the variation of the maximum tension difference of the cable with angle γ. Moreover, we can also note that the tension difference of the lower cable decreases from 371 to 328 kN.

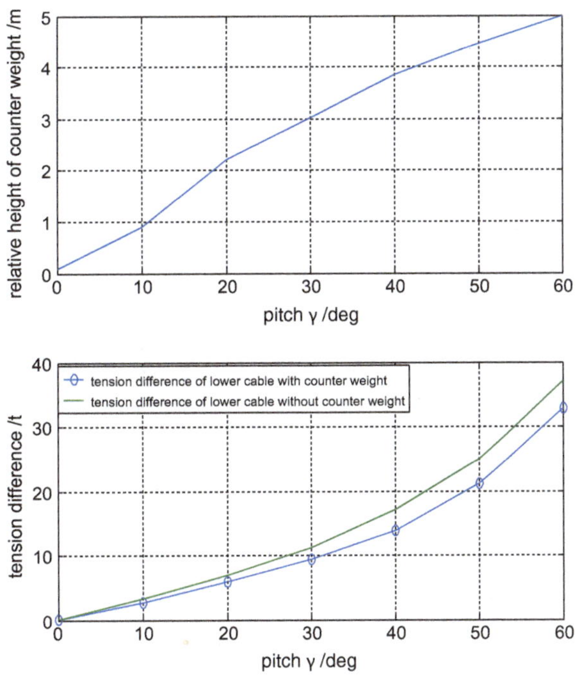

Fig. 2.10 Clogs are added on cable towers. **a** variation of bar length with γ, **b** variation of cable tension difference with γ

2.2.3 Experimental Results and Discussion of Enhancing the Stiffness of the Feed Cable-Suspended Structure

In order to enhance the system stiffness, especially, twisting stiffness, Paper [19] proposes a method by adding stable cables with useful results. Considering the length of paper, the following design process and results of experiment are given only.

To evaluate the performance of the proposed method, the LT 5-m model is selected for implementation. As shown in Fig. 2.11, when the cabin is located at the position (0, 0, 110) (cm), a rope is drawn along the tangent direction of the cabin. Then, the rope is cut off suddenly. At the same time three laser ranging equipments record the vibration of the cabin. For each cables there is a clog with 3 N connected at the free end. The weights of the cabin and tube with water are 60 and 40 N, respectively.

The vibration of the cabin along *y*-direction with 6-cable and 9-cable is shown in Fig. 2.11. It can be observed that the application of the stable cables introduces significant amount of passive damping. The time to be stable 6 and 9 cables are about 4 and 30 s, reduced almost 8 times. As for the amplitude, 5 times are obtained.

Actually, the above experiments have been done repeatedly in addition to adding the tube with liquid. Typical responses are given in Fig. 2.12, and compared with other three cases. It can be noted from Fig. 2.12 that the tube with liquid is not significant effectiveness. Hence, the nine-cable structure is positive.

The experiments shown in Figs. 2.11 and 2.12 are carried out when cabin is fixed at the given position. It is help to see the benefit from 3-added downward stable cables. However, it is not enough to evaluate the structural behavior comprehensive. Dynamic property of the structure is probably much more important. For the sake of this, the following dynamic experiment with LT5-m is done. During

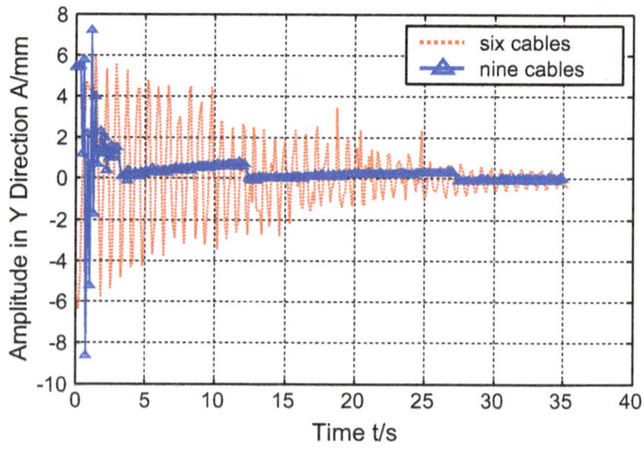

Fig. 2.11 Comparison of vibration in y direction

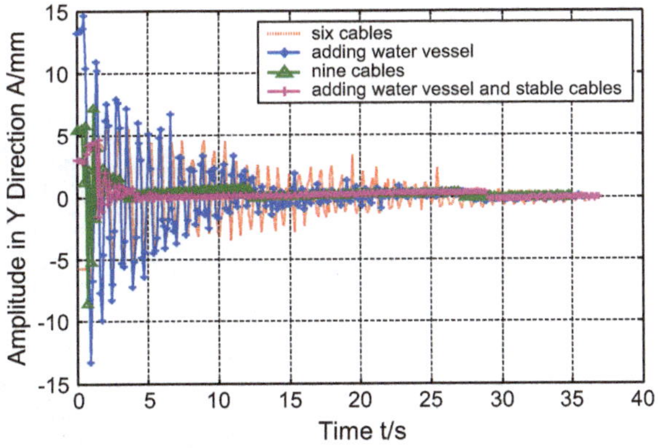

Fig. 2.12 Comparison of vibration in y direction

experiment, the cabin is asked to move along the given trajectory. The artificial wind forces, through a fan with 5 kW, on the structure with a mean velocity of 17 m/s is provided. A sample of the wind velocity persists 50 s, then, the fan is cut off and vibration recording is continued 50 s more. Vibration responses of the 6 cables and 9 cables are compared in Fig. 2.13. Obviously, the incensement in dynamic stiffness is significant by adding three downward stable cables.

For further observing the variety of the twisting stiffness, three moments in x, y and z directions (Fig. 2.14) are acted on the cabin. Figures 2.15 and 2.16 show the experimental results in the case of 6 and 9 cables. The clog is changed from 3, 5 and to 9 N. The incensement in the twisting stiffness is considerably. On the other hand, as far as the 6 cables are concerned, we do not worry about the stiffness in the x and y directions, while worrying about the stiffness about z direction. The experimental results in Fig. 2.15 demonstrate the conclusion. In addition, through optimizing the weight of clog, we examine the vibration response. Figure 2.16 shows the experimental results. It is observed from Fig. 2.16 that the stiffness about Z_1 with 9 cables is in the same level as the stiffness about X_1 and Y_1 of 6 cables.

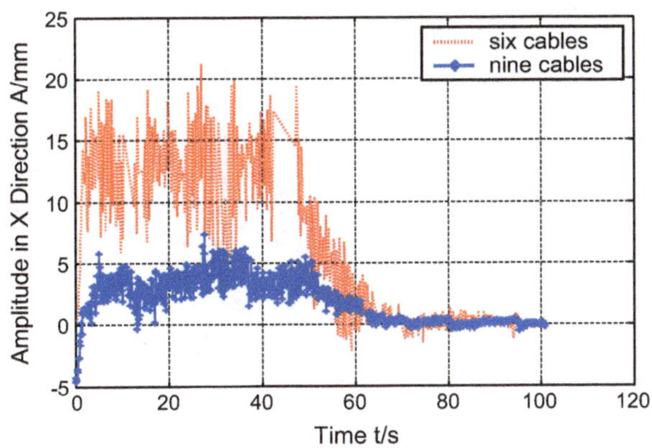

Fig. 2.13 Comparison of vibration in x direction

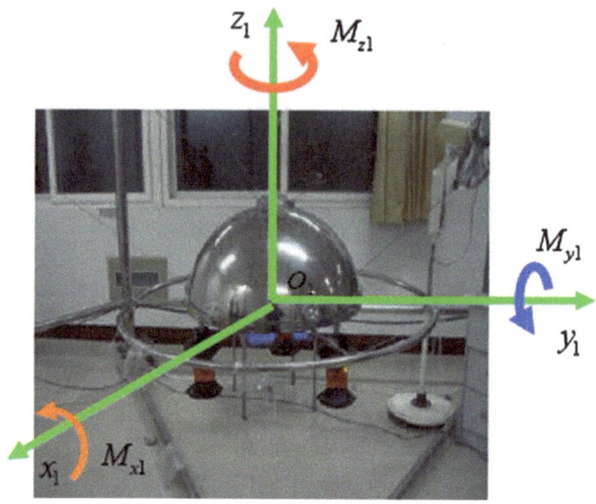

Fig. 2.14 Schematic representation of the cabin

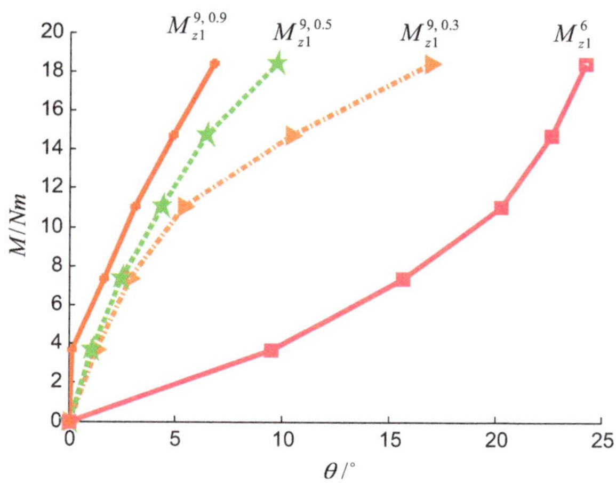

Fig. 2.15 Experimental results of twisting stiffness

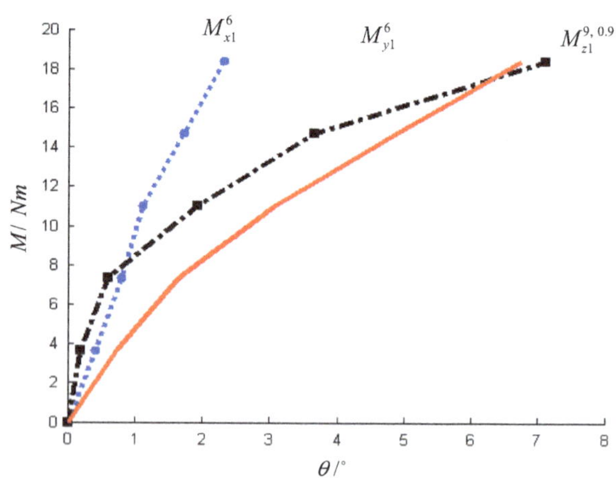

Fig. 2.16 Experimental results of system stiffness

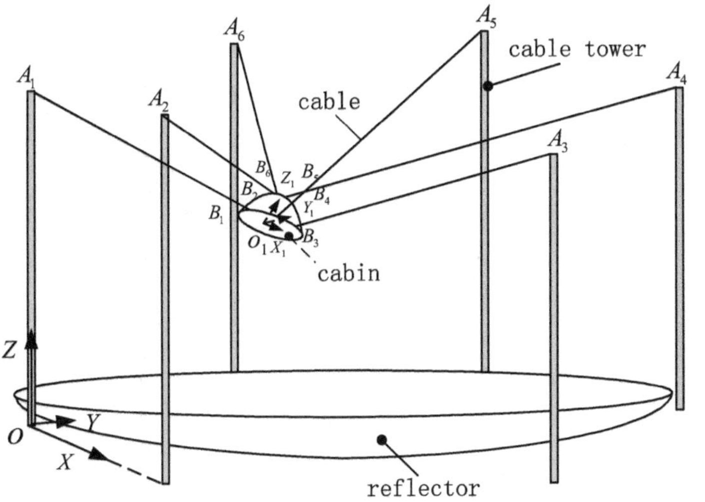

Fig. 2.17 Schematic diagram of CPR for LT

2.3 Dynamic Modeling of a Cable-Suspended Parallel Robot with Non-negligible Cable Mass

2.3.1 *System Dynamic Model*

A. **Model definition and kinematics**

A simple schematic of the CPR for FAST 5-m representing the coordinate systems is shown in Fig. 2.2. The CPR is considered that is kinematically and statically determined. In the design, there is a cable tower/winch pair at each vertex of a regular hexagon of radius a, which actuate six cables that are linked to a cabin. Therefore, the cabin can translate and rotate in the inertial frame. Let the origin of the inertial frame $OXYZ$ be the bottom of tower A_1. The origin of $O_1X_1Y_1Z_1$ (cabin coordinate system) is in the center of the bottom of the cabin as shown in Fig. 2.18.

In Fig. 2.18, towers A_1–A_6 are distributed evenly on a circle with the diameter of D. The cables A_1B_1, A_3B_3, and A_5B_5 are connected from the bottom of the cabin to the towers A_1, A_3, and A_5, respectively. The other three cables A_2B_2, A_4B_4, and A_6B_6 are connected from the towers A_2, A_4, and A_6 to the top points B_2, B_4, and B_6 of the cabin as shown in Fig. 2.18. Angle η between O_1B_2 and O_1B_1 indicates the relative position relationship of spherical joint B_1 and spherical joint B_2. It can be noted from Fig. 2.18 that the coordinates of B_2, B_4, and B_6 have relationships with η.

The coordinates of the joint A_i on the tower i with respect to the frame $OXYZ$ can be described as

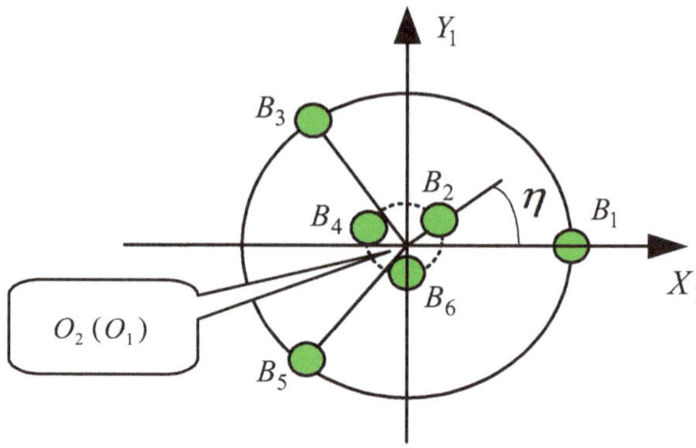

Fig. 2.18 Connection between the cables and the cabin

$$\left.\begin{array}{ll}\mathbf{r}_{A1} = (0 \quad 0 \quad h)^{\mathrm{T}}, & \mathbf{r}_{A2} = (a \quad 0 \quad h)^{\mathrm{T}} \\ \mathbf{r}_{A3} = (\frac{3}{2}a \quad \frac{\sqrt{3}}{2}a \quad h)^{\mathrm{T}}, & \mathbf{r}_{A4} = (a \quad \sqrt{3}a \quad h)^{\mathrm{T}} \\ \mathbf{r}_{A5} = (a \quad \sqrt{3}a \quad h)^{\mathrm{T}}, & \mathbf{r}_{A6} = (-\frac{1}{2}a \, \frac{\sqrt{3}}{2}a\, h)^{\mathrm{T}}\end{array}\right\} \tag{2.10}$$

where h is the height of tower and a is the side length between the tower A_1 and tower A_2 in regular hexagon. The length of the cable can be changed by rotating the winch to reel the cable in or let it out. The cabin is assumed to be a rigid body.

We proceed to consider the vector diagram for an ith cable. The position of frame $OXYZ$ is represented by vector $\mathbf{r}_{Bi} = [x_{Bi}\; y_{Bi}\; z_{Bi}]^{\mathrm{T}}$, which contains the Cartesian coordinates x, y, z of the origin of frame $O_1X_1Y_1Z_1$ with respect to frame $OXYZ$. The length vector $\mathbf{l}_i$ expressed with respect to frame $OXYZ$ can be computed by

$$l_i = r_{Bi} - r_{Ai} \quad (i = 1, 2, \ldots 6) \tag{2.11}$$

where, $r_{Bi} = r_{O1} + R \cdot r_{Bi1} \quad (i = 1, 2, \ldots 6)$ $\mathbf{r}_{O1}$ is the position vector of the origin of $O_1X_1Y_1Z_1$ with respect to $OXYZ$; $\mathbf{r}_{Bi1}$ is the position vector of the point B_i related to $O_1X_1Y_1Z_1$. $\mathbf{r}_{Bi1}$ can be expressed as

$$\left.\begin{array}{l}r_{B11} = [r, 0, 0]^T \\ r_{B31} = [r\,\cos(\theta), r\,\sin(\theta), 0]^T \\ r_{B51} = [r\,\cos(2\theta), r\,\sin(2\theta), 0]^T\end{array}\right\} \tag{2.12}$$

$$\left.\begin{array}{l}r_{B21} = \left[r_1\cos(\eta)\,, r_1\sin(\eta)\,, \sqrt{r^2 - (r_1)^2}\right]^T \\ r_{B41} = \left[r_1\cos(\eta+\theta)\,, r_1\sin(\eta+\theta)\,, \sqrt{r^2 - (r_1)^2}\right]^T \\ r_{B61} = \left[r_1\cos(\eta+2\theta)\,, r_1\sin(\eta+2\theta)\,, \sqrt{r^2 - (r_1)^2}\right]^T\end{array}\right\} \tag{2.13}$$

In which, r and r_1 represent the radii of the points B_1, B_3, B_5 and the points B_2, B_4, B_6 of the cabin, respectively. $\theta = 120°$ as shown in Fig. 2.18.

$\boldsymbol{R}$ is the orientation matrix of the cabin with respect to $OXYZ$. By introducing the roll, pitch, and yaw angles into R, the matrix becomes,

$$\mathbf{R} = \begin{bmatrix} c\alpha c\beta & c\alpha s\beta s\gamma - s\alpha c\gamma & c\alpha s\beta c\gamma + s\alpha s\gamma \\ s\alpha c\beta & s\alpha s\beta s\gamma + c\alpha c\gamma & s\alpha s\beta c\gamma - c\alpha s\gamma \\ -s\beta & c\beta s\gamma & c\beta c\gamma \end{bmatrix} \tag{2.14}$$

where roll is the rotation about the fixed axis Z by α, pitch is the rotation about the fixed axis Y by β, and yaw is the rotation about the fixed axis X by γ; c represents cosine function; and s represents sine function.

The Jacobian matrix of CPR is defined as the relation between the velocity of the cabin and the velocity of the driven cable. We specify the velocity of the cable-driven as $\dot{\boldsymbol{l}} = [\dot{l}_1\,\dot{l}_2\,\dot{l}_3\,\dot{l}_4\,\dot{l}_5\,\dot{l}_6]^T$ (m/s), define the velocity of the center of the cabin as $\mathbf{v} = (\dot{x}, \dot{y}, \dot{z})^{\mathrm{T}}$ (m/s), and express the angular velocity of the cabin with respect to three axes Z, Y and X as $\mathbf{w} = (\dot{\alpha}, \dot{\beta}, \dot{\gamma})^{\mathrm{T}}$ (rad/s), so the velocity of the cabin can be described as $\dot{\mathbf{r}}_{O1} = [\mathbf{vw}]^{\mathrm{T}} = [\dot{x}\,\dot{y}\,\dot{z}\,\dot{\alpha}\,\dot{\beta}\,\dot{\gamma}]^{\mathrm{T}}$. The relation between $\dot{\mathbf{l}}_i$ and $\dot{\mathbf{r}}_{O1}$ can be defined as [20]

$$\dot{\mathbf{l}} = \mathbf{J} \cdot \dot{\mathbf{r}}_{O1} \tag{2.15}$$

where the Jacobian matrix $\mathbf{J}$ is

$$\mathbf{J} = \begin{bmatrix} \frac{\partial l_i}{\partial x} & \frac{\partial l_i}{\partial y} & \frac{\partial l_i}{\partial z} & \frac{\partial l_i}{\partial \alpha} & \frac{\partial l_i}{\partial \beta} & \frac{\partial l_i}{\partial \gamma} \end{bmatrix} \quad (i = 1, 2, \ldots, 6) \tag{2.16}$$

For simplicity, we use the symbolism to develop the Jacobian matrix. The formula can be obtained.

Thus, we can compute the cable-driven lengths, i.e., norms of $\mathbf{l}_i$, from the given positions and orientations of the cabin. This problem is called an inverse kinematics problem of a CPR. The forward kinematics problem is the opposite of the inverse kinematics, i.e., to obtain the positions and orientations from the given cable-driven lengths.

B. **Dynamic equations of the cabin**

The general dynamic equations of motion can be obtained from the Lagrangian formulation. The generalized coordinates is $\mathbf{q} = [x, y, z, \alpha, \beta, \gamma]^{\mathrm{T}}$. Lagrange's equation can be written in the form of potential energy V, kinetic energy T, and generalized forces or torques in the following form,

$$\frac{d}{dt}\left(\frac{\partial T(\mathbf{q}, \dot{\mathbf{q}})}{\partial \dot{\mathbf{q}}}\right) - \frac{\partial T(\mathbf{q}, \dot{\mathbf{q}})}{\partial \mathbf{q}} + \frac{\partial V(\mathbf{q})}{\partial \mathbf{q}} = \boldsymbol{\tau} \tag{2.17}$$

This results in a kinetic energy of the cabin, which can be written in Cartesian coordinates as,

$$T = \frac{1}{2} m \mathbf{v}^{\mathrm{T}} \mathbf{v} + \frac{1}{2} \mathbf{w}^{\mathrm{T}} \mathbf{R} \mathbf{I}_{O1} \mathbf{R}^{\mathrm{T}} \mathbf{w} \tag{2.18}$$

or,

$$\begin{aligned} T = {} & \frac{1}{2} m(\dot{x}^2 + \dot{y}^2 + \dot{z}^2) + \frac{1}{2} [I_{xx}(\dot{\alpha} c\alpha c\beta + \dot{\beta} s\alpha c\beta - \dot{\gamma} s\beta)^2 \\ & + I_{yy}((c\alpha s\beta s\gamma - s\alpha c\gamma)\dot{\alpha} + (c\alpha c\gamma + s\alpha s\beta s\gamma)\dot{\beta} + \dot{\gamma} c\beta s\gamma)^2 \\ & + I_{zz}((c\alpha s\beta c\gamma + s\alpha s\gamma)\dot{\alpha} + (s\alpha s\beta c\gamma - c\alpha s\gamma)\dot{\beta} + \dot{\gamma} c\beta c\gamma)^2] \end{aligned} \tag{2.19}$$

where $\mathbf{v} = [\dot{x}\,\dot{y}\,\dot{z}]^{\mathrm{T}}; \mathbf{w} = [\dot{\alpha}\,\dot{\beta}\,\dot{\gamma}]^{\mathrm{T}}$; m and $\mathbf{I}_{O1}$ is the mass and the moment of inertia of the cabin, respectively. $\mathbf{I}_{O1}$ is expressed as

$$\mathbf{I}_{O1} = \begin{bmatrix} I_{xx} & -I_{xy} & -I_{xz} \\ -I_{yx} & I_{yy} & -I_{yz} \\ -I_{zx} & -I_{zy} & I_{zz} \end{bmatrix} \tag{2.20}$$

In this study, each cable is assumed to be a force element. Therefore, the potential energy of the system is due only to gravitational forces. The potential energy takes the form:

$$V = mgz \tag{2.21}$$

where g is the acceleration due to gravity.

It is noted that there is a relation between externally applied wrench on the cabin and the cable tensions required to keep the system in equilibrium, this relationship [10] is

$$[F_x, F_y, F_z, M_{\partial}, M_{\beta}, M_{\gamma}]^{\mathrm{T}} = -\mathbf{J}^{\mathrm{T}} \mathbf{u} \tag{2.22}$$

where the external force on the cabin at the reference point is F_x, F_y, F_z and the external moment on the cabin body is given by M_{α}, M_{β}, M_{γ}; $\mathbf{u}$ is the vector of cable-driven tensions.

Through a series of transformations, substitutions and simplifying the resulting expression, here, we write the equations of motion in terms of $\mathbf{q}$ generalized coordinates as the following general form

$$\mathbf{M}(\mathbf{q})\ddot{\mathbf{q}} + \mathbf{C}(\mathbf{q}, \dot{\mathbf{q}})\dot{\mathbf{q}} + \mathbf{G}(\mathbf{q}) + \boldsymbol{\tau}_d = -\mathbf{J}^{\mathrm{T}} \mathbf{u} \tag{2.23}$$

where $\mathbf{M}(\mathbf{q})$ is the inertia matrix of the system, $\mathbf{C}(\mathbf{q}, \dot{\mathbf{q}})$ is the vector of Coriolis and centripetal terms, $\mathbf{G}(\mathbf{q})$ is the vector of gravity terms, $\boldsymbol{\tau}_d$ is the vector of external disturbance terms (e.g., random wind, etc.). Here, $\mathbf{u}$ is the vector of cable-driven tensions T_{Ji} for $i = 1, 2, \ldots, 6$. In the following discussion, we will briefly describe how T_{Ji} is derived.

2.3.2 Mechanics Analysis of the Cable

Six flexible cables are used to drive the cabin to implement high-precision trajectory tracking. Due to the eigenfrequency of the cable-cabin system is very low, the response of the cabin is very slow in the presence of external disturbances. The quasi- dynamics analysis is made by substituting static tension of cable at a static position for the dynamic force of the cable at the same position. When the cabin operates at the practically observed speed (0.5–2 cm/s), the difference between static tension and the dynamic force is very small [21]. However, this difference has little influence on the calculation of the cable length, so the dynamic characteristic of cables can be neglected. Owing to six identical cables, we only perform mechanics analysis of a cable.

The first challenge for analyzing a cable-suspended system is to mathematically describe the static displacement (or the shape) of the cables under the influence of gravity [22]. To compute the static displacement of a cable, we will assume for the moment that the unstressed length of the cable, L_{u}, are known. The procedure in this section determines the complete geometry of the cable, end forces from a given strained or unstrained cable length and the positions of cable ends by using Newton-Raphson method.

Considering an elastic cable element stretched in vertical plane, with unstressed length L_u, corresponding stressed length L, elastic modulus E, cross sectional area A, and self-weight per unit length q_0, as shown in Fig. 2.19, the rigorous relationship between cable projection and end force of the cable is subjected to [23]

$$L^2 = h^2 + \frac{l^2}{\eta^2} \sin \boldsymbol{h}^2 \eta \tag{2.24}$$

where

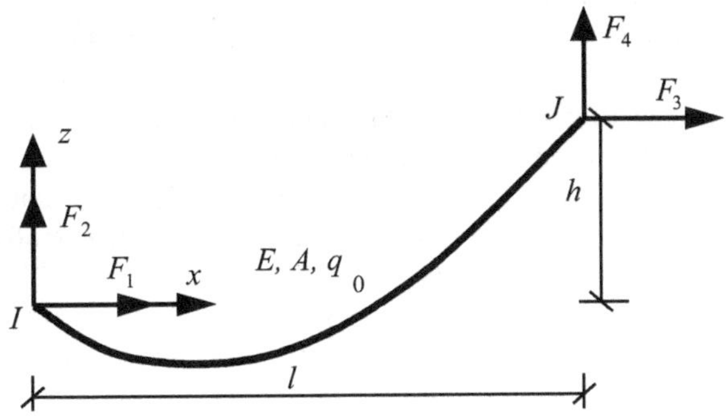

Fig. 2.19 Elastic catenary cable element

$$\eta = \frac{q_0 l}{2F_1} \tag{2.25}$$

$$L_u = L - \frac{1}{2EAq_0}\left(F_4 T_J + F_2 T_I + F_1^2 \ln\frac{F_4 + T_J}{T_I - F_2}\right) \tag{2.26}$$

and

$$F_2 = \frac{q_0}{2}\left[L - h\frac{\cosh\eta}{\sinh\eta}\right] \tag{2.27}$$

$$l = -F_1\left[\frac{L_u}{EA} + \frac{1}{q_0}\ln\frac{F_4 + T_J}{T_I - F_2}\right] \tag{2.28}$$

$$h = \frac{1}{2EAq_0}\left(T_J^2 - T_I^2\right) + \frac{T_J - T_I}{q_0} \tag{2.29}$$

The variables F_1, F_2, F_3, F_4, T_{Ii} and T_{Ji} (for $i = 1, 2, \ldots, 6$) are subjected to the following equations

$$\left.\begin{array}{ll} F_3 = -F_1, & F_2 + F_4 = q_0 L_u \\ T_{Ii} = \left(F_1^2 + F_2^2\right)^{1/2}, & T_{Ji} = \left(F_3^2 + F_4^2\right)^{1/2} \end{array}\right\} \tag{2.30}$$

There exists an unique solution for the cable-driven tension using (2.13), for a given motion of the cabin, as long as $\mathbf{J}$ is not singular. In this case, the solution of (2.13) can be written as

$$\mathbf{u} = -\left(\mathbf{J}^{\mathrm{T}}\right)^{-1}\left[\mathbf{M}(\mathbf{q})\ddot{\mathbf{q}} + \mathbf{C}(\mathbf{q}, \dot{\mathbf{q}})\dot{\mathbf{q}} + \mathbf{G}(\mathbf{q}) + \boldsymbol{\tau}_d\right] \tag{2.31}$$

2.4 Conclusions

In this chapter, design and optimization of the feed cable-suspended structures is presented. And the counterweight is added to the top of cabin, and its position varies with orientation in time, which is beneficial to the further reduction of cable tension difference, economizing energy and big orientation. In addition, to enhance the twisting stiffness of the cable-suspended structure, a method by adding stable cables is proposed. The experimental results have demonstrated the effectiveness and feasibility of the proposed method. Then, when the cabin locates at the pose angle of 0° and 20°, respectively, analytical calculations are conducted in three cases, and the antenna pattern is simulated with GRASP 8.0 (see Fig. 2.18). The numerical results show the stable cables have little influence on the electronic performance. Then, the inverse dynamics problem of the CPR with non-negligible cable mass is formulated by the Lagrange formulation. modeling and analysis of the cable-supporting system including servomechanism dynamics for super antenna is presented. Subsequently, to improve the precision and performance of our model, the dynamic analysis of the cabin including servomechanism dynamics without regard to cable mass is established in detail for the controller design in next chapter.

References

1. Duan BY, Qiu YY, Zhang FS et al (2009) On design and experiment of the feed cable-suspended structure for super antenna. Mechatronics 19(4):503–509
2. Nan R (2006) Five hundred meter aperture spherical radio telescope (FAST). Sci China Ser G 49(2):129–148
3. Duan BY (1999) A new design project of the line feed structure for large spherical radio telescope and its nonlinear dynamic analysis. Mechatronics 9(1):53–64
4. Qiu YH (1998) A novel design for a giant Arecibo-type spherical radio telescope with an active main reflector. Mon Not R Astron Soc 301(3):827–830
5. Su YX, Duan BY (2000) The application of the Stewart platform in large spherical radio telescopes. J Robotic Syst 17(7):375–383
6. Luo YF, Deng CG, Li GQ et al (2001) Structural analysis of FAST reflector supporting system. Astrophys Space Sci 278(1–2):231–236
7. Albus J, Bostelman R, Dagalakis N (1992) The NIST robocrane. J Robot Syst 10(5)
8. Shiang WJ, Cannon D, Gorman J (1999) Dynamic analysis of the cable array robotic crane. 1999 IEEE international conference on robotics and automation, 1999. Proceedings. vol 4. pp 2495–2500
9. Yamamoto M, Yanai N, Mohri A (2004) Trajectory control of incompletely restrained parallel-wire-suspended mechanism based on inverse dynamics. IEEE Trans Rob 20(5):840–850
10. Oh S, Agrawal SK (2005) A reference governor-based controller for a cable robot under input constraints. IEEE Trans Control Syst Technol 13(4):639–645
11. Kamman JW, Huston RL (2001) Multibody dynamics modeling of variable length cable systems. Multibody SysDyn 5(3):211–221
12. Craig JJ (2005) Introduction to robotics: mechanics and control. Pearson Prentice Hall, Upper Saddle River
13. Anupoju CM, Su CY, Oya M (2005) Adaptive motion tracking control of uncertain nonholonomic mechanical systems including actuator dynamics. IEE Proc Control Theor Appl 152(5):575–580
14. Ming A, Higuchi T (1994) Study on multiple degree-of-freedom positioning mechanism using wires. II: Development of a planar completely restrained positioning mechanism. Int J Jpn Soc Precis Eng 28(3):235–242
15. Zi B, Duan BY, Du JL et al (2008) Dynamic modeling and active control of a cable-suspended parallel robot. Mechatronics 18(1):1–12
16. Oh SR, Agrawal SK (2005) Cable suspended planar robots with redundant cables: controllers with positive tensions. IEEE Trans Rob 21(3):457–465
17. Hajela P (1990) Genetic search-an approach to the nonconvex optimization problem. AIAA J 28(7):1205–1210
18. Kaveh A, Rahami H (2006) Nonlinear analysis and optimal design of structures via force method and genetic algorithm. Comput Struct 84(12):770–778
19. Qiu YY, Duan BY, Wei Q et al (2002) Optimal distribution of the cable tensions and structural vibration control of the cable-cabin flexible structure. Struct Eng Mech 14(1):39–56

20. Guo Q, Liu Y, Liu Y (2002) Local structurization kinematic decoupling of six-leg virtual-axis NC machine tool. IEEE/ASME Trans Mechatron 7(4):515–518
21. Sun X (2001) A new project of the large spherical radio telescope suspend cable feed supporting system for nonlinear statics, kinematics and dynamics. Ph. D. Dissertation, Xidian University
22. Kozak K, Zhou Q, Wang J (2006) Static analysis of cable-driven manipulators with non-negligible cable mass. IEEE Trans Rob 22(3):425–433
23. Karoumi R (1999) Some modeling aspects in the nonlinear finite element analysis of cable supported bridges. Comput Struct 71(4):397–412

Chapter 3
Trajectory Tracking Control of the Feed Cable-Suspended Structure for Super Antenna

3.1 Preamble

In view of a control aspect, recently, there has been an increasing interest in the CPR, and different control schemes, derived from nonlinear control theory, classical control techniques have been proposed with various theoretical and practical contributions being made [1–4]. From an engineering viewpoint, the major focus of this investigation is one position control method that seem well suited to the CPR. It is well known that up until now, a conventional proportional-integral-derivative (PID) controller is most widely used in industry due to its simple control structure, ease of design, and inexpensive cost. However, the PID controller cannot yield a good control performance if a controlled object is highly nonlinear and uncertain. Therefore, we would like to keep the advantages of the PID controller while designing a new controller for an enhanced control performance. Fuzzy controllers using fuzzy if-then rules are able to effectively incorporate nonlinear properties and unmodeled effects into its basic design [5]. The idea of fuzzy set and fuzzy control is introduced by Zadeh in an attempt to control systems that are structurally difficult to model [6]. Since Mamdani did the first fuzzy control application [7], fuzzy control is being increasingly applied to many systems with nonlinearity and uncertainty [8]. Especially, the most successful fuzzy logic controllers applied into industrial plants are designed by control engineers. Recently, fuzzy-logic and conventional-technique are combined to design fuzzy logic controller [9, 10], and the control technology based on fuzzy logic is being increasingly applied to systems with nonlinearity and uncertainty. Lin et al. [11] developed an integrated fuzzy-logic-based missile guidance law against high speed target. Siripongwutikorn et al. [12] proposed an augmented Fuzzy control algorithm for a quantitative packet loss rate guarantee to aggregate traffic in packet switched networks. Hwang and Kim [13] employed an adaptive fuzzy logic method to provide the solution to the control of an electrically driven robot. In this chapter, a fuzzy plus

B. Zi and S. Qian, *Design, Analysis and Control of Cable-Suspended Parallel Robots and Its Applications*, DOI 10.1007/978-981-10-1753-7_3

proportional-integral control (FPPIC) method of CPR is proposed to track a given trajectory based on dynamic models that established by different methods.

As we analyzed in Chap. 2, the cables that used to replace rigid links can gives CPRs several advantages such as very large workspace and high payload-to-weight ratios. However, due to the characteristics of the cable, the parallel cable structure also introduces many new challenges in the control of CPRs. The non-rigid cables which can only bear unidirectional force make the robot more sensitive to the external disturbance force. Thus, a sophisticated control scheme is required for an enhanced control performance. However, from the kinematic and dynamic models of CPRs we have established in Chap. 2, we know that the traditional control theories for rigid links parallel robot cannot be used directly for CPRs and they must be modified to provide positive tension for the cables. In this chapter the fuzzy plus proportional-integral control (FPPIC) method, which can enhance the control performance for steady state errors, is utilized to control the wind-induced vibration of the trajectory tracking of the feed. Moreover, the simulation and experiment are respectively deployed on the 5-m and 50-m scaled model to test the different dynamic modeling present in former chapter and justify the efficiency of purposed control strategy.

3.2 Fuzzy Plus Proportional-Integral Controller Design

A robot control scheme stems from two frameworks. One is to design a control on the workspace and the other is on the linkspace. The linkspace control is a conventional control, and is a kind of tracking control to follow the desired link length computed from the position command of the mobile plate by inverse kinematics. Normally, most controllers in applications are based on the linkspace coordinates, which they consider only an approximated manipulator model [14]. From an engineering viewpoint, considering a control aspect, a PID control has been employed in industry due to its simple control structure, ease of design, and inexpensive cost. But it does not always guarantee a high performance for a parallel robot. Therefore, we would like to keep the advantages of the PID controller while designing a new controller for an enhanced control performance. Fuzzy logic has been considered to be an efficient and effective tool to manage system uncertainties since the seminal paper of Zadeh. Among the many applications of fuzzy logic, control design appears to be the one that has attracted the most attention in the past two decades [15]. In this chapter, a fuzzy plus proportional-integral controller combining PI control with fuzzy logic control is developed for more effective and robust performance in the presence of random wind and other disturbances. Its viability for practical applications has been demonstrated by simulation and experiments.

The CPR control system of FAST consists of six cable-driven alternating current (AC) subservo system, and the control mode of the subservosystem is Insulated

Gate Bipolar Transistor (IGBT) Modem Pulse-Width Modulated (PWM) sine wave control.

For the high requirement of real-time and synchronization of CPR, we adopt a centralized control scheme, which consists of a main computer (M-C), a measure computer (Measure-C) and the multi-movement-node based on servodrive. In real-time motion, the real position and orientation of the center of the feed cabin can be detected by three laser ranging equipments under the control of Measure-C, together with the desired trajectory are fed into M-C, and M-C calculates the lengths of six cables in each time slice based on the inverse dynamic model, and then converts the length change value of cable into the number of pulse. Thus each servo control system can realize corresponding movement via multi-axis movement controller. That is, the length of each cable at every sampling time can be adjusted by high power servomotor through a favorable control method to guarantee the high-precision trajectory tracking.

Fuzzy logic systems belong to the category of computational intelligence techniques. One advantage of the fuzzy logic over other forms of knowledge-based controllers lies in the interpolative nature of the fuzzy control rules. In a fuzzy logic controller, the dynamic behavior of a fuzzy system is characterized by a set of linguistic description rules based on expert knowledge usually of the form: IF (a set of conditions are satisfied) THEN (a set of consequence can be inferred). Figure 3.1 shows a basic configuration of a fuzzy logic system. The inputs, which are the state variables, characterize the key features of the system, and the output control variable influence the states of the system. The basic consideration is aimed at stabilizing the rotor angles of the machines and also damping the oscillation of the generator speed to zero [16]

The CPR for FAST is a nonlinear system with a large envelope of control space. One common task in controlling the CPR system is to demand the cabin to follow a given reference trajectory. System dynamics in tracking control may change significantly. Therefore, the fuzzy control is a convenient choice because it is possible to finely tune the control actions in a nonlinear fashion for different tracking error ranges [17]. Hence, in order to improve its control performance, in some nonlinear systems and in the case of strong disturbances, fuzzy and PID controllers are used in parallel [17, 18]. In view of the characteristics of nonlinearity for CPR, a proportional-integral controller needs to be added in parallel to the fuzzy controller for the tracking control of the CPR in order to improve the steady-state error performance.

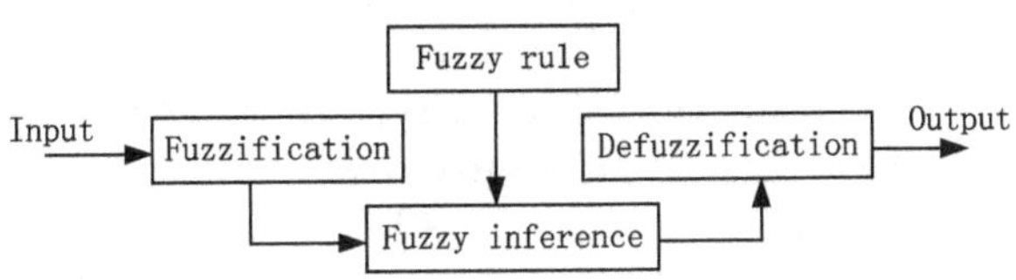

Fig. 3.1 Basic configuration of a fuzzy logic system

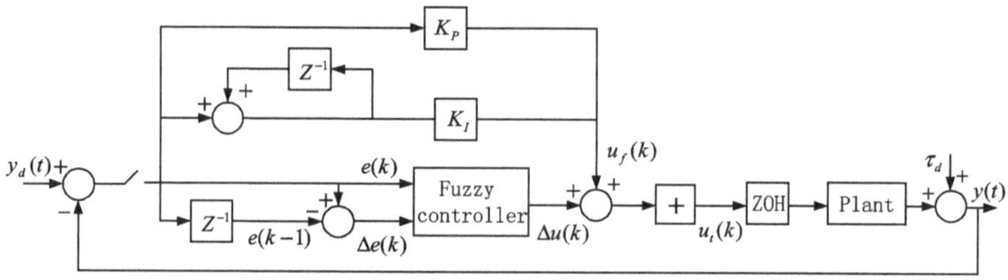

Fig. 3.2 Block diagram of the closed-loop fuzzy plus proportional-integral control system

The proposed controller consists of fuzzy logic control and proportional-integral control algorithms. A proportional-integral controller is added in parallel to enhance the control performance for steady state errors, a block diagram as shown in Fig. 3.2.

Here, K_P and K_I are called proportional gain and integral gain, respectively. τ_d is the external disturbance. The two inputs are the error $e(k)$ and the change of the error $\Delta e(k)$. The output is the change in the control action $\Delta u(k)$

$$e(k) = y_d(k) - y(k) \tag{3.1}$$

$$\Delta e(k) = e(k) - e(k-1) \tag{3.2}$$

where $y_d(k)$ is the desired output at time instant k, $y(k)$ is the actual output, $e(k)$ is the error and $\Delta e(k)$ is the change of the error.

The change of control action $\Delta u(k)$ is calculated based on the two inputs using a fuzzy inference mechanism. $u_f(k)$ is the proportional-integral control signal, and the total control signal is $u_t(k) = u_f(k) + \Delta u(k)$. The fuzzy controller is implemented in a discrete-time form using the zero-order-hold (ZOH), which holds its input signal for a specified sampling period to get a continuous signal out. The inputs to the controller system, $e(k)$, and $\Delta e(k)$, are actual values in form of "crisp" numbers. Fuzzification converts the numerical value into a linguistic variable which can be understood by the fuzzy control system. Singleton fuzzification, which transforms a crisp value into a fuzzy singleton value, is adopted in this fuzzy control scheme. Triangular type membership functions are defined over the range of input and output space, which linguistically describe the variable's universe of discourse, due to their ease in real-time hardware implementation [19]. All membership functions of the fuzzy logic control inputs, $e(k)$, $\Delta e(k)$, and the output, $\Delta u(k)$, are defined on the common normalized domain $[-1, 1]$ as shown in Fig. 3.3. Each of the fuzzy logic reasoning inputs, $e(k)$ and $\Delta e(k)$, and the output, $\Delta u(k)$, is assumed to take seven linguistic sets defined as negative big (NB), negative medium (NM), negative small (NS), zero (ZO), positive small (PS), positive medium (PM) and positive big (PB). Thus, a complete set of 49 fuzzy rules is given in Table 3.1 based on expert knowledge and testing. For example, if the output is falling far away from the desired trajectory, a large control signal that pulls the output toward the desired

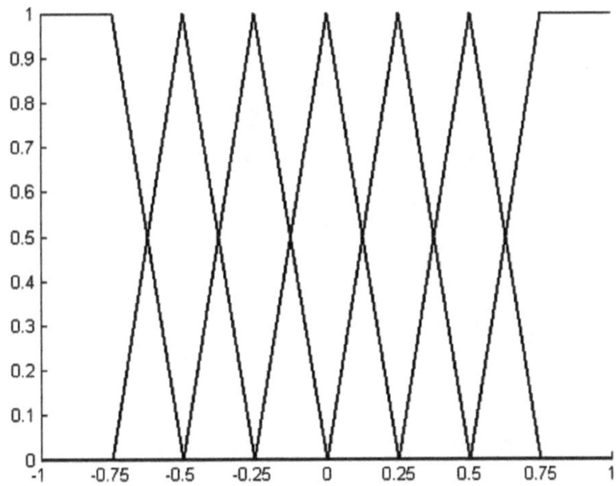

Fig. 3.3 Membership functions of e, Δe and Δu

Table 3.1 Rules for the fuzzy tracking system

Δu		Δe						
		NB	NM	NS	ZO	PS	PM	PB
	NB	NB	NB	NB	NB	NM	NS	ZO
	NM	NB	NB	NB	NM	NS	ZO	PS
	NS	NB	NM	NM	NS	ZO	PS	PM
e	ZO	NB	NM	NS	ZO	PS	PM	PB
	PS	NM	NS	ZO	PS	PM	PB	PB
	PM	NS	ZO	PS	PM	PB	PB	PB
	PB	ZO	PS	PM	PB	PB	PB	PB

trajectory is expected, whereas a small control signal is required when the output is near and approaching the desired trajectory. For both the input and output variables membership functions are defined to be symmetric, equi-spaced with an equal area so that each set can be described by its central value. The max-min (Mamdani type) inference is used to generate the best possible conclusion. This type of inference is computationally easy and effective; thus it is appropriate for real-time control applications. The crisp control command is calculated here using the center-of-gravity (COG) defuzzification [20]. The criterion provides the defuzzified output with better continuity.

In fuzzy control systems, the performance of the fuzzy controller, which is dependent on the decision of fuzzy control rules, influences that of the controlled system greatly. In order to adjust the control rules conveniently, a weighting factor is introduced to the fuzzy controller. It can be expressed as

$$\left.\begin{array}{l} \Delta U(k) = -\langle aE(k) + (1-a)\Delta E(k)\rangle \quad a = (0,\,1) \\ \Delta u(k) = b\Delta U(k) \end{array}\right\} \tag{3.3}$$

where $\langle\rangle$ is the operator, denoting a minimum integer which has the same sign as its content, but the absolute value is required. $E(k)$ is the fuzzy value of the error of input variable, $\Delta E(k)$ is the fuzzy value of the change in error, a is the weighting factor, b is the proportional factor, and $\Delta U(k)$ and $\Delta u(k)$ are the fuzzy and non-fuzzy outputs of the fuzzy controller.

Through adjusting the weighting factor the weighted degree to error and change in error can be changed. If the order of the controlled system is low, the weighted value to error should be larger than that to change in error. On the contrary, if the order of the controlled system is high, the weighted value to change in error should be larger than that to error [21].

3.3 Simulation Results and Analysis of Dynamic Model and Trajectory Tracking Controller

In this Section, the simulations and experiments are respectively deployed on the 5 and 50-m scaled model to test the dynamic model purposed in Chap. 2 and FPPIC controller purposed in Sect. 3.2.

3.3.1 *Case 1: 5-m Scaled Model of FAST*

Here, we provide the examples of simulation and experiment to justify the dynamic modeling for control and to test our proposed method. The model parameters of FAST 5-m used in the simulation and experiment are given in Table 3.2. Moment of inertia of CPR in cabin frame (in kg m^2 unit):

$$\mathbf{I}_{O1} = \begin{bmatrix} 0.782 & 0 & 0 \\ 0 & 0.782 & 0 \\ 0 & 0 & 0.293 \end{bmatrix}$$

A. Simulation example studies

Example 3.1 aims to justify the dynamic modeling for control, while example 3.2 demonstrates the effectiveness of the FPPIC active control method through simulations on the comparison between the conventional PID and the proposed FPPIC method.

Table 3.2 Parameters of CPR

m(kg)	a(m)	D(m)	h(m)	r(m)	r_1(m)	η(°)	q_0(N/m)	E(GPa)	A(m^2)	g(m/s^2)
20	2.5	5	2.5	0.21	0.015	8	2.57	28	2.788×10^{-5}	9.8

Example 3.1 The algorithm for the inverse dynamics of CPR without τ_d described in the foregoing statement has been implemented in a program written in MATLAB and C, separately. The program has been developed to plan straight-line and circular paths. We use the simulation programs for solving the inverse dynamics. That is, the given data values representing the task space are passed through the mathematical equation representing the inverse dynamics problem solution [i.e., f: $(x, y, z, \alpha, \beta, \gamma) \to (l_1, l_2, l_3, l_4, l_5, l_6)$] to obtain the corresponding data values for the cable space. By inverting these data values [i.e., f: $(l_1, l_2, l_3, l_4, l_5, l_6) \to (x, y, z, \alpha, \beta, \gamma)$], the data values corresponding the cable space become the input values, and task-space values become the output to the forward dynamics model. The trajectory planning scheme described above has been selected just for simplicity. In fact, any scheme for planning the trajectory in the Cartesian space can be used with equal ease based on the established dynamic formulation. Considering the length of the chapter, the simulation result of a circle case is presented only. In the simulation of tracking a moving target with constant velocity $\mathrm{v} = 0.01\ \mathrm{m/s}$ and constant pose (α, β and γ are all assumed to be 0°), the initial point is $(2.45,\ 2.5 \times \cos(\pi/6), 1.3)\ (\mathrm{m})$. The geometric path is the circle with the radius of 0.6 m parallel to the x-y plane.

$$\begin{cases} x = 1.85 + 0.6\sin(0.0167t) \\ y = 2.5\cos(\pi/6) + 0.6\cos(0.0167t) \\ z = 1.3 \end{cases} \tag{3.4}$$

The trajectory tracking of the center of the cabin is shown in Fig. 3.4. From Fig. 3.4, the formulation follows the planned trajectory relatively well. Figure 3.5 illustrates the changes in length of the cables. L_i, for $i = 1, 2, \ldots, 6$, denotes the length of cable between A_i and B_i. Owing to the coordinates of the joint A_1, A_2 and A_3 is just symmetric to that of the point A_4, A_5 and A_6, as shown in Fig. 2.19, the length of cable vary symmetricly. Obviously, the plots shown in Fig. 3.5 indicate the changes in length of the cables are valid. From Fig. 3.6, it is known that the

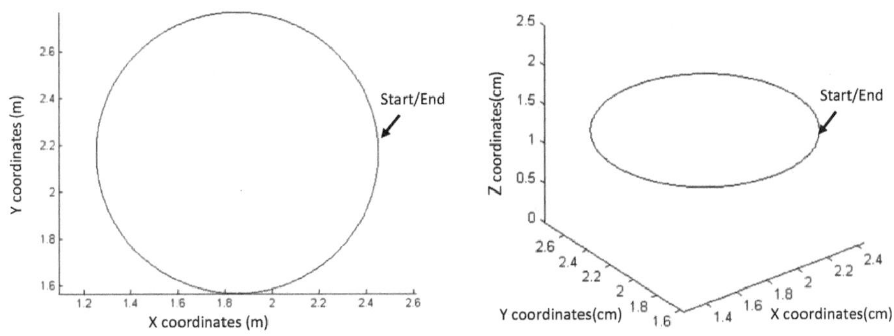

Fig. 3.4 Following trajectory of the circle motion

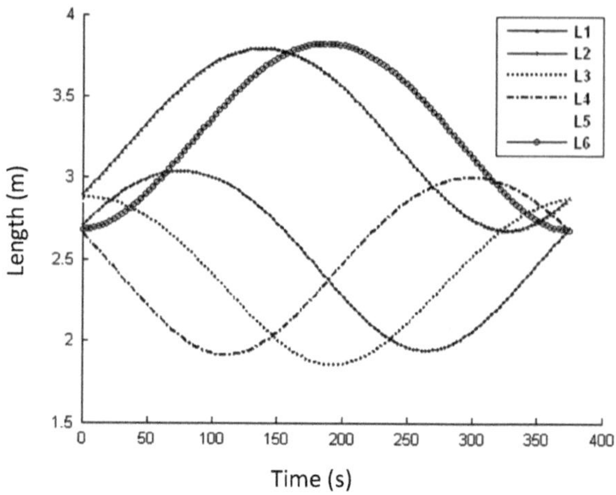

Fig. 3.5 Following trajectory of the circle motion

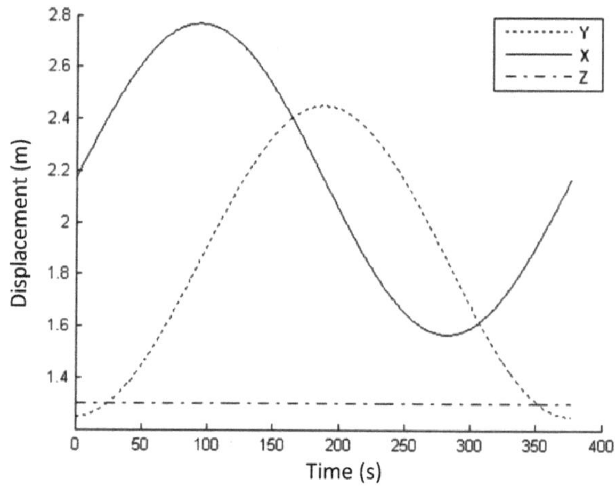

Fig. 3.6 Tracking displacement of the center of cabin

tracking displacements of the center of cabin in X, Y and Z directions accord with the desired motion trajectory. Figure 3.7 represents the cabin upon which the forces are exerted in all directions. The forces actuating on the cabin in the direction of X and Y are equal to the centripetal forces in the direction of X and Y as the cabin moves along the circle, respectively. On the other hand, the force actuating on the cabin in the direction of Z equals the gravity of the cabin.

From the above simulation results, it may be concluded that we justify the dynamic modeling for control.

Example 3.2 Because the CPR is suspended in the space, the wind disturbance has significant effect on the tracking precision. Therefore, the simulation is carried out based on the trajectory tracking with the random wind τ_d. According to the historical maximum wind speed and the simulation of the maximum wind-induced

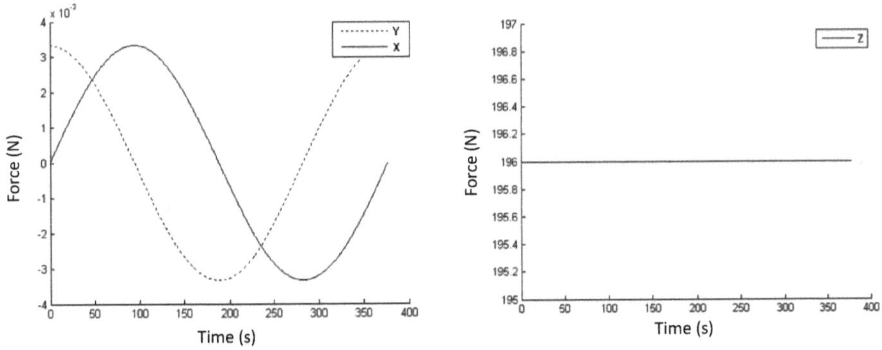

Fig. 3.7 The forces actuating on the cabin in X, Y and Z directions

displacement of the CPR, random wind τ_d, which is a disturbance at the velocity of 17 m/s, is modeled in a program written in MATLAB.

Along-wind velocity is considered as an ergodic stationary random process and samples of the wind velocity related to time are generated by the following equation [22]

$$V_j(x,y,z,t) = \overline{V}_j(z) + \sum_{m=1}^{j}\sum_{l=1}^{N} \left|H_{jm}(\omega_l)\right|\sqrt{2\Delta\omega}\cos\left[\omega_l \cdot t + \psi_{jm}(\omega_l) + \theta_{ml}\right], \quad (j = 1,2,\ldots n) \tag{3.5}$$

In which $\overline{V}_j(z)$ is the mean wind velocity at jth height; n and N denote the numbers of divided segments uniformly along the structural height and in frequency band of the wind velocity spectrum, respectively; $\Delta\omega$ is the frequency increment; $\psi_{jm}(\omega_l)$ refers to phase angle related to two points at different heights; θ_{ml} is random variable ranging from 0 to 2π. In this section, the wind-induced vibration caused by the along-wind velocity is considered especially. Further, the wind pressure can be simulated as follows,

$$w(x,y,z,t) = \frac{1}{2}\rho V^2(x,y,z,t) = \frac{1}{2}\rho\left[\overline{V}(z) + v(x,y,z,t)\right]^2 \tag{3.6}$$

where $v(x,y,z,t)$ is the fluctuating velocity and $\overline{V}(z)$ is the mean velocity at height z satisfying the exponential law,

$$\overline{V}(z) = \overline{V}_{10}(z/10)^{\alpha} \tag{3.7}$$

where $\overline{V}_{10}$ indicates the mean velocity at 10 meters high; α is coefficient related to the ground roughness.

As a comparison, a conventional PID controller is also integrated to the system in the form of

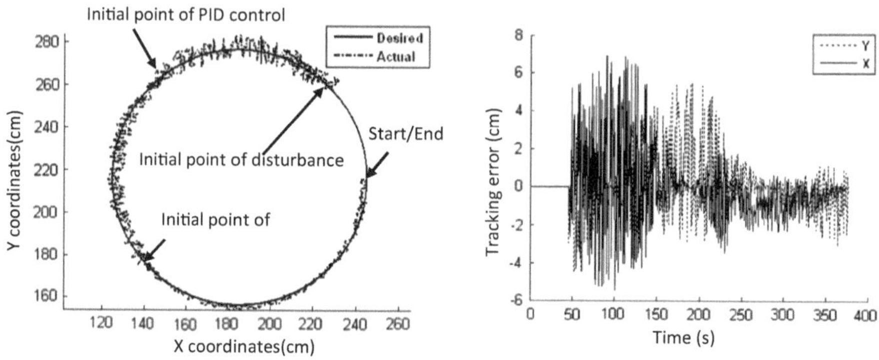

Fig. 3.8 The tracking trajectory and tracking errors in the direction of X and Y

$$u(k) = k_p(e(k) + k_i \sum_{j=0}^{k} e(j)T + k_d \frac{e(k) - e(k-1)}{T} \tag{3.8}$$

where k_p, k_i and k_d represent the proportional, integral and derivative gains, respectively, and T is the sampling period.

Because of the same selected servomotor systems, the parameters for all the six control subsystem are determined as identical primarily, and then make the necessary small tuning according to the respective simulation results. After several simulations, the sampling period is determined as $T = 0.02$ s; the parameters for the PID and the FPPIC can be determined as, $k_p = 21.5$, $k_i = 1.4$ and $k_d = 0.46$; $a = 0.34$, $b = 1.7$, $K_P = 12.5$, $K_I = 0.23$.

To see the performance of the proposed fuzzy plus proportional-integral controller, Fig. 3.8 shows the actual motions in the x-y plane tracked by the conventional PID and FPPIC methods, and the tracking errors in the x and y direction. The arrow in the Fig. 3.8 indicates switching state of controllers and disturbance. Figure 3.8 shows that the active control strategy can effectively attenuate the wind-induced vibration, and improve the tracking accuracy of the feed cabin. We observe that the maximum and average errors of the FPPIC method are smaller than those of the conventional PID controller. From the above simulation results, it may be concluded that the FPPIC strategy can achieve a favourable control performance and has high robustness.

B. Experimental results

The proposed algorithm is verified by experiments using the CPR control system in Fig. 3.9. Two different control schemes are implemented: conventional PID control and FPPIC. Figure 3.10 shows the hardware configuration of the experimental system. The extendable actuator is comprised of an AC servomotor and drive unit, worm-gear speed reducer (reduction ratio: 1:60), and bobbin. The servomotor and driver used is YASKAWA $\sum$-L series (low inertial moment)

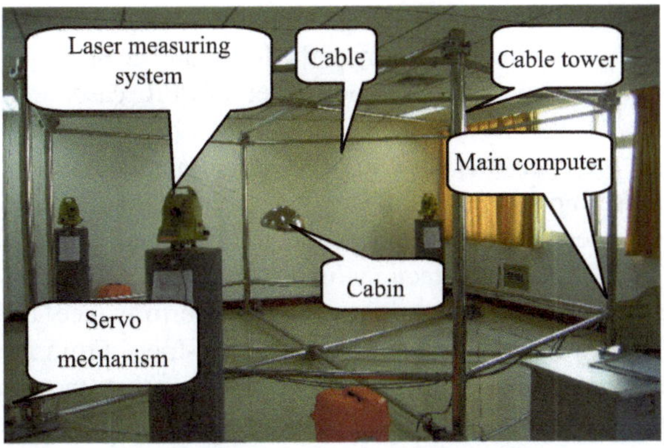

Fig. 3.9 The model of feed support for 5-m FAST

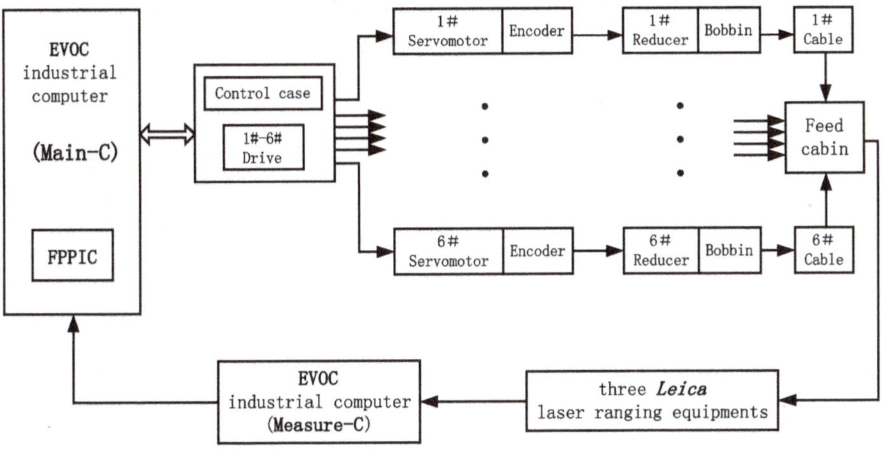

Fig. 3.10 The hardware configuration of experimental system

SGML-04AF14 AC servomotor with an incremental encoder (1024 P/R) and the matching SGDL-04AP driver. The motor parameters are: rated power 0.4 kW, rated voltage 200 V, rated torque 1.27 Nm, rated speed 3000 r/min, motor moment of inertia 1.91×10^{-5} kg m^2. The servo drive was set in the position mode.

The FPPIC algorithm and the PID method were implemented in C on an EVOC industrial computer with a Pentium 4–2.4 GH$_{Z,}$ i.e., M-C. The position and orientation of the center of the feed cabin is measured using three Leica TCA1800 laser ranging equipments (made in Switzerland) under the control of a Measure-C.

A series of experiments with FAST 5-m was performed with the same reference trajectory as proposed in Sect. 4.1. During experiment, the cabin is asked to move along the given trajectory. The artificial wind forces, through a fan with 5 kw, on the structure with a mean velocity of 17 m/s is produced. For real-time, the parameters of the controller, obtained by simulation, are manually adjusted on the FAST 5-m experimental setup to minimize the following error. Experimental results are presented for three cases: an open-loop case; a FPPIC case; and a conventional PID control case.

We begin with the open-loop case to justify the dynamic modeling. In this case, the number of pulse inputs is calculated using inverse dynamics. Figures 3.11 and 3.12 show the desired trajectory and actual trajectory of the circle motion and tracking errors in X, Y and Z directions under open-loop.

Figures 3.13 and 3.14 illustrate the tracking performance and errors using conventional PID and FPPIC methods under closed-loop. From the results of the proposed method as shown in Fig. 3.14, it is clear that the tracking error is improved and is smaller than the PID control. In addition, it is also shown that the computational and experimental performance values are in good agreement. The simulation and experimental results together highlight the effectiveness of the proposed control framework. However, the trajectory following errors in the experiments are poor in contrast to in the simulation. During the experimental tests, there are mostly due to problems in practical implementation such as nonlinear

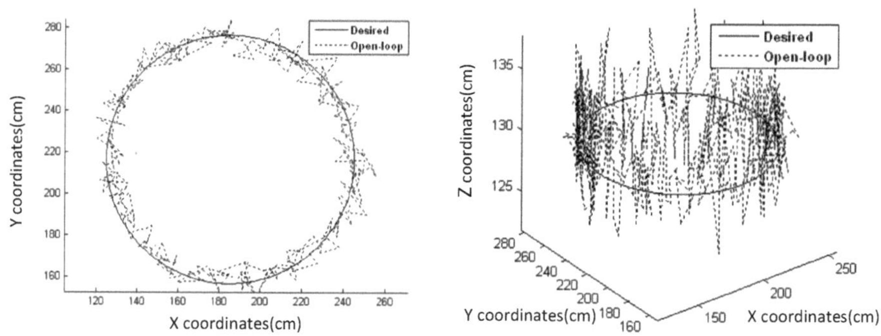

Fig. 3.11 Desired trajectory and actual trajectory of the circle motion under open-loop

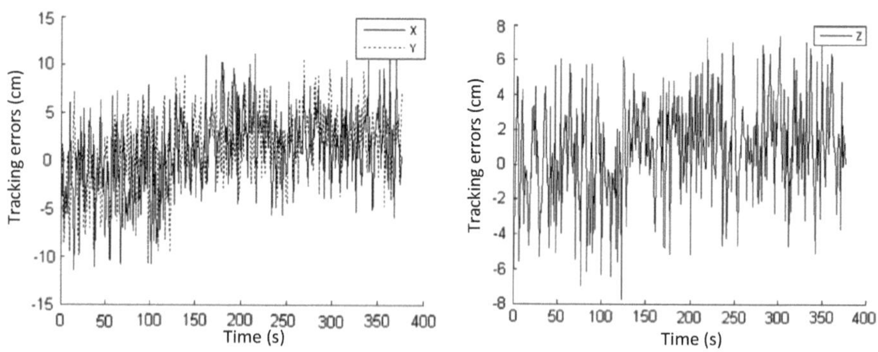

Fig. 3.12 Tracking errors in *X*, *Y* and *Z* directions under open-loop

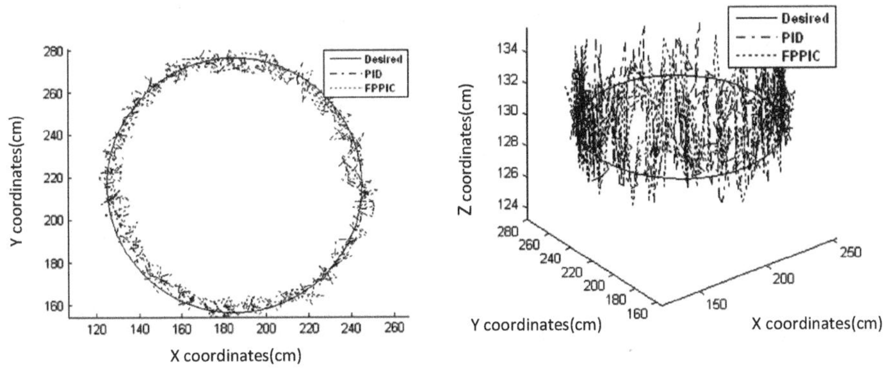

Fig. 3.13 Desired trajectory and actual trajectory of the circle motion under open-loop

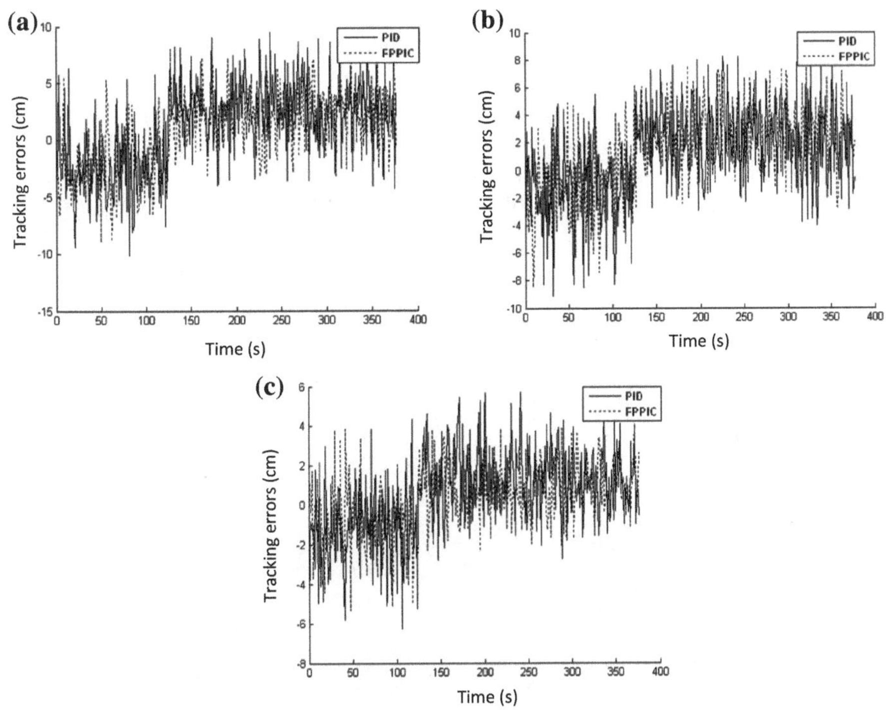

Fig. 3.14 Tracking errors in *X*, *Y* and *Z* directions under open-loop

effects (backlash and friction etc.), (dynamic) uncertainties, cross coupling and other imperfections of such a nonlinear system. On the other hand, the other major assumption we have made in our model is the gearing, the shafts, and the bearings are not flexible. In reality, all of these elements have finite stiffness. However, for purposes of control-system analysis and design, we neglect these effects and use a simpler dynamic model in the simulation.

3.3.2 Case 2: 50-m Scaled Model of FAST

As for the outer LT scaled model of 50-m was built in Xidian University (see Fig. 3.15). According to the notation introduced before, the model parameters of LT 50-m used in the simulation studies are given in Table 3.3. Moment of inertia of CPR in cabin frame (in kg m^2 unit):

$$\mathbf{I}_{O1} = \begin{bmatrix} 288 & 0 & 0 \\ 0 & 288 & 0 \\ 0 & 0 & 115 \end{bmatrix}$$

The algorithm for the inverse dynamics of CPR without $\boldsymbol{\tau}_d$ described in the foregoing statement has been implemented in a program written in MATLAB. The program has been developed to plan straight-line and circular paths. We use the simulation programs for solving the inverse dynamics. That is, the given data values representing the task space are passed through the mathematical equation representing the inverse dynamics problem solution to obtain the corresponding data

Fig. 3.15 The model of feed cable-supporting for LT 50-m

Table 3.3 Model parameters of LT 50-m

m(kg)	a(m)	D(m)	h(m)	r(m)	r_1(m)	η(°)	g(m/s^2)
720	25	50	22	1.33	0.08	50	9.8

values for the cable space. By inverting these data values, the data values corresponding the cable space become the input values, and task-space values become the output to the forward dynamics model. The trajectory planning scheme described above has been selected just for simplicity. In fact, any scheme for planning the trajectory in the Cartesian space can be used with equal ease based on the established dynamic formulation. Considering the length of chapter, the following result of simulation is given only a circle motion case. In the simulation of tracking a moving target with constant velocity v = 0.01 m/s and constant pose (α, β and γ are all assumed to be 0°), the initial point is $(8, 25\cos(\pi/6), 14)$ (m). The geometric path is the circle with the radius of 4.5 m parallel to the x-y plane.

$$\begin{cases} x = 12.5 + 4.5\sin(0.0022t) \\ y = 25\cos(\pi/6) + 4.5\cos(0.0022t) \\ z = 14 \end{cases} \tag{3.9}$$

The trajectory tracking of the center of the cabin is shown in Fig. 3.16. From Fig. 3.16, the formulation tracks the planned trajectory relatively well. Figure 3.17 shows the changes in length of the cables. L_i, for $i = 1, 2, \ldots, 6$, denotes the length of cable between A_i and B_i. Owing to the coordinates of the joint A_1, A_2 and A_3 is just symmetric to that of the point A_4, A_5 and A_6, as shown in Fig. 2.19, the length of cable vary symmetricly. Obviously, the plots shown in Fig. 3.17 indicate the changes in length of the cables are valid. From Fig. 3.18, it is known that the tracking displacements of the center of cabin in X, Y and Z directions accord with the desired motion trajectory.

Figure 3.19 represents the cabin upon which the forces are exerted in all directions. The forces actuating on the cabin in the direction of X and Y equal the centripetal forces in the direction of X and Y as the cabin moves along the circle, respectively. On the other hand, the force actuating on the cabin in the direction of Z equals the gravity of the cabin.

From the above simulation results, it may be concluded that we justify the dynamic modeling for control.

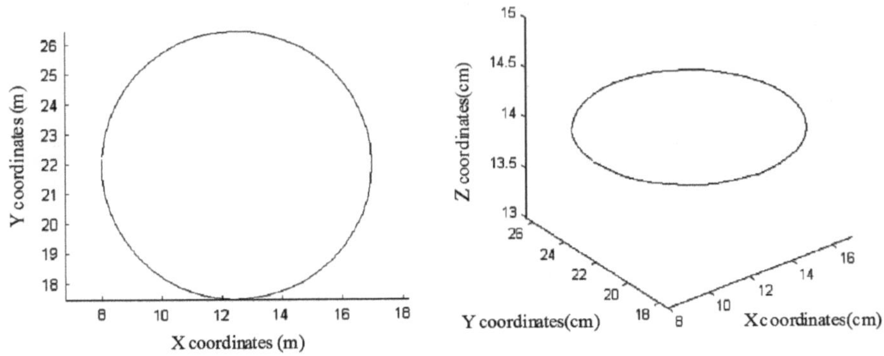

Fig. 3.16 Following trajectory of the circle motion

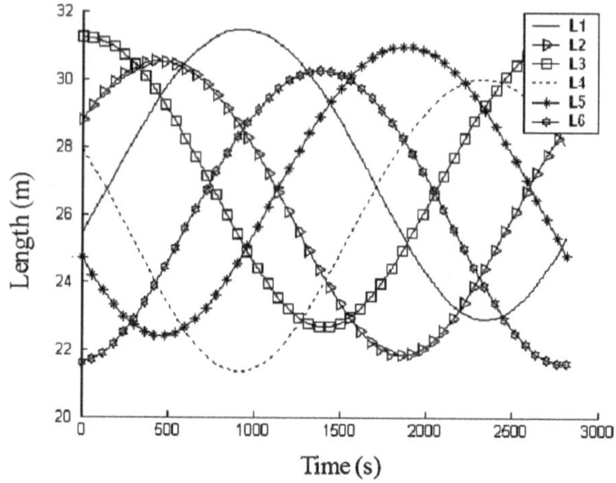

Fig. 3.17 Change in length of cable

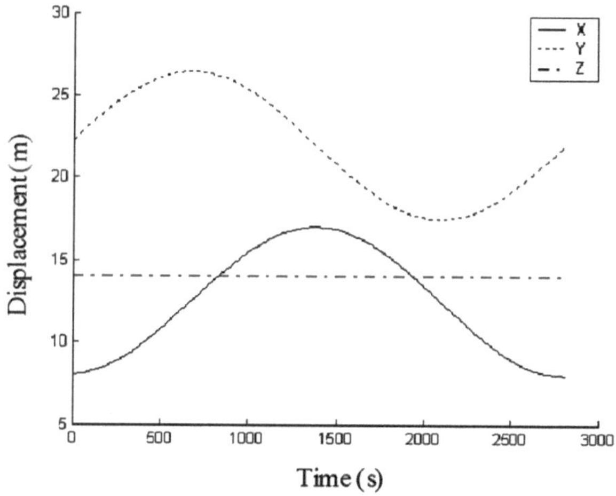

Fig. 3.18 Tracking displacement of the center of cabin in X, Y and Z directions

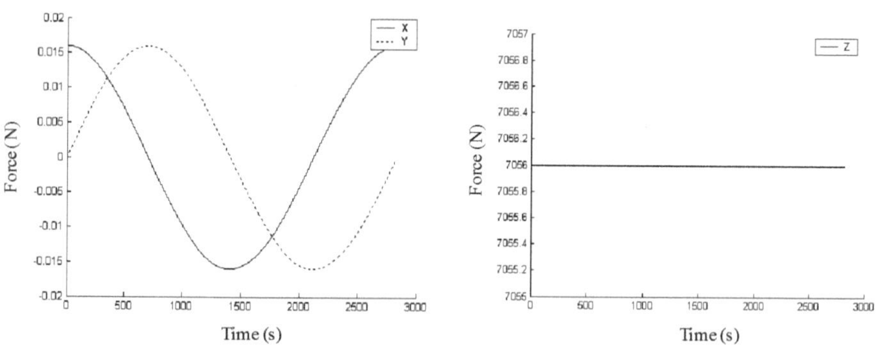

Fig. 3.19 The forces actuating on the cabin in X, Y and Z directions

A. Simulation of wind-induced vibration control of the cable-cabin structure in stationary position

At any point within the workspace of the cabin, all six parameters x, y, z, α, β, γ of the cabin are known. The six cables are controlled with six computer-controlled servomotors.

Each cable is divided into a group of cable bars. The frame structure of the cabin consists of 146 aluminum bar elements and 128 steel beam elements, respectively. If all six servomotors are braked, the cable-cabin mechanism will become a structure. When the origin O_1 of local coordinate system of the cabin is at point (330.5, 928.7, 1610.5) (cm) and the orientation angles $\alpha = -80°37'$, $\beta = 32°59'$ and $\gamma = 25°24'$, the cabin reaches a boundary point of its workspace. At such position and orientation, the structure is unsymmetrical badly, so the structural stiffness will be quite weakness. Choosing the point as a typical point, the wind-induced vibration of the structure is simulated and compared under no control and active control.

Dynamic differential equation of the cable-cabin structure can be mathematically stated as [22],

$$\mathbf{M}\ddot{\delta} + \mathbf{C}\dot{\delta} + \mathbf{K}\delta = \mathbf{F} \tag{3.10}$$

Where δ and F denote the displacement of the structural nodes and the load acting on the structure; M, C and K are the mass matrix, damp matrix and stiffness matrix, respectively. Especially $K = K_0 + K_\sigma + K_L$. K_0 is the linear stiffness matrix; K_σ is the initial stress matrix or geometry stiffness matrix and K_L is the initial displacement matrix. $C = 2M\left[\sum_{i=1}^{N} \omega_i \xi_i \delta_{0i} \delta_{0i}^T\right] M$. ω_i, ξ_i and δ_{0i} are the natural frequency, damp ratio and a normal modal vector, respectively. Since the structure contains cables, many low natural frequencies exist. Frequency-domain analysis is suit for linear structure. To make a nonlinear analysis precisely, time-domain analysis has to be made. Along-wind velocity is considered as an ergodic stationary random process and samples of the wind velocity related to time are generated by the Eq. 3.5.

Exerting wind forces on the structure with a mean velocity of 17 m/s which is the largest one in the site. For a sample of the wind velocity persisting 60 s, the nonlinear dynamic Eq. (3.10) is integrated directly by Newmark-β method.

Among all the cabin nodes, the suspension point B_3 (see Fig. 2.20) has the largest vibration displacement. Figure 3.20a–c shows the time history of displacement of the point B_3 under no control, ADRC and FPPIC, respectively. From Fig. 3.20a–c, the structure with ADRC and FPPIC decreases the maximum displacement of the point B_3 from 24 to 13 cm, and from 24 to 9 cm, respectively. Because the wind velocity blows in the direction of axis X, the displacement along axis X is the largest one.

Active vibration control simulation results, showing a 40–60% reduction of vibration amplitudes, validate the effectiveness of the control strategy.

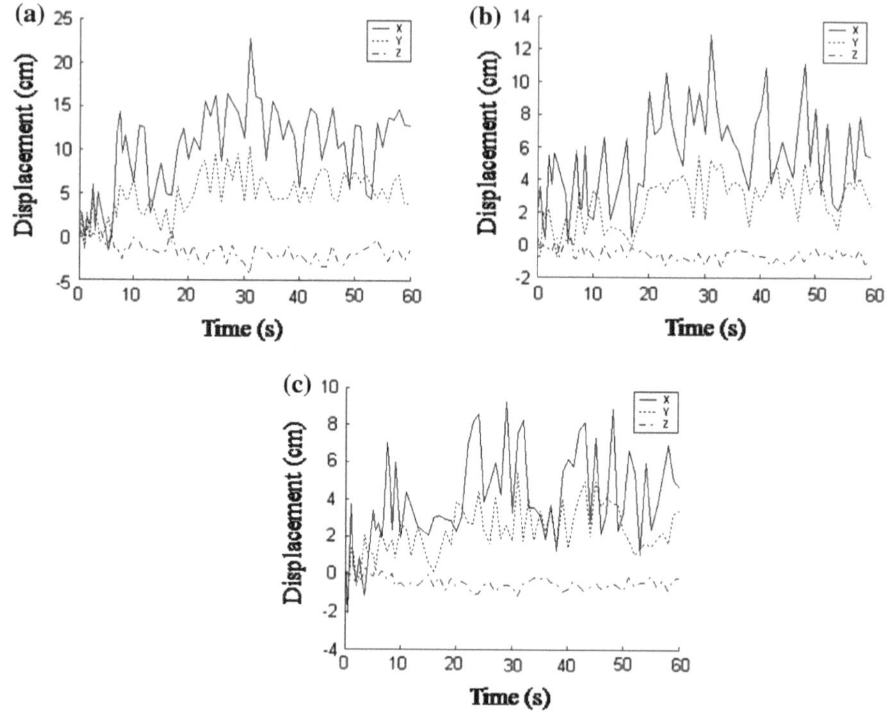

Fig. 3.20 Time history of displacement of the point B_3 in X, Y and Z directions **a** Without control. **b** With ADRC control. **c** With FPPIC control

B. Simulation in trajectory-tracking control of the CPR

In this subsection, several simulations in trajectory-tracking motion have been performed to assess the behavior of the CPR driven by six AC servomotors with the FPPIC and the ADRC. In the simulations, we applied the trajectory mentioned in Sect. 3.2. The control configuration of LT 50-m scaled mode is shown in Fig. 3.21. The position and orientation of the center of the feed cabin is measured using three Leica TCA1800 laser ranging equipments (made in Switzerland) under the control of a Measure-C.

For comparison, a discrete-time ADRC is also used for the CPR motion control system. The ADRC consists of a tracking differentiator (TD) in the feedforward path, an extended state observer (ESO), and a nonlinear proportional derivative control in the feedback path. The TD solves the difficulties posed by low-order reference trajectories which are quantized at the sensor resolution, and the ESO provides the estimate of the unmeasured system's state and the real action of the unknown disturbances only based on a measurement output of the system. The schematic diagram of the ADRC control is shown in Fig. 3.22. For more details on the physical construction and specification of the ADRC, refer to [23, 24].

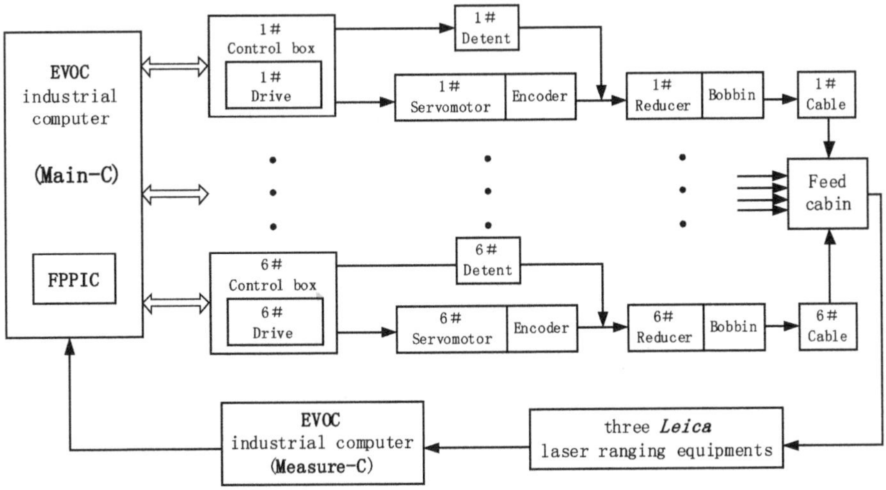

Fig. 3.21 The control configuration of LT 50-m scaled model

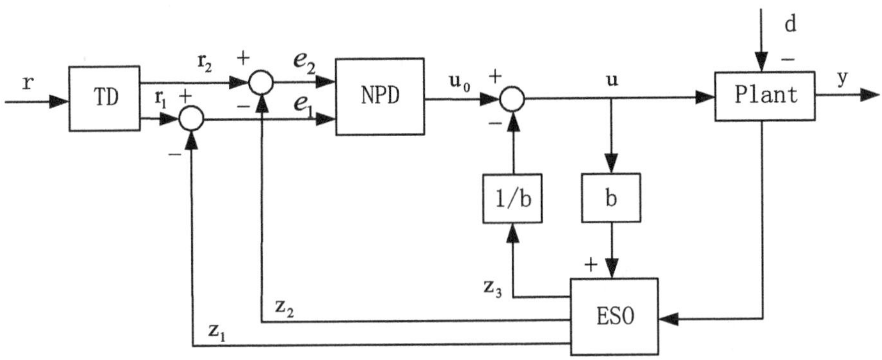

Fig. 3.22 Schematic diagram of ADRC

In our simulations (without regard to the friction term), T is chosen to be 2 ms. Because of the same selected servomotor systems, the parameters for all the six control subsystem are determined as identical primarily, and then make the necessary small tuning according to the respective simulation results. After several simulations, the parameters of the FPPIC can be determined as $a = 0.34$, $b = 2.6$, $K_P = 8.5$, $K_I = 0.23$. Combining the practical operating condition and the results presented in [23, 24], the parameters of the ADRC were selected as follows: $\delta = 329$, $\alpha_1 = 0.48$, $\alpha_2 = 0.23$, $\delta_1 = 2.6 \times 10^{-4}$, $\beta_1 = 79$, $\beta_2 = 85$, $\beta_3 = 185$, $k_p = 39$, $k_d = 1.2$, $\alpha_{01} = 0.69$, $\alpha_{02} = 1.61$, $\delta_0 = 0.03$, $b_0 = 0.15$.

For comparison, Fig. 3.23 shows the desired trajectory (dashdotted line) and actual motions under ADRC (real line) and FPPIC (dashed line). The corresponding tracking errors in X, Y and Z directions are shown in Fig. 3.24a–c (only one eighth

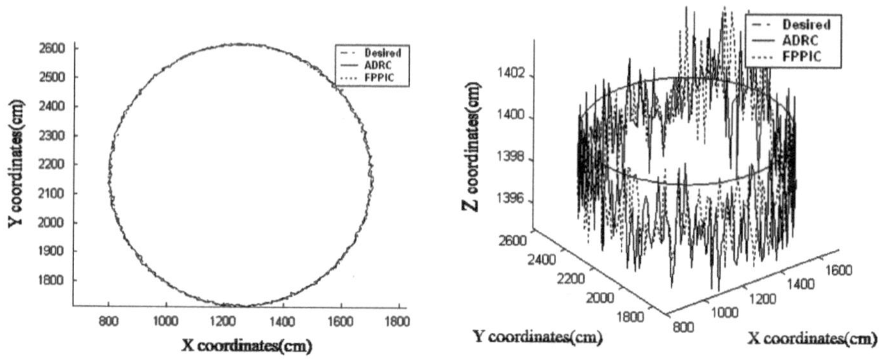

Fig. 3.23 Desired trajectory and actual trajectory of the circle motion under closed-loop

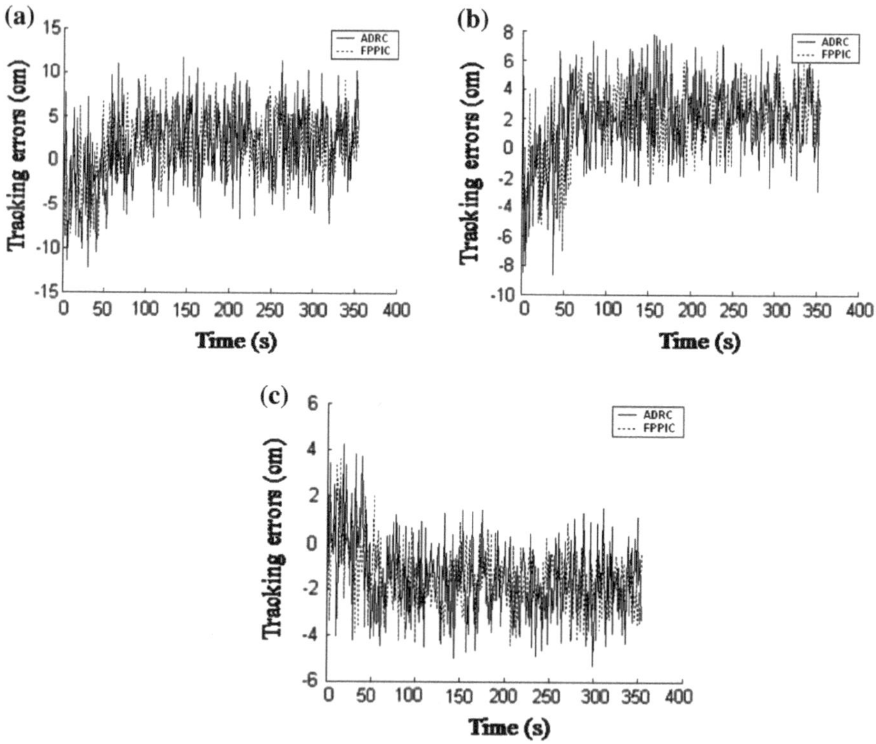

Fig. 3.24 Tracking errors in X(a), Y(b) and Z(c) directions under closed-loop

path is given). From Fig. 3.18a, the tracking errors in X direction with ADRC and FPPIC hold in the range of ± 15 and ± 10 cm, respectively. Similar to the Fig. 3.19a, in Fig. 3.19b, c, the tracking errors in Y and Z direction with ADRC and FPPIC hold in the range of ± 8 and ± 5 cm, respectively. Refer to Figs. 3.19a–c, the simulation results have less error for the FPPIC compared to the ADRC. Moreover, we can see that the robustness of the proposed method is nearly the same as that of the ADRC method. From the above simulation results, it may be concluded that the FPPIC strategy can achieve a favourable control performance and has high robustness.

3.4 Conclusions

In this chapter, the Simulations and experiments on the CPR of FAST model of 5 and 50-m is carry out to justify the dynamic model of super antenna that established in Chap. 2. In view of random wind forces acted on the cabin are simulated based on the characters of the structure, a fuzzy plus proportional-integral controller is designed to control the wind-induced vibration of the trajectory tracking of the feed. Then, the examples of simulation and experiment are employed to justify the dynamic modeling for control and to test the proposed controller. Specifically, the simulation and experiments of dynamic model with non-negligible cable mass and FPPIC method is carry out on the 5-m scaled model, and the simulation and experiments of dynamic model including servomechanism dynamics and FPPIC method for trajectory tracking control is carry out on the 5-m scaled model. Moreover, for comparison, a discrete-time automatic disturbances rejection controller (ADRC) is also used for the CPR motion control system. Two algorithms are compared with respect to performance of the closed-loop system. The results illustrate that the satisfactory performance of the proposed control scheme as compared with a discrete-time automatic disturbances rejection controller even in the presence of internal model uncertainties in both the cabin and servomechanism dynamics and external disturbances.

Some intrinsic limitations of the system are also outlined. In particular, a dynamic characteristic of the cable is negligible. Our aim was to perform dynamic analysis of the cabin including servomechanism dynamics without regard to cable mass and develop a fuzzy plus proportional-integral controller design methodology and hence, modeling errors were assumed to be tolerated. The consideration of system modeling errors (cable dynamic modeling) is a topic for further research and is currently under investigation. In addition, more work is in progress to implement real-time experiments.

References

1. Yamamoto M, Yanai N, Mohri A (1999) Inverse dynamics and control of crane-type manipulator. Intelligent robots and systems, 1999, IROS'99, Proceedings, 1999 IEEE/RSJ international conference on, IEEE, vol 2, pp 1228–1233
2. Yanai N, Yamamoto M, Mohri A (2002) Feedback control for wire-suspended mechanism with exact linearization. Intelligent robots and systems, 2002, IEEE/RSJ international conference on, IEEE, vol 3, pp 2213–2218

3. Zi B, Duan BY, Du JL et al (2008) Dynamic modeling and active control of a cable-suspended parallel robot. Mechatronics 18(1):1–12
4. Heyden T, Woernle C (2006) Dynamics and flatness-based control of a kinematically undetermined cable suspension manipulator. Multibody SysDyn 16(2):155–177
5. Sanchez E, Flores V (2006) Real-time underactuated robot swing-up via fuzzy PI + PD control. J Intel Fuzzy Syst 17(1):1–13
6. Zadeh LA (1965) Fuzzy sets. Inf Control 8(3):338–353
7. Mamdani EH (1974) Application of fuzzy algorithms for control of simple dynamic plant. Electr Eng Proc Inst 121(12):1585–1588
8. Lee CC (1990) Fuzzy logic in control systems: fuzzy logic controller. II. IEEE Trans Syst Man Cybernet 20(2):419–435
9. Cho HJ, Cho KB, Wang BH (1997) Fuzzy-PID hybrid control: automatic rule generation using genetic algorithms. Fuzzy Sets Syst 92(3):305–316
10. Li W, Chang XG, Wahl FM et al (1999) Technical note hybrid fuzzy P + ID control of manipulators under uncertainty. Mechatronics 9(3):301–315
11. Lin CL, Hung HZ, Chen YY et al (2004) Development of an integrated fuzzy-logic-based missile guidance law against high speed target. IEEE Trans Fuzzy Syst 12(2):157–169
12. Siripongwutikorn P, Banerjee S, Tipper D (2005) Fuzzy-based adaptive bandwidth control for loss guarantees. IEEE Trans Neural Netw 16(5):1147–1162
13. Hwang JP, Kim E (2006) Robust tracking control of an electrically driven robot: adaptive fuzzy logic approach. IEEE Trans Fuzzy Syst 14(2):232–247
14. Kim DH, Kang JY, Lee KI (2000) Robust tracking control design for a 6 DOF parallel manipulator. J Robotic Syst 17(10):527–547
15. Chang WJ, Sun CC, Chung HY (2004) Fuzzy controller design for discrete controllability canonical Takagi-Sugeno fuzzy systems. IEE Proc Control Theory Appl 151(3):319–328
16. Rubaai A, Ofoli A (2003) Multi-layer fuzzy controller for control of power networks. Industry applications conference, 2003, 38th IAS annual meeting, Conference record of the, IEEE, vol 1. pp 277–284
17. Onat M, Dogruel M (2004) Fuzzy plus integral control of the effluent turbidity in direct filtration. IEEE Trans Control Syst Technol 12(1):65–74
18. Reznik L, Ghanayem O, Bourmistrov A (2000) PID plus fuzzy controller structures as a design base for industrial applications. Eng Appl Artif Intell 13(4):419–430
19. Peyravi H, Khoei A, Hadidi K (2002) Design of an analog CMOS fuzzy logic controller chip. Fuzzy Sets Syst 132(2):245–260
20. Yang SX, Li H, Meng MQH et al (2004) An embedded fuzzy controller for a behavior-based mobile robot with guaranteed performance. IEEE Trans Fuzzy Syst 12(4):436–446
21. Tian L, Collins C (2004) A dynamic recurrent neural network-based controller for a rigid–flexible manipulator system. Mechatronics 14(5):471–490
22. Zi B, Zhu Z, Du J (2011) Analysis and control of the cable-supporting system including actuator dynamics. Control Eng Pract 19(5):491–501
23. Su YX, Duan BY, Zheng CH et al (2004) Disturbance-rejection high-precision motion control of a Stewart platform. IEEE Trans Control Syst Technol 12(3):364–374
24. Su YX, Zheng CH, Duan BY (2005) Automatic disturbances rejection controller for precise motion control of permanent-magnet synchronous motors. IEEE Trans Industr Electron 52(3):814–823

Chapter 4
Integrated Mechanism Design and Control of the Hybrid-Driven Based Cable-Suspended Parallel Robots

4.1 Preamble

In recent years, the CPRs are finding increased use in a wide variety of modern engineering applications. Therefore, they are required not only for operations with high accuracy, high payload and low cost, but also for output with greater flexibility, which can change the law of output motion quickly and conveniently [1]. Nevertheless, the existing research on the drive system of the CPRs usually uses the low power controllable motor. The low power controllable motor can not directly drive the high-loading CPRs due to the restrictions of power and torque. Hence, it is necessary to carry out for a new-type drive system for the CPRs. Planar two degrees of freedom five-bar mechanisms have been widely studied since it is simple and the actuators can be fixed at the base, which reduces the inertia of the mobile body. The existing hybrid-driven planar 2-DOF mechanisms (HDPM) are the well-known five-bar mechanism whose drive system consists of a constant velocity (CV) motor and a servomotor, which is also a closed-loop linkage manipulator. The CV motor provides main power and motion required; however, it lacks flexibility. On the other hand, the servomotor acts as a motion regulation device which suffers from the limited power capacity and high cost. The HDPM can, therefore, take the advantage of the complementary characteristics of both motors to generate a programmable range of highly nonlinear output motions with high power capacities at low costs. In this investigation, the HDPM is introduced into a completely restrained CPR with 3 DOFs in place of the servomotor as a power source of the CPR system on the basis of theories of mechanism structure synthesis.

The hybrid-driven based cable-suspended parallel robots (HDCPR) is a nonlinear multivariable dynamic system and suffers from structured and unstructured uncertainties, such as payload variation, external disturbances. In addition, because of the changes in cable length and joint force, the CV motor of the HDPM will bring in the velocity fluctuation, which can not be attenuated by the CV motor itself, due to the lack of a control mechanism in the CV motor, yet be propagated to

B. Zi and S. Qian, *Design, Analysis and Control of Cable-Suspended Parallel Robots and Its Applications*, DOI 10.1007/978-981-10-1753-7_4

the servomotor, and further to the end-effector of the HDCPR. Therefore, it is of interest to investigate the control of the HDCPR, and the control problem of the multivariable systems becomes more challenging [2, 3]. Clearly, it is difficult to achieve a good enough tracking accuracy using standard PID controllers [4]. The adaptive control can cope with parameter uncertainties include the link length, mass, inertia, etc. [5]. However, it requires extensive computation, and will not guarantee the estimated parameters of the manipulators converge to their true value, and the tracking errors would repeatedly be brought into the system when the manipulators repeat their task [6].

In order to achieve a high-precision performance, the HDCPR control system must effectively and accurately manipulate the motion trajectory of the end-effector. Since the HDCPR is generally used in repetitive task, one should take advantage of the fact that the reference trajectory is repeated over a given operation time. Iterative learning control (ILC) is a well-established technique that presents itself as the most suitable method to improve on repetitive tasks without excessive requirements on sensor-feedback quality or control-loop bandwidth [7]. Specifically, ILC is a simple and effective technique of tracking control aiming at improving system tracking performance from trial to trial in a repetitive mode [8]. The basic principle of ILC is to use information collected from previous executions of the same task repetitively to form the control action for the current operation so as to improve tracking performance from iteration to iteration [9]. For this kind of control scheme, it has a lot of applications such as robot arm manipulators, assembly line tasks, chemical batch process, reliability testing rigs, continuous casting process, and so on [10, 11].

Optimization has been playing an utmost important role in the motion control and mechanism design to effectively carry out the design for the mechatronic systems [12, 13]. Optimization design of multi-domain engineering systems can be rather complex and it requires an integrated and concurrent approach in order to obtain the optimal results [14]. It is not difficult to understand that the performance of the mechatronic systems not only relies on their controller design, but also on the design of their mechanical structure [15, 16], and many researchers have presented integrated optimization design of the mechatronic systems [17–21]. Recently, some global search approaches, such as genetic algorithm (GA), particle swarm optimization technique and differential evolution, have been increasingly used to carry out the optimization solution of the mechanisms, as they are more likely to converge to the global optimum [22]. The GAs have been growing in popularity over the last few years as more and more researchers discover the benefits of their adaptive search [23]. Their performance is superior to that of classical techniques [24]. This section uses GAs to optimize the HDCPR due to their good adequacy with the complicated multiobjective optimization problem.

This chapter is organized as following. In Sect. 4.2, the dynamic model of the HDCPR is developed on the basis of Lagrange method. Then, an adaptive iterative learning control strategy is designed for the high-precision trajectory tracking. Furthermore, the stability of the controller is proved by means of Lyapunov function. In order to improve the dynamic performance of the HDCPR system, a methodology of simultaneous optimal design of mechanism and control for the HDCPR is presented in Sect. 4.3. The dynamic modeling of the HDCPR is performed based on Newton-Euler method, and the workspace of the manipulator is also analyzed in Sect. 4.4.

4.2 Design, Dynamics and Workspace of a Hybrid-Driven Based Cable Parallel Manipulator

4.2.1 Mechanism Description

In this research, for the purpose of analytical modeling and numerical analysis, the three-dimensional design model of the completely restrained HDCPR with three translational motions is taken as an example (see Fig. 4.1). The HDCPR suspends an end-effector by four cables and restrains all motion degrees of freedom for the object using the cable force when the end-effector moves within the workspace. For each cable, one end is connected to the end-effector, the other one rolls through a pulley fixed on the top of the relative cable tower rack and then is fed into the HDPM. The HDCPR comprises of two modules: (1) the CPR consisting of four-cable tower racks, four cables, pulley struts, pulleys, girder, cargo (i.e. end-effector); (2) four groups of HDPMs containing three-phase asynchronous motors, servomotors, reducers and double crank five-bar linkage; the asynchronous motors are connected using the pulley transmission mechanisms, while the servomotors and the reducers are linked by couplings. At the same time, the pulley transmission mechanisms and reducers are joined to the double crank planar

Fig. 4.1 Three-dimensional model of the HDCPR

five-bar linkage. The double crank planar five-bar linkage is made up of a large crank disk, a long connecting rod, a small crank disk, and a short connecting rod. Both long and short connecting rods are articulated with each other and attached to the cable. Moreover, each tower rack of the CPR is equipped with a cable guide pulley. The mechanical configuration of the HDCPR is illustrated in Fig. 4.2.

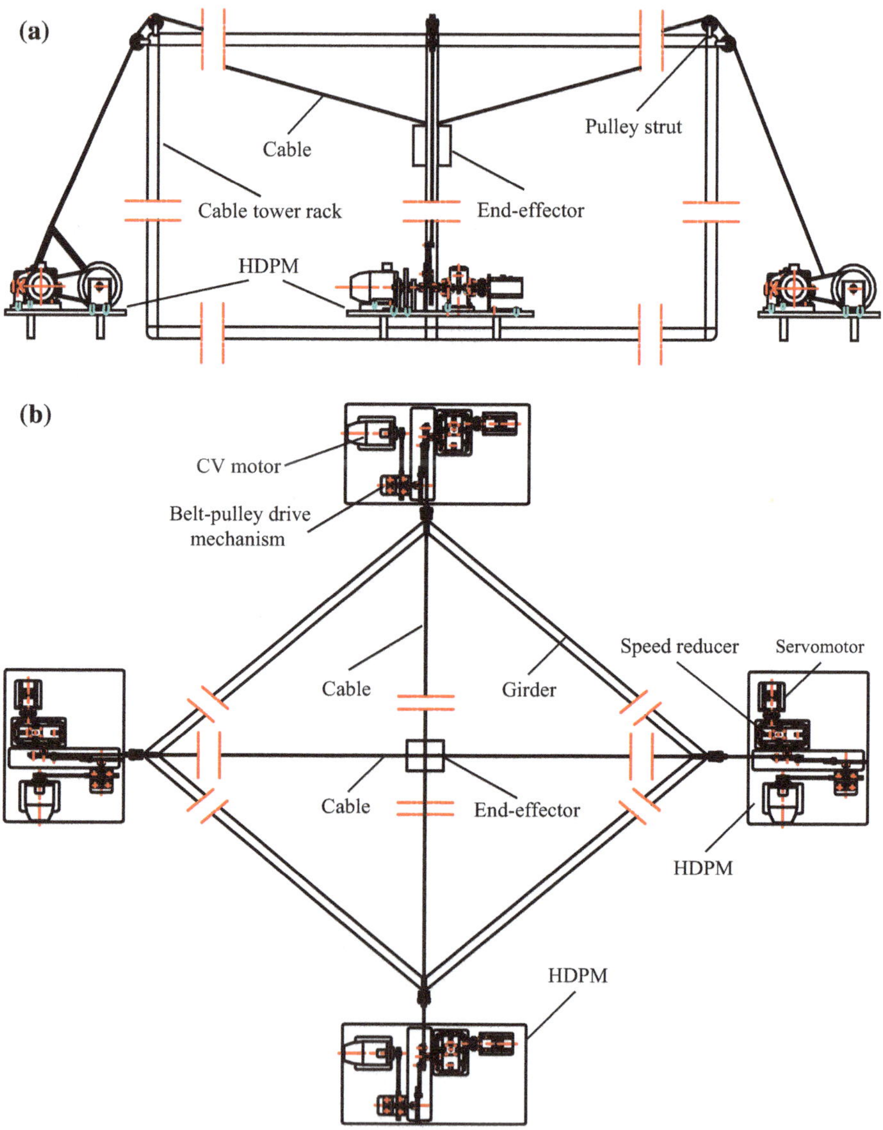

Fig. 4.2 Mechanical configuration of the HDCPR: **a** front view, **b** top view

The working process of the HDCPR is as follows: the CV three-phase asynchronous motors and servomotors are power sources. On the one hand, the servomotor and the reducer are linked by the coupling, then the servomotors through a small crank disk are connected with a short rod connection thus adjusting the output motion of the HDPM. On the other hand, each CV three-phase asynchronous motor is connected with the pulley transmission mechanisms, then it links to a long connecting rod by a large crank disk, thus providing the main power for the HDCPR. These two types of input motions are hinged through the short and long connecting rod, so that the power distribution and other characteristics of the HDPM are improved while ensuring the output motions. Four groups of the HDPM with the same structure produce rotary motion, thus the four cables can be driven in order to realize the output motion trajectory of the end-effector.

4.2.2 Dynamic Modeling

A simple schematic sketch of the HDCPR structure model with the associated coordinate systems are depicted in Fig. 4.3. At the bottom of one cable tower rack a global coordinate system O(XYZ) is established. The end-effector has location coordinates G(x, y, z). The distance between each cable tower rack top $M_i(x_i, y_i, z_i)$ and the end-effector is $l_i(i = 1, 2, 3, 4)$. The four cable tower racks have same height and are arrayed in a rectangle on the ground, where deformation is ignored. In order to simplify the model, the cables are treated as a massless body with no deformation, with length s. The bottom of the HDPM corresponds to the joint A as the origin, where a local coordinate system $A(x'y'z')$ is established.

Link AB driven by a CV motor and link ED driven by a servomotor are the two driving links (Fig. 4.2a), the joint C is the output. There are two independent constraints in the HDPM, which can be used to derive the kinematic equation of the HDPM:

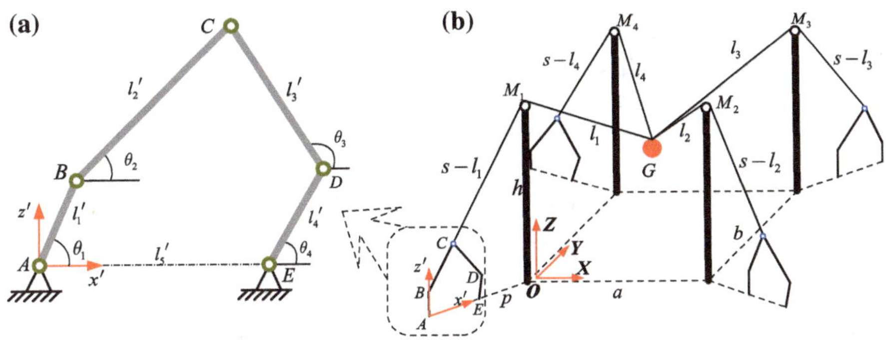

Fig. 4.3 Structure model **a** of the HDPM and schematic sketch **b** of the HDCPR

$$\begin{cases} l'_1 \cos\theta_1 + l'_2 \cos\theta_2 - l'_3 \cos\theta_3 - l'_4 \cos\theta_4 - l'_5 = 0 \\ l'_1 \sin\theta_1 + l'_2 \sin\theta_2 - l'_3 \sin\theta_3 - l'_4 \sin\theta_4 = 0 \end{cases} \tag{4.1}$$

In Eq. (4.1) θ_1 and θ_4 are known. Consequently θ_2 and θ_3 can be calculated as

$$\theta_3(\theta_1, \theta_4) = \arccos\left(\frac{C}{\sqrt{A^2 + B^2}}\right) + \arctan\left(\frac{\mu}{\lambda}\right) \tag{4.2}$$

$$\theta_2(\theta_1, \theta_4) = \arctan\left(\frac{\mu + l'_3 \sin\theta_3}{\lambda + l'_3 \cos\theta_3}\right) \tag{4.3}$$

where $A = 2l'_3\lambda, B = 2l'_3\mu, C = l'^2_2 - l'^2_3 - \lambda^2 - \mu^2, \lambda = l'_4 \cos\theta_4 - l'_1 \cos\theta_1 + l'_5,$ $\mu = l'_4 \sin\theta_4 - l'_1 \sin\theta_1.$

Let

$$\begin{cases} x'_C = l'_1 \cos\theta_1 + l'_2 \cos\theta_2 = l'_5 + l'_3 \cos\theta_3 + l'_4 \cos\theta_4 \\ z'_C = l'_1 \sin\theta_1 + l'_2 \sin\theta_2 = l'_3 \sin\theta_3 + l'_4 \sin\theta_4 \end{cases} \tag{4.4}$$

be the coordinates of joint C in the local coordinate system. According to the geometric relationship of the HDCPR, l1 can be derived as

$$l_1 = s - \sqrt{(p + l'_5 - x'_C)^2 + (h - z'_C)^2} \tag{4.5}$$

where p represents the distance $\overline{EO}$ and s is the total length of cable. Similarly, l_2, l_3 and l_4 can be derived, respectively.

The relationships between the cable length li and the end-effector location G(x, y, z), can be easily obtained as follows:

$$\begin{cases} l_1 = \sqrt{x^2 + y^2 + (z - h)^2} \\ l_2 = \sqrt{(x - a)^2 + y^2 + (z - h)^2} \\ l_3 = \sqrt{(x - a)^2 + (y - b)^2 + (z - h)^2} \\ l_4 = \sqrt{x^2 + (y - b)^2 + (z - h)^2} \end{cases} \tag{4.6}$$

resulting in

$$\begin{cases} x = \frac{l_1^2 - l_2^2 + a^2}{2a} \\ y = \frac{l_1^2 - l_4^2 + b^2}{2b} \\ z = h - \sqrt{l_1^2 - x^2 - y^2} \end{cases} \tag{4.7}$$

Newton's law for the end-effector of the HDCPR results in

$$m\begin{bmatrix}\ddot{x}\\ \ddot{y}\\ \ddot{z}\end{bmatrix} = \sum_{i=1}^{4}\overrightarrow{T_i} + \begin{bmatrix}0\\ 0\\ -mg\end{bmatrix} \tag{4.8}$$

where T_i(i = 1, 2, 3 ,4) is a tension cable force, i.e., the driving force of cable i exerted on the end-effector for the HDCPR, and $\overrightarrow{T_i}$ is force vector.

Thus the dynamic model can be expressed as follows

$$N_{3\times4}T_{4\times1} = M_{3\times1} \tag{4.9}$$

$$N_{3\times4} = \begin{bmatrix}\frac{-x}{l_1} & \frac{a-x}{l_2} & \frac{a-x}{l_3} & \frac{-x}{l_4}\\ \frac{-y}{l_1} & \frac{-y}{l_2} & \frac{b-y}{l_3} & \frac{b-y}{l_4}\\ \frac{h-z}{l_1} & \frac{h-z}{l_2} & \frac{h-z}{l_3} & \frac{h-z}{l_4}\end{bmatrix}$$

where $N_{3\times4}$ is the matrix of coordinate transformer, $T_{4\times1} = \begin{bmatrix}T_1 & T_2 & T_3 & T_4\end{bmatrix}^{\mathrm{T}}$ is the vector of cable tension, $M_{3\times1} = \begin{bmatrix}m\ddot{x} & m\ddot{y} & mg + m\ddot{z}\end{bmatrix}^{\mathrm{T}}$ is the vector of inertia force, m is mass of the end-effector, and g the gravitational acceleration.

The dynamic model of the HDCPR is presented in two parts. The first is directed to the structural model (CPR) above, the other one to the actuator dynamics (servomechanism). Applying the Newton-Euler equation establishes the dynamic model of the HDPM. The force analysis of each link of the HDPM is shown in Fig. 4.4.

τ_1 and τ_4 represent the driving torque of the CV motor on link AB and the servomotor on link ED, respectively. $\beta = \arctan\left(\frac{h-l'_1\sin\theta_1 - l'_2\sin\theta_2}{p+l'_5-l'_1\cos\theta_1 - l'_2\cos\theta_2}\right)$ is the angle between cable tension and x'-coordinate.

By the force analysis above, the Newton-Euler formulation of the links can be listed as follows:

Link AB:

$$\begin{cases} F_{Ax'} - F_{Bx'} = m_1\ddot{x}'_1 \\ F_{Az'} - m_1 g - F_{Bz'} = m_1\ddot{z}'_1 \\ \tau_1 + F_{Ax'}r_1\sin\theta_1 - F_{Az'}r_1\cos\theta_1 + F_{Bx'}\left(l'_1 - r_1\right)\sin\theta_1 - F_{Bz'}\left(l'_1 - r_1\right)\cos\theta_1 = J_1\ddot{\theta}_1 \end{cases} \tag{4.10}$$

Link BC:

$$\begin{cases} T_1\cos\beta + F_{Bx'} - F_{Cx'} = m_2\ddot{x}'_2 \\ T_1\sin\beta + F_{Bz'} - F_{Cz'} - m_2 g = m_2\ddot{z}'_2 \\ F_{Bx'}r_2\sin\theta_2 - F_{Bz'}r_2\cos\theta_2 + \left(F_{Cx'} - T_1\cos\beta\right)\left(l'_2 - r_2\right)\sin\theta_2 \\ \quad + \left(T_1\sin\beta - F_{Cz'}\right)\left(l'_2 - r_2\right)\cos\theta_2 = J_2\ddot{\theta}_2 \end{cases} \tag{4.11}$$

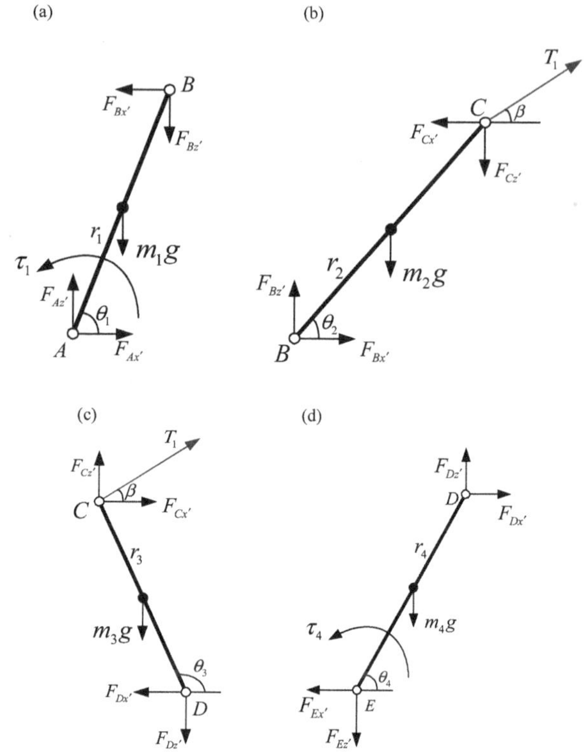

Fig. 4.4 Free body diagrams of the HDPM: **a** link *AB*, **b** link *BC*, **c** link *CD*, **d** link *ED*

Link CD:

$$\begin{cases} T_1 \cos\beta + F_{Cx'} - F_{Dx'} = m_3 \ddot{x}_3' \\ T_1 \sin\beta + F_{Cz'} - m_3 g - F_{Dz'} = m_3 \ddot{z}_3' \\ (F_{Cz'} + T_1 \sin\beta) r_3 \cos\theta_3 - (F_{Cx'} + T_1 \cos\beta) r_3 \sin\theta_3 - F_{Dx'}\left(l_3' - r_3\right) \sin\theta_3 \\ + F_{Dz'}\left(l_3' - r_3\right) \cos\theta_3 = J_3 \ddot{\theta}_3 \end{cases} \tag{4.12}$$

Link ED:

$$\begin{cases} F_{Dx'} - F_{Ex'} = m_4 \ddot{x}_4' \\ F_{Dz'} - m_4 g - F_{Ez'} = m_4 \ddot{z}_4' \\ \tau_4 - F_{Dx'} r_4 \sin\theta_4 + F_{Dz'} r_4 \cos\theta_4 - F_{Ex'}\left(l_4' - r_4\right) \sin\theta_4 + F_{Ez'}\left(l_4' - r_4\right) \cos\theta_4 = J_4 \ddot{\theta}_4 \end{cases} \tag{4.13}$$

where m_j, l'_j, J_j, r_j, $\left(x'_j, z'_j\right)$ $(j = 1, 2, 3, 4)$ represent the mass, length, inertia, distance between the mass centre and the joint, and coordinates of the mass centre of each link, respectively.

The equation can be summarized as

$$A_{12\times14}H_{14\times1} = B_{12\times1} \tag{4.14}$$

where

$$A_{12\times24} = \begin{bmatrix}
1 & 0 & -1 & 0 & 0 & 0 & 0 & 0 & 0 & 0 & 0 & 0 & 0 & 0 \\
0 & 1 & 0 & -1 & 0 & 0 & 0 & 0 & 0 & 0 & 0 & 0 & 0 & 0 \\
r_1s_1 & -r_1c_1 & (l'_1-r_1)s_1 & (r_1-l'_1)c_1 & 0 & 0 & 0 & 0 & 0 & 0 & 0 & 0 & 0 & 0 \\
0 & 0 & 1 & 0 & -1 & 0 & 1 & 0 & 0 & 0 & 0 & 0 & 0 & 0 \\
0 & 0 & 0 & 1 & 0 & -1 & 0 & 1 & 0 & 0 & 0 & 0 & 0 & 0 \\
0 & 0 & r_2s_2 & -r_2c_2 & (l'_2-r_2)s_2 & (r_2-l'_2)c_2 & (r_2-l'_2)s_2 & (l'_2-r_2)c_2 & 0 & 0 & 0 & 0 & 0 & 0 \\
0 & 0 & 0 & 0 & 1 & 0 & 1 & 0 & -1 & 0 & 0 & 0 & 0 & 0 \\
0 & 0 & 0 & 0 & 0 & 1 & 0 & 1 & 0 & -1 & 0 & 0 & 0 & 0 \\
0 & 0 & 0 & 0 & -r_3s_3 & r_3c_3 & -r_3s_3 & r_3c_3 & (r_3-l'_3)s_3 & (l'_3-r_3)c_3 & 0 & 0 & 0 & 0 \\
0 & 0 & 0 & 0 & 0 & 0 & 0 & 0 & 1 & 0 & -1 & 0 & 0 & 0 \\
0 & 0 & 0 & 0 & 0 & 0 & 0 & 0 & 0 & 1 & 0 & -1 & 0 & 0 \\
0 & 0 & 0 & 0 & 0 & 0 & 0 & 0 & -r_4s_4 & r_4c_4 & (r_4-l'_4)s_4 & (l'_4-r_4)c_4 & 0 & 1
\end{bmatrix}$$

is the matrix of force transformer, $s_i = \sin\theta_i, c_i = \cos\theta_i$.

$$H_{14\times1} = [F_{Ax'}, F_{Az'}, F_{Bx'}, F_{Bz'}, F_{Cx'}, F_{Cz'}, T_1\cos\beta, T_1\sin\beta, F_{Dx'}, F_{Dz'}, F_{Ex'}, F_{Ez'}, \tau_1, \tau_4]^{\mathrm{T}}$$

is the vector of resultant and active force.

$$\begin{aligned} B_{12\times1} = {} & [m_1\ddot{x}'_1, m_1g + m_1\ddot{z}'_1, J_1\ddot{\theta}_1, m_2\ddot{x}'_2, m_2g + m_2\ddot{z}'_2, J_2\ddot{\theta}_2, m_3\ddot{x}'_3, m_3g \\ & + m_3\ddot{z}'_3, J_3\ddot{\theta}_3, m_4\ddot{x}'_4, m_4g + m_4\ddot{z}'_4, J_4\ddot{\theta}_4]^{\mathrm{T}} \end{aligned}$$

is the vector of inertia force.

Taking into account Eqs. (4.9) and (4.14), the dynamic model of the HDCPR is derived from the dynamic equations of the CPR and four groups of the HDPM, as follows:

$$\begin{cases} N_{3\times4}T_{4\times1} = M_{3\times1} \\ A^i_{12\times14}H^i_{14\times1} = B^i_{12\times1}\ (i = 1, 2, 3, 4) \end{cases} \tag{4.15}$$

4.2.3 Workspace Analysis

The workspace of the HDCPR is characterized as the set of points where the end-effector can be positioned while all cables are in tension ($T_i > 0$). At each point within the possible workspace, Eq. (4.9) describing the force in each cable is used to see if tension is obtainable. However, it is not sufficient to obtain the actual workspace of the HDCPR if it only depends on the tension condition. Accordingly, when analyzing the workspace of the HDCPR, one should also consider the following constraints such as the workspace of the HDPM and the motion range of the end-effector:

1. The workspace of the HDPM. According to Eq. (4.7), the set of points (x, y, z) where the end-effector can be positioned are influenced by cable length l_i. Cable length l_i is calculated by the coordinates joint C $\left(x'_C, z'_C\right)$ from Eq. (4.5). Thus, the workspace of the joint C in x'-z' plane is one of the constraints of the workspace of the HDCPR, and it can be expressed as (only the first group of the HDPM)

$$\begin{cases} |l'_1 - l'_2| < L_{AC} < l'_1 + l'_2 \\ |l'_3 - l'_4| < L_{CE} < l'_3 + l'_4 \end{cases} \tag{4.16}$$

   where $L_{AC} = \sqrt{x'^2_C + z'^2_C}$, $L_{CE} = \sqrt{(x'_C - l'_5)^2 + z'^2_C}$
2. The motion range of the end-effector for the HDCPR can be written in the following form:

$$\begin{cases} 0 < x = \frac{l_1^2 - l_2^2 + a^2}{2a} < a \\ 0 < y = \frac{l_1^2 - l_4^2 + b^2}{2b} < b \\ 0 < z = h - \sqrt{l_1^2 - x^2 - y^2} < h \end{cases} \tag{4.17}$$

A general numerical workspace generation approach is employed here. The possible motion range of joint C in the HDPM is first discretized into a number of points. According to the Eq. (4.16), a series of the cable length l_i can be calculated. Then, from Eq. (4.17), various combinations of l_i can calculate the end-effector coordinates (x, y, z). Thus, a series of points in Cartesian coordinate system are determined. At the end, the points are checked and those satisfying the tension condition ($T_i > 0$) will generate the workspace. The flowchart of generating the workspace of the HDCPR is shown in Fig. 4.5.

4.2.4 Results and Discussion

Simulation studies were performed with the software named MATLAB 2010. A three-dimensional simulation model of the HDCPR has been established, where the four groups of HDPMs have the same structure parameters and symmetry in the three-dimensional space. The parameters of the HDCPR are listed in Table 4.1. All computations were carried out on a computer with a 2.99-GHz CPU and 2.00 GB memory; the circle trajectory took about 9.67 s, while the line segment trajectory requires only 2.93 s.

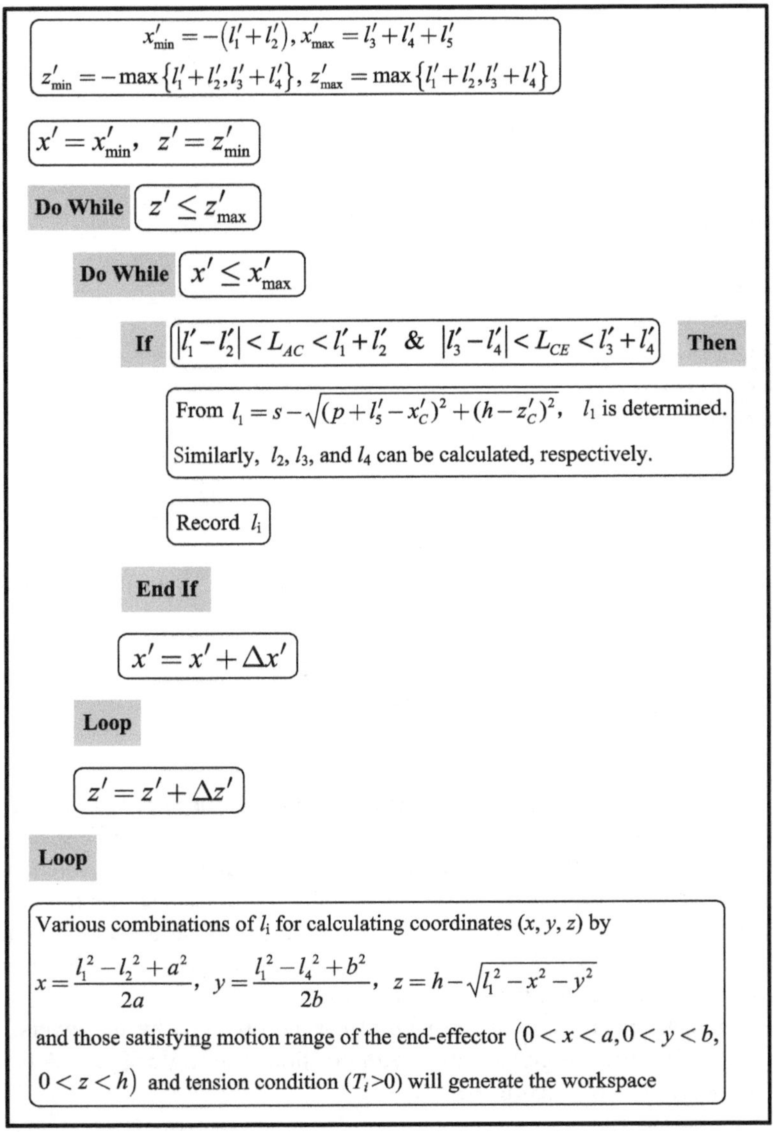

Fig. 4.5 Flowchart of generating the workspace of the HDCPR

4.2.4.1 Dynamic Simulation

This section presents two motion cases of the end-effector for dynamic simulation.

Case 1: **Circle trajectory**

Table 4.1 Parameters of the HDCPR

System parameters	Value
Mass of the end-effector m	20 kg
Acceleration due to gravity g	9.81 m/s^2
Height of the cable tower rack h	1 m
Side length of four towers distributed in rectangle a	1 m
Side length of four towers distributed in rectangle b	1 m
Length of the cable s	2.3 m
Length of the *EO* p	1 m
Length of the link *AB* l'_1	0.2 m
Length of the link *BC* l'_2	0.5 m
Length of the link *CD* l'_3	0.5 m
Length of the link *ED* l'_4	0.28 m
Length of the link *AE* l'_5	0.51 m
Density of the links	7.85×10^{-3} kg/m^3
Cross sectional area of the links	2×10^{-4} m^2

The equation of the circle with a radius of 0.25 m parallel to the X-Y plane in global coordinate system is:

$$\begin{cases} x = 0.25\cos t + 0.5 \\ y = 0.25\sin t + 0.5 \\ z = 0.5 \end{cases} \tag{4.18}$$

where $0 \le t \le 6.2(s)$. The end-effector moves along the trajectory with the constant velocity $v = 0.25\,\mathrm{m/s}$ as shown in Fig. 4.6.

Figure 4.7 shows cable motion in the process of simulation. The cable lengths l_i are generated by Eq. (4.6) and vary symmetrically. Figure 4.8 shows the curves of cable tension exerted on the end-effector related to the circle trajectory. Since the end-effector moves on a horizontal circle trajectory, the changes of cable tensions in Fig. 4.8 are reasonable and the transition is smooth.

The kinematic parameters of the HDPM can be obtained by the inverse kinematics of the HDCPR, for given end-effector position. The angle of joint A is driven by a CV motor and assumed to be

$$\theta_1 = \omega t \tag{4.19}$$

where $\omega = 1$ rad/s. According to Fig. 4.3b the following geometrical relation is verified:

$$(l'_5 + p - l'_1\cos\theta_1 - l'_2\cos\theta_2)^2 + (h - l'_1\sin\theta_1 - l'_2\sin\theta_2)^2 = (s - l_i)^2 \tag{4.20}$$

where l_1 is known from Eq. (4.6). Combining Eqs. (4.19) and (4.20), θ_2 can be calculated as

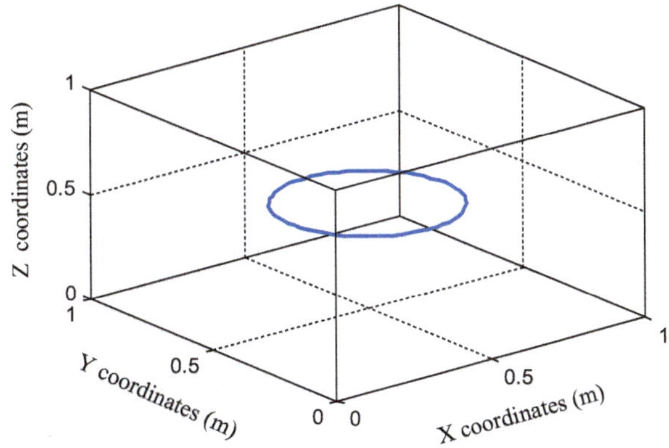

Fig. 4.6 Following trajectory of the circle motion

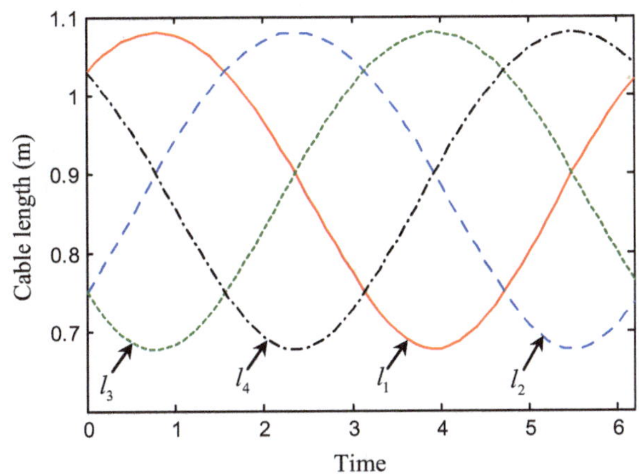

Fig. 4.7 Time behavior of cable length for Case 1

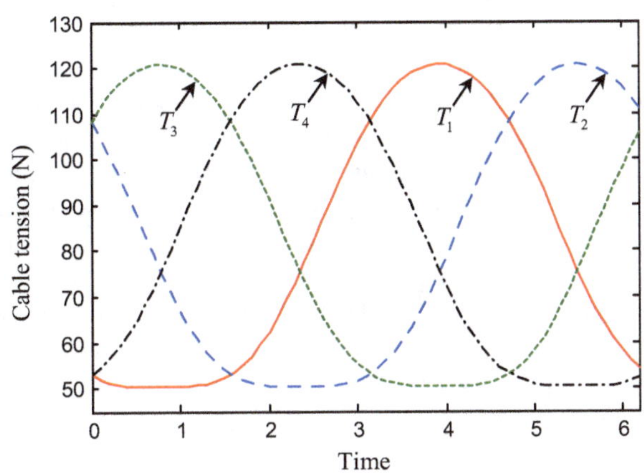

Fig. 4.8 Curves of cable tension for Case 1

$$\theta_2 = 2\arctan\left(\frac{E_1 - \sqrt{F_1^2 + E_1^2 - G_1^2}}{F_1 - G_1}\right) \tag{4.21}$$

where

$$E_1 = 2l_2'(l_1' \sin\theta_1 - h), F_1 = 2l_2'(l_1' \cos\theta_1 - l_5' + p),$$

$$G_1 = (l_5' + p) - 2_1'\left(h \sin\theta_1 + (l_5' + p)\cos\theta_1\right) + l_1'^2 + l_2'^2 + h^2 - (s - l_1)^2.$$

From Eqs. (4.4), (4.19) and (4.21), the coordinates of joint C $\left(x_C', z_C'\right)$ can be determined as

$$\begin{cases} x_C' = l_1' \cos\theta_1 + l_2' \cos\theta_2 \\ z_C' = l_1' \sin\theta_1 + l_2' \sin\theta_2 \end{cases} \tag{4.22}$$

According to Eq. (4.4), the coordinates of joint C $\left(x_C', z_C'\right)$ can be rewritten as

$$\begin{cases} x_C' = l_5' + l_3' \cos\theta_3 + l_4' \cos\theta_4 \\ z_C' = l_3' \sin\theta_3 + l_4' \sin\theta_4 \end{cases} \tag{4.23}$$

Solving Eq. (4.23), one gets:

$$\theta_4 = 2\arctan\left(\frac{E_2 - \sqrt{F_2^2 + E_2^2 - G_2^2}}{F_2 - G_2}\right) \tag{4.24}$$

$$\theta_3 = \arcsin\left(\frac{z_C' - l_4' \sin\theta_4}{l_3'}\right) \tag{4.25}$$

where $E_2 = -2z_C' l_4', F_2 = 2l_4' l_5' - 2x_C' l_4'$ and $G_2 = (x_C' - l_5')^2 + z_C'^2 + l_4'^2 - l_3'^2$.

According to the inverse kinematics analysis of the HDCPR above, one can calculate the angular variables for all links of the HDPM as the end-effector moves along the circle. Figure 4.9 shows the angle curves of the links of the HDPM. It can be noted that the angular variable law for link AB is a line, while for the other links are curves. This is in line with the motion law of the HDPM. Moreover, angular curves of all the links maintain good continuity, and the numerical values are reasonable. The angular velocity curves and angular acceleration curves of links can be generated by take the first and second time derivative of the angular change ($d\theta_i/dt$, $d^2\theta_i/dt^2$). The coordinates of the joint C in local x'-z' plane can be computed by Eq. (4.22). The running trajectory of joint C is depicted in Fig. 4.10.

Figure 4.11 depicts the resulting force curves of the joints of the HDPM. From it one can see that, the resultant force curves of the joints are smooth, which is consistent with the desired motion law.

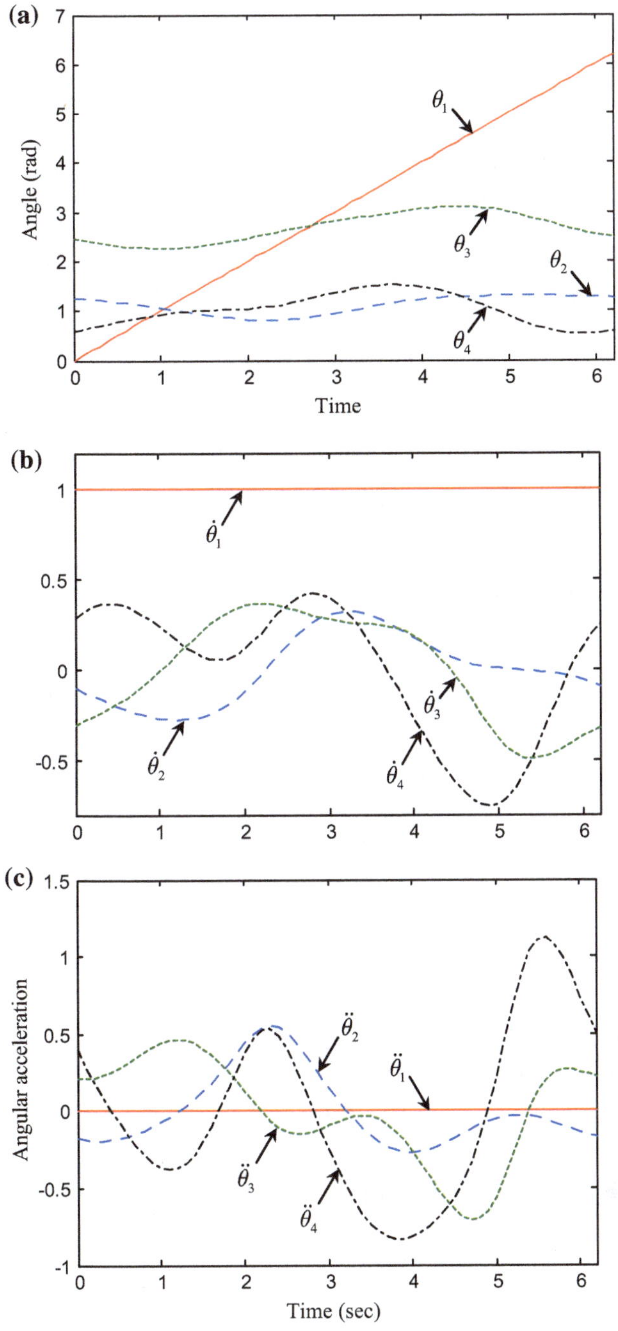

Fig. 4.9 Curves of the links motion parameters of the HDPM for Case 1 **a** angle, **b** angular velocity, **c** angular acceleration

If the cables are driven only by coils and servomotors, which is a classical approach for the conventional CPR, the range and speed of obtainable output motions may be limited by certain servomotor power capacity and high cost. This problem can be solved by using hybrid-driven machines instead of coils and servomotors. For the hybrid-driven machines, the CV motor provides the majority of power supply to drive the mechanism's motion. The servomotor is real-time controllable and off-line programmable. Therefore, the hybrid-driven mechanisms take the advantage of the complementary characteristics of two types of motors to generate programmable output motions with high power capacities in low cost.

The CV motor power and servomotor power of the HDPM can be calculated by:

$$\begin{cases} P_{CV1} = \tau_1 \times \dot{\theta}_1 \\ P_{SV1} = \tau_4 \times \dot{\theta}_4 \end{cases} \tag{4.26}$$

where P_{CV1} is the CV motor power of the first group of the HDPM, P_{SV1} is the servomotor power of the first group of the HDPM. Similarly, the other motor

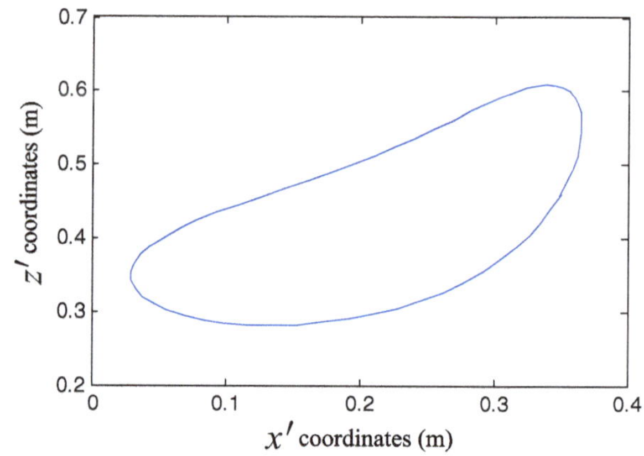

Fig. 4.10 Trajectory of joint C in x'-z' plane for Case 1

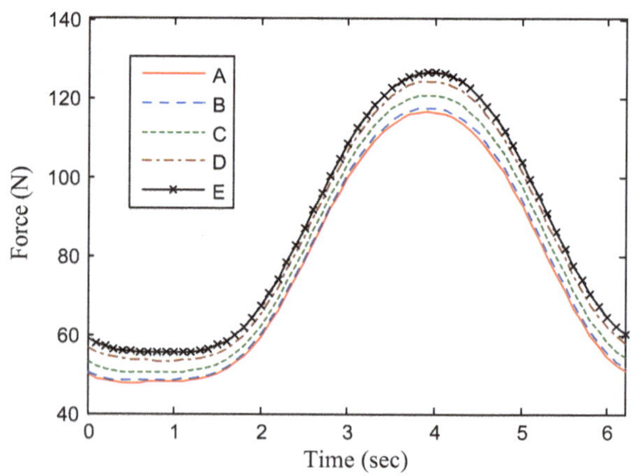

Fig. 4.11 Resultant force curves of the joints of the HDPM for Case 1

power,P_{CV2}, P_{SV2}, P_{CV3}, P_{SV3}, P_{CV4}, P_{SV4}, of the HDPM can be derived, respectively.

Figure 4.12 shows motor absolute value power curves of four groups of the HDPMs of the HDCPR. From Fig. 4.12, for each of the HDPM, the range of the CV motor power (P_{CVi}) is larger than the servomotor power (P_{SVi}). It is illustrated also that the CV motors undertake the majority of power for the HDCPR system.

As the motion circular trajectory of the end-effector is the same, and as the structural parameters of the classical approach for the conventional CPR are the same, (cables are driven only by coils and servomotors), the servomotor power of the conventional CPR can be obtained by:

$$P_i = T_i \times \dot{l}_i \ (i = 1, 2, 3, 4) \tag{4.27}$$

where P_i is the i-th servomotor power, T_i represents the i-th cable tension, $\dot{l}_i$ denotes the i-th cable velocity.

Figure 4.13 shows the servomotor absolute value power curves of the classical approach for the conventional CPR. For comparison, absolute value power curves of the first group of HDPM of the HDCPR and the first servomotor of the conventional CPR are given in Fig. 4.14. As it is seen, large range of servomotor power |P—1| of the conventional CPR may be decreased by using the HDPM.

In order to investigate the power property of the HDCPR further, the following simulation is performed as the CV motors of the four groups of the HDPMs of the HDCPR run with 1.1 rad/s (ω = 1.1 rad/s). Figure 4.15 displays curves of the link angle and velocity of the HDPM. Angular curves of all links maintain good continuity, and the numerical values are reasonable. Figure 4.16 shows absolute value

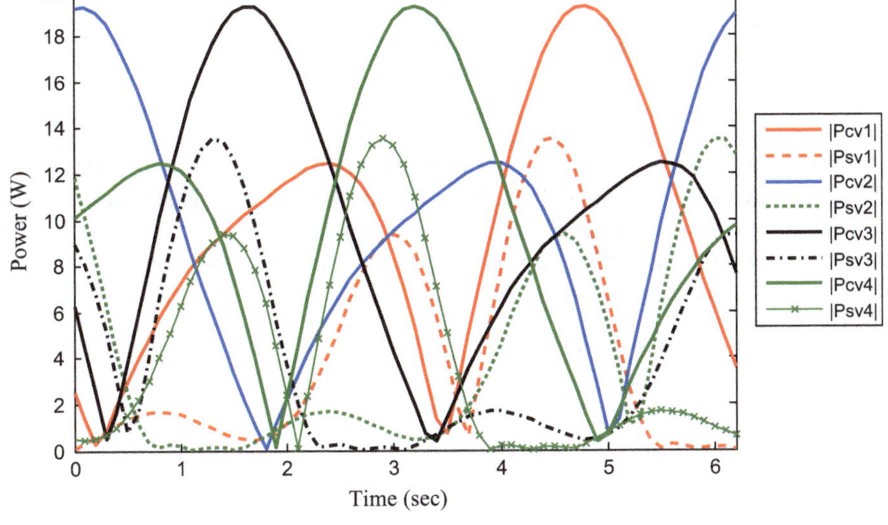

Fig. 4.12 Motor absolute value power curves of the four groups of HDPMs of the HDCPR

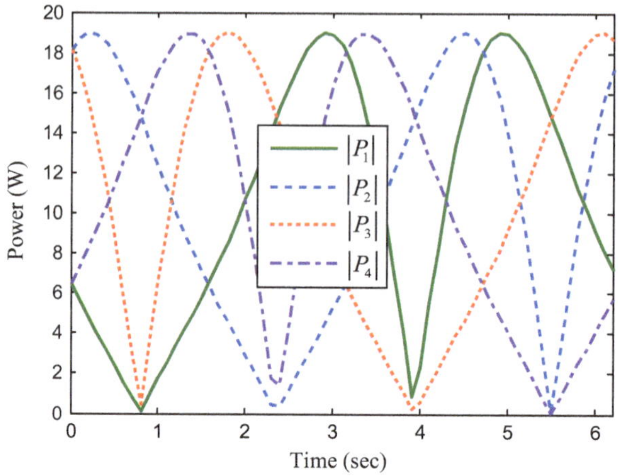

Fig. 4.13 Servomotor absolute value power curves of the conventional CPM

Fig. 4.14 Comparison of absolute value power curves between the HDCPR and conventional CPR

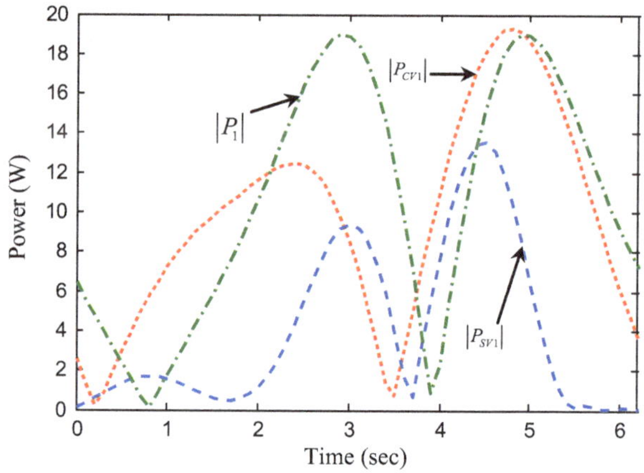

power curves of four groups of the HDPMs of the HDCPR. Figure 4.17 illustrates absolute value power curves of the first group of HDPM of the HDCPR and the first servomotor of the conventional CPR. From the simulations, we can see that the CV motors provide the majority of power supply and the servomotors mainly act as a motion regulation device as depicted in Figs. 4.12, 4.14, 4.16 and 4.17.

Case 2: **Line segment trajectory**

The equation of the line in global coordinate system is:

$$\begin{cases} x = t/6 + 0.25 \\ y = t/12 + 1/3 \\ z = t/6 + 0.25 \end{cases} \tag{4.28}$$

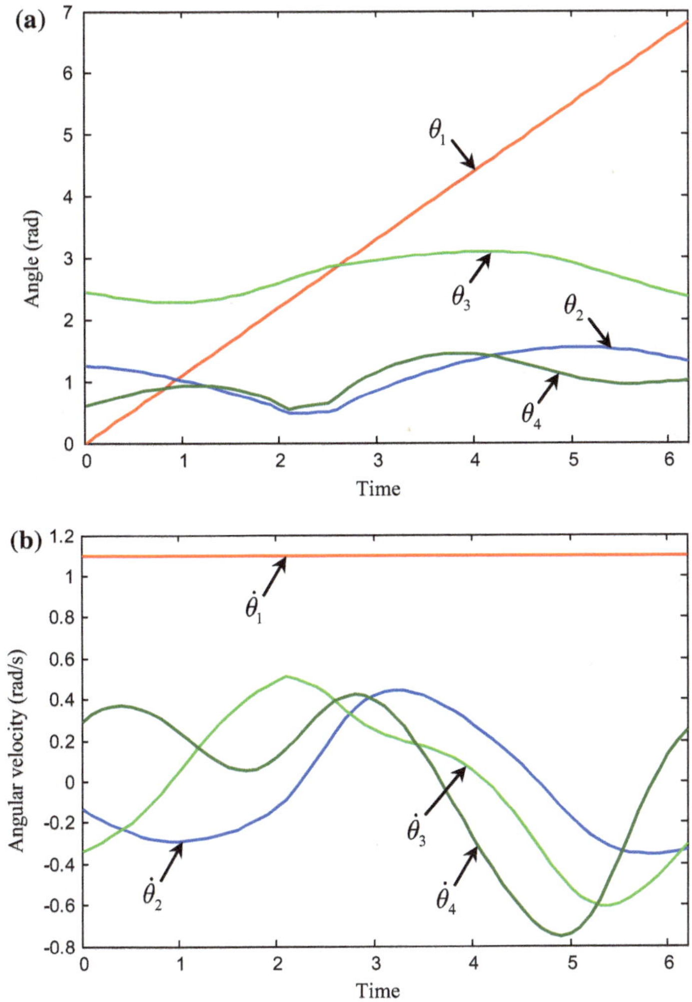

Fig. 4.15 Curves of the links motion parameters of the HDPM as ω = 1.1 rad/s: **a** angle, **b** angular velocity

where $0 < t < 2(s)$. From the Eq. (4.28), it can be noted that the center of gravity of the end-effector moves that the geometric path is line segment with the constant velocity $v = 0.25\,\mathrm{m/s}$, the initial point is (0.25, 1/3, 0.25) (m). The trajectory tracking of the end-effector related to Case 2 is shown in Fig. 4.18. From this figure the formulation tracks the planned trajectory relatively well.

Figure 4.19 shows the changes of cables lengths. As can be seen, the variations of the curves of cables lengths are in accordance with the reference path (Eq. 4.28) of the end-effector.

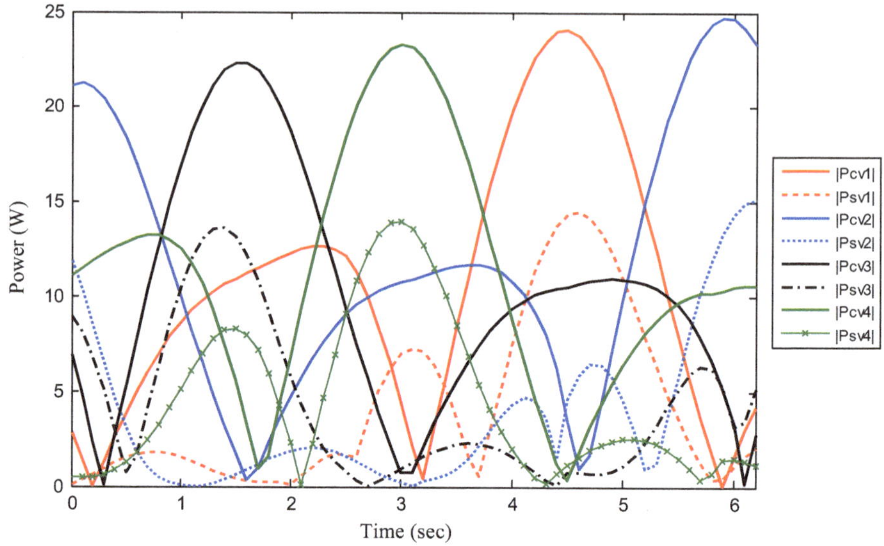

Fig. 4.16 Motor absolute value power curves of the four groups of HDPMs of the HDCPR as ω = 1.1 rad/s

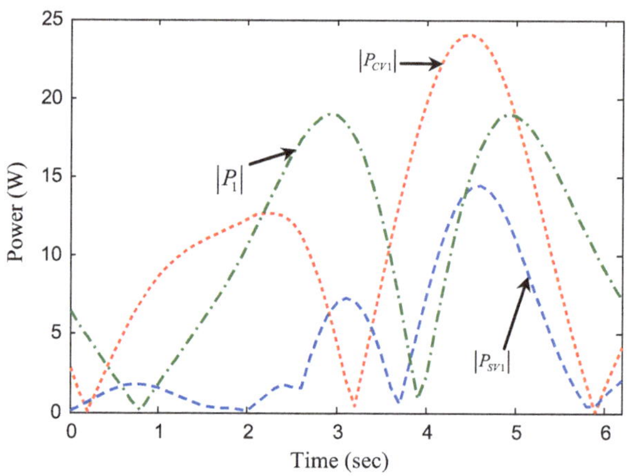

Fig. 4.17 Comparison of absolute value power curves between the HDCPR as ω = 1.1 rad/s and conventional CPR

Figure 4.20 depicts the curves of cable tensions for Case 2. One can see that the curves are smooth, which is the ideal data for the HDCPR system operates steadily.

The angular variables of links of the HDPM for Case 2 can be obtained by the inverse kinematics analysis of the HDCPR. The angle curves of the HDPM links are shown in Fig. 4.21a. The corresponding angular velocity and angular

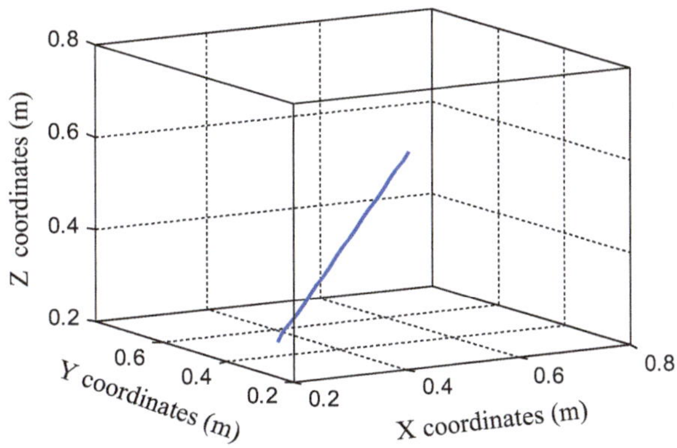

Fig. 4.18 Following trajectory of the line segment motion

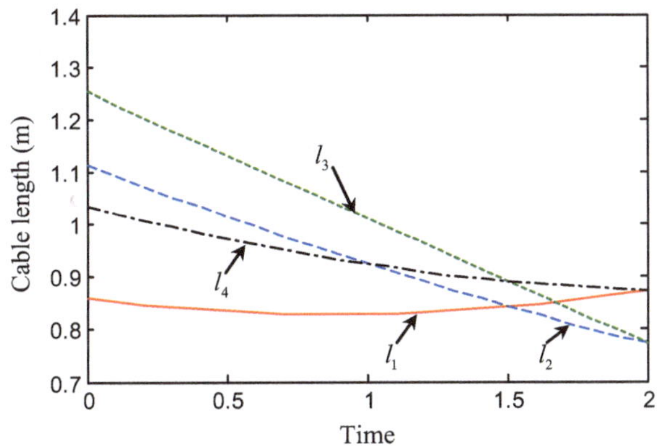

Fig. 4.19 Curves of cable length for Case 2

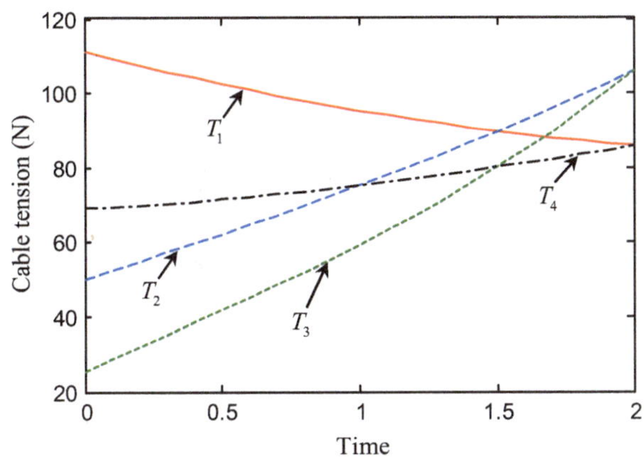

Fig. 4.20 Curves of cable tension for Case 2

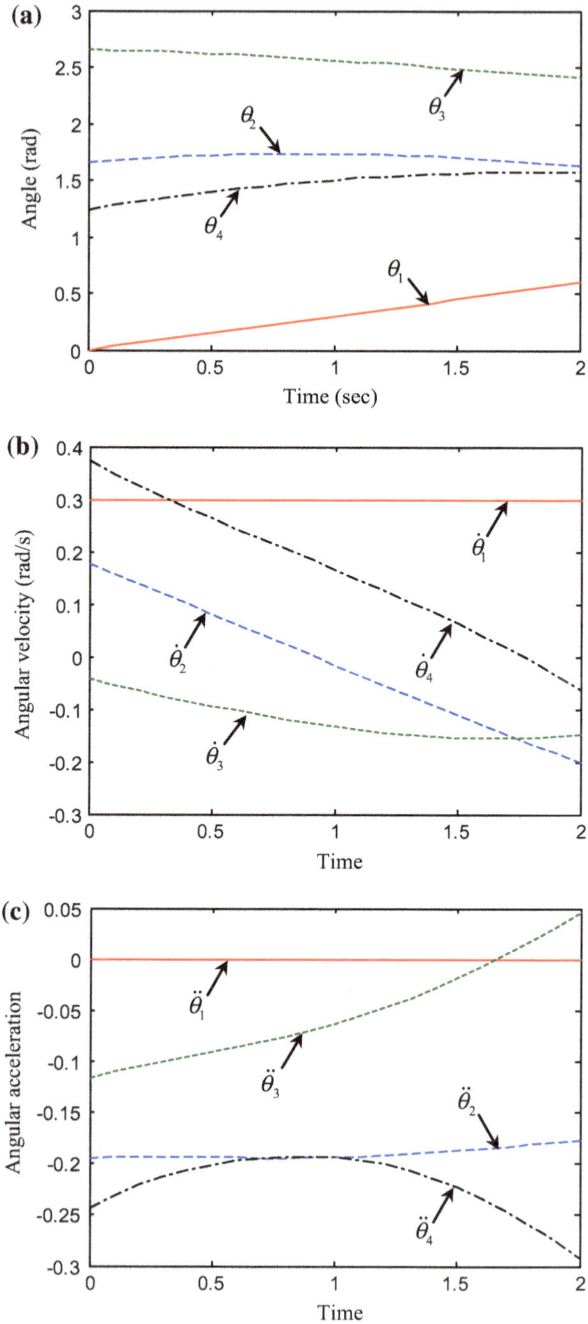

Fig. 4.21 Curves of the links motion parameters of the HDPM for Case 2: **a** angle, **b** angular velocity, **c** angular acceleration

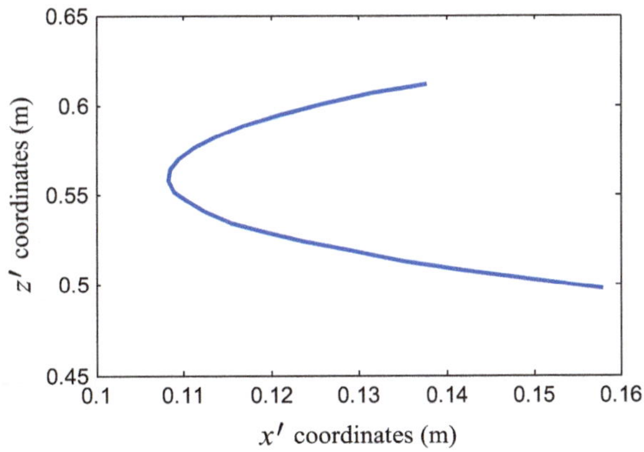

Fig. 4.22 Running trajectory of the joint C in x'-z' plane for Case 2

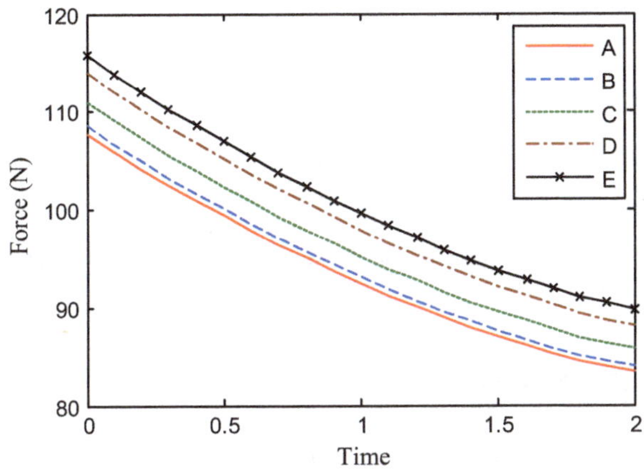

Fig. 4.23 Resultant force curves of the joints of the HDPM for Case 2

acceleration are shown in Fig. 4.21b, c. Figure 4.22 shows the running trajectory of the joint C in local $x'o'z'$ plane.

The resulting force curves of the joints of the HDPM for Case 2 can be solved by the dynamics modeling of the HDCPR in Sect. 4.3, as shown in Fig. 4.23. It can be noted that the resulting force curves of the joints have the same trend, and change smoothly. Comparison of absolute value power curves of the HDPMs in the HDCPR and the servomotors of the conventional CPR for Case 2 are shown in Fig. 4.24. Referring to Fig. 4.24, it can be seen that the CV motors provide the majority of power supply and the servomotors mainly act as a motion regulation device.

From the results and discussions, illustrative simulation studies highlight its performances, which lay a foundation for the further research on optimization and real-time control.

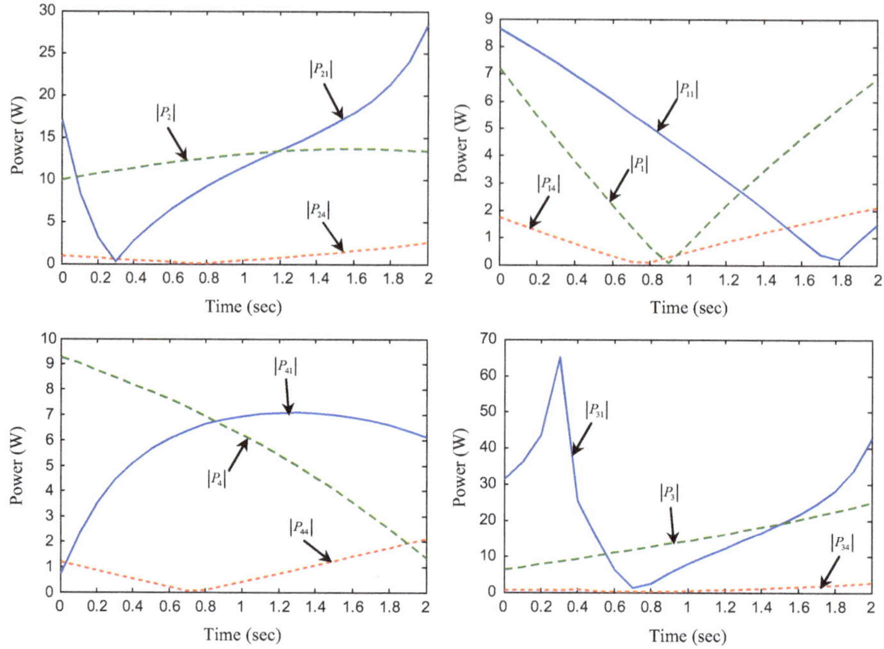

Fig. 4.24 Comparison of absolute value power curves between the HDCPR and conventional CPR for Case 2

4.2.4.2 Workspace Simulation

The workspace of joint C is determined by geometric constraints (4.16) given in Fig. 4.25. The simulation results of the reachable 3D workspace for the HDCPR and the projections of workspace onto the X-Y plane and X-Z plane are shown in Fig. 4.26. The whole workspace is approximately an upside-down cone. It should be noted that most of the workspace volume is concentrated in the upper part of the tower cube, which is important for trajectory planning of the end-effector.

4.3 Dynamic Modeling and Iterative Learning Control of a Novel Hybrid-Driven Based Three-Cable Parallel Manipulator

In this section, in view of the nonlinear, time-varying characteristics and repetitive unknown disturbances of the HDCPR system, a control strategy incorporates adaptive traditional feedback PD control with iterative learning control is designed for the high-precision trajectory tracking of the HDCPR control system.

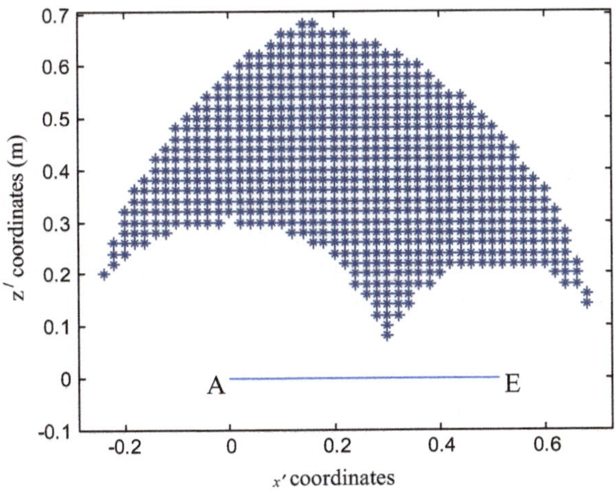

Fig. 4.25 Workspace of joint C in x'-z' plane

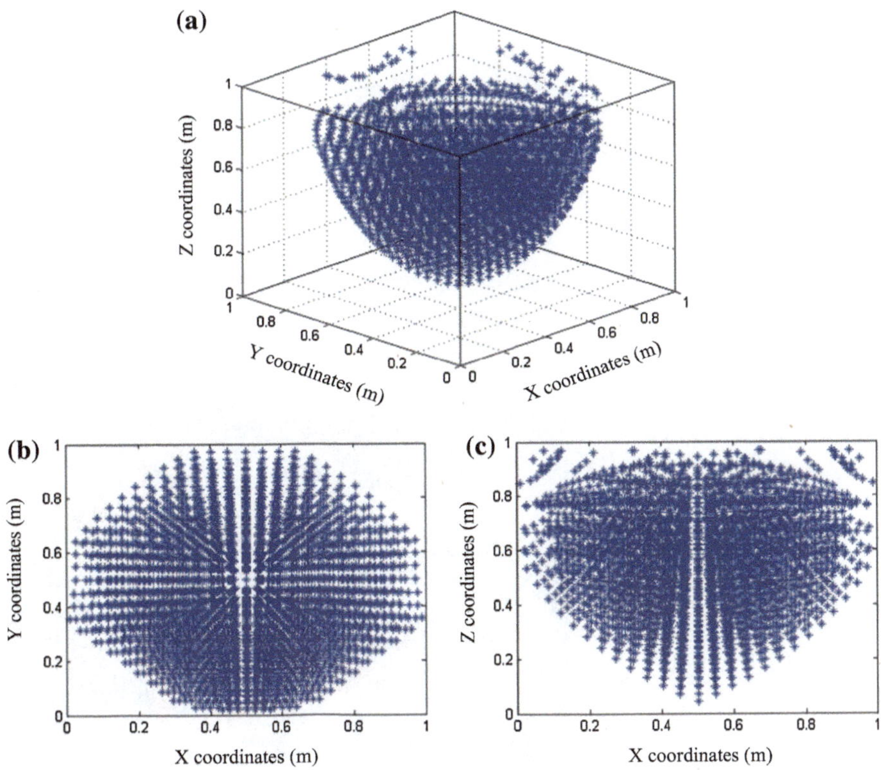

Fig. 4.26 Workspace of the HDCPR (**a**) and projections of workspace onto *X*-*Y* plane (**b**) and *X*-*Z* plane (**c**)

4.3.1 System Description

In this research, for the purpose of numerical analysis, the three-dimensional design model of the incompletely restrained HDCPR with three translational motions is taken as an example (see Fig. 4.27). The HDCPR suspends an end-effector by three cables and restrains all motion degrees of freedom for the object using the cables and gravitational force when the end-effector moves within the workspace. For each cable, one end is connected to the end-effector, the other one rolls through a pulley fixed on the top of the relative cable tower rack and then is fed into the HDPM. The HDCPR comprises of two modules: (1) the CPR consisting of three-cable tower racks, three cables, pulley struts, pulleys, girder, cargo (i.e. end-effector); (2) three groups of HDPMs containing three-phase asynchronous motors, servo-motors, reducers and double crank five-bar linkage; the asynchronous motors are connected using the pulley transmission mechanisms, while the servomotors and the reducers are linked by couplings. At the same time, the pulley transmission mechanisms and reducers are joined to the double crank planar five-bar linkage. The double crank planar five-bar linkage is made up of a large crank disk, a long connecting rod, a small crank disk, and a short connecting rod. Both long and short connecting rods are articulated with each other and attached to the cable. Moreover, each tower rack of the CPR is equipped with a cable guide pulley.

A simple schematic sketch of the HDCPR structure model with the associated coordinate systems are depicted in Fig. 4.28. At the bottom of one cable tower rack, a global coordinate system O(XYZ) is established. The end-effector has location

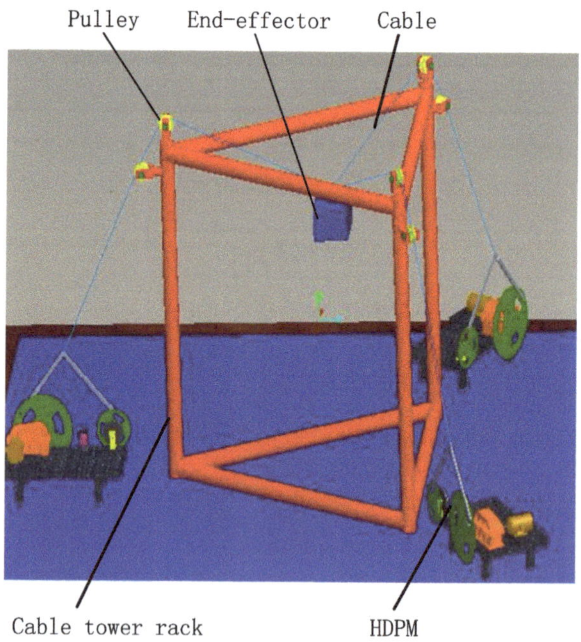

Fig. 4.27 Three-dimensional model of the HDCPR

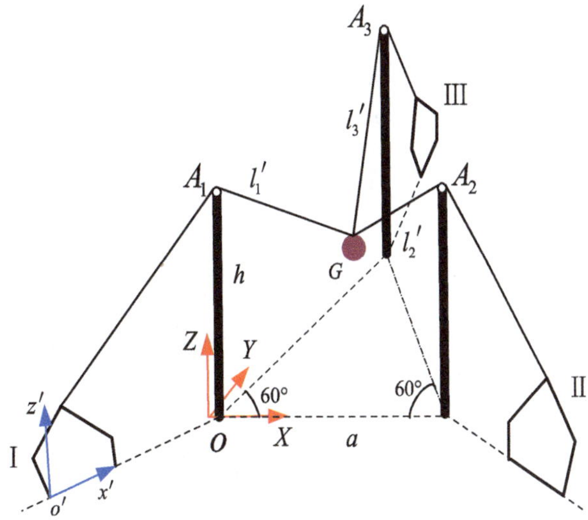

Fig. 4.28 Schematic sketch of the HDCPR

coordinates G(X, Y, Z), with mass m. The distance between each cable tower rack top $A_i(Xi, Yi, Z_i)$ and the end-effector is l_i' (i = 1, 2, 3). The three cable tower racks have same height and are arrayed in a triangle on the ground, when deformation is ignored. In order to simplify the model, the cables are treated as a massless body with no deformation. At the bottom of the HDPM, a local coordinate system $(x'o'z')$ is established.

4.3.2 Dynamic Modeling

Mathematical dynamic models of the HDCPR system are essential for good control design and analysis. In order to obtain a sufficient representation of the dynamical HDCPR behavior, a rigorous model of the HDCPR is derived using the method of Lagrange method. Figure 4.29 displays the structure diagram of the ith group of the HDPM and cable. Link AE is fixed on the ground. Link AB driven by a CV motor and link DE driven by a servomotor are two driving links. Link BC and link CD are driven by the two driving links, so that the joint C is the output. This section assumes that the links have the same density, and the centroid is in the center of links. The length of the links are l_1, l_2, l_3, l_4, l_5, the angel between links and x'-axis are $\theta_{i1}, \theta_{i2}, \theta_{i3}, \theta_{i4}, 0°$, the mass of links are m_1, m_2, m_3, m_4, the rotational inertia of links are J_1, J_2, J_3, J_4, the mass of the end-effector is m, respectively.

Angular displacement θ_{i1} and θ_{i4} of the two driving links are selected to be as generalized coordinates, and the angular velocity of all links can be expressed as

$$\dot{\theta}_{ij} = \frac{\partial \theta_{ij}}{\partial \theta_{i1}} \dot{\theta}_{i1} + \frac{\partial \theta_{ij}}{\partial \theta_{i4}} \dot{\theta}_{i4} \quad \mathrm{j} = 1, 2, 3, 4 \tag{4.29}$$

The centroid velocity of each link in x' and z' directions can be expressed as follows

$$\begin{cases} \dot{x}'_{ij} = \frac{\partial x'_{ij}}{\partial \theta_{i1}} \dot{\theta}_{i1} + \frac{\partial x'_{ij}}{\partial \theta_{i4}} \dot{\theta}_{i4} \\ \dot{z}'_{ij} = \frac{\partial z'_{ij}}{\partial \theta_{i1}} \dot{\theta}_{i1} + \frac{\partial z'_{ij}}{\partial \theta_{i4}} \dot{\theta}_{i4} \end{cases} \quad \mathrm{j} = 1, 2, 3, 4 \tag{4.30}$$

From Eqs. (4.29) and (4.30), one can derive the kinetic energy of the ith group of the HDPM

$$\begin{aligned} E_{li} &= \frac{1}{2} \sum_{j=1}^{4} \left[m_j \left(\dot{x}'^2_{ij} + \dot{z}'^2_{ij} \right) + J_j \dot{\theta}^2_{ij} \right] \\ &= \frac{1}{2} d_{i1} \dot{\theta}^2_{i1} + d_{i2} \dot{\theta}_{i1} \dot{\theta}_{i4} + \frac{1}{2} d_{i3} \dot{\theta}^2_{i4} \end{aligned} \tag{4.31}$$

where

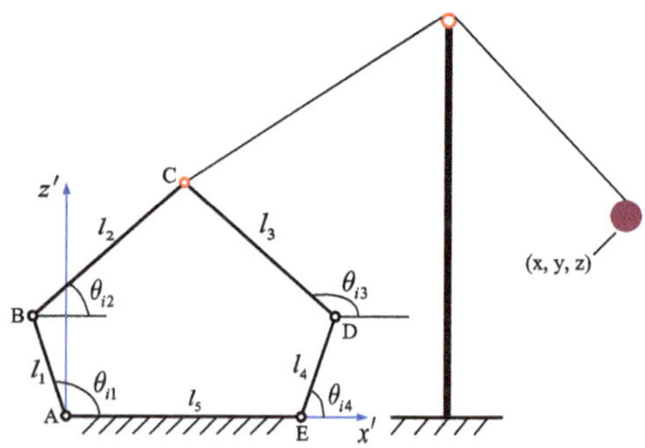

Fig. 4.29 Structure diagram of the *i*th group of the HDPM and cable

$$d_{i1} = \sum_{j=1}^{4}\left[m_j\left(\frac{\partial x'_{ij}}{\partial \theta_{i1}}\right)^2 + m_j\left(\frac{\partial z'_{ij}}{\partial \theta_{i1}}\right)^2 + J_j\left(\frac{\partial \theta_{ij}}{\partial \theta_{i1}}\right)^2\right]$$

$$d_{i2} = \sum_{j=1}^{4}\left[m_j\left(\frac{\partial x'_{ij}}{\partial \theta_{i1}}\right)\left(\frac{\partial x'_{ij}}{\partial \theta_{i4}}\right) + m_j\left(\frac{\partial z'_{ij}}{\partial \theta_{i1}}\right)\left(\frac{\partial z'_{ij}}{\partial \theta_{i4}}\right) + J_j\left(\frac{\partial \theta_{ij}}{\partial \theta_{i1}}\right)\left(\frac{\partial \theta_{ij}}{\partial \theta_{i4}}\right)\right]$$

$$d_{i3} = \sum_{j=1}^{4}\left[m_j\left(\frac{\partial x'_{ij}}{\partial \theta_{i4}}\right)^2 + m_j\left(\frac{\partial z'_{ij}}{\partial \theta_{i4}}\right)^2 + J_j\left(\frac{\partial \theta_{ij}}{\partial \theta_{i4}}\right)^2\right]$$

Then, the sum of kinetic energy of three groups of the HDPM is the following

$$E_l = \sum_{i=1}^{3} E_{li} \tag{4.32}$$

The velocity of the end-effector of the HDCPR in X, Y, and Z directions can be written as

$$\begin{cases} \dot{V}_X = \sum_{i=1}^{3}\left(\frac{\partial X}{\partial \theta_{i1}}\dot{\theta}_{i1} + \frac{\partial X}{\partial \theta_{i4}}\dot{\theta}_{i4}\right) \\ \dot{V}_Y = \sum_{i=1}^{3}\left(\frac{\partial Y}{\partial \theta_{i1}}\dot{\theta}_{i1} + \frac{\partial Y}{\partial \theta_{i4}}\dot{\theta}_{i4}\right) \\ \dot{V}_Z = \sum_{i=1}^{3}\left(\frac{\partial Z}{\partial \theta_{i1}}\dot{\theta}_{i1} + \frac{\partial Z}{\partial \theta_{i4}}\dot{\theta}_{i4}\right) \end{cases} \tag{4.33}$$

Thus, the following kinetic energy of the end-effector of the HDCPR will be obtained

$$\begin{aligned} \tilde{E} &= \frac{1}{2}m\left(\dot{V}_X^2 + \dot{V}_Y^2 + \dot{V}_Z^2\right) \\ &= \sum_{i=1}^{3}\left(\frac{1}{2}m\tilde{d}_{i1}\dot{\theta}_{i1}^2 + \frac{1}{2}m\tilde{d}_{i3}\dot{\theta}_{i4}^2\right) + m\tilde{d}\dot{\theta}_{i1}\dot{\theta}_{i4} \end{aligned} \tag{4.34}$$

where

$$\tilde{d}_{i1} = \left(\frac{\partial X}{\partial \theta_{i1}}\right)^2 + \left(\frac{\partial Y}{\partial \theta_{i1}}\right)^2 + \left(\frac{\partial Z}{\partial \theta_{i1}}\right)^2$$

$$\tilde{d}_{i3} = \left(\frac{\partial X}{\partial \theta_{i4}}\right)^2 + \left(\frac{\partial Y}{\partial \theta_{i4}}\right)^2 + \left(\frac{\partial Z}{\partial \theta_{i4}}\right)^2$$

$$\tilde{d} = \sum_{i=1}^{3}\frac{\partial X}{\partial \theta_{i1}}\sum_{i=1}^{3}\frac{\partial X}{\partial \theta_{i4}} + \sum_{i=1}^{3}\frac{\partial Y}{\partial \theta_{i1}}\sum_{i=1}^{3}\frac{\partial Y}{\partial \theta_{i4}} + \sum_{i=1}^{3}\frac{\partial Z}{\partial \theta_{i1}}\sum_{i=1}^{3}\frac{\partial Z}{\partial \theta_{i4}}$$

In view of Eqs. (4.32) and (4.34), the total kinetic energy of the HDCPR system can be generated as

$$E = E_l + \tilde{E} \tag{4.35}$$

The potential energy of the HDCPR system can be expressed as

$$\tilde{G} = m_1 g \sum_{i=1}^{3} z'_{i1} + m_2 g \sum_{i=1}^{3} z'_{i2} + m_3 g \sum_{i=1}^{3} z'_{i3} + m_4 g \sum_{i=1}^{3} z'_{i4} + mgZ \tag{4.36}$$

Using Lagrange's equation:

$$\frac{\partial}{\partial t}\left(\frac{\partial E}{\partial \dot{\theta}}\right) - \frac{\partial E}{\partial \theta} + \frac{\partial \tilde{G}}{\partial \theta} = F \tag{4.37}$$

where $\theta = (\theta_{i1}\ \theta_{i4})^{\mathrm{T}} (\mathrm{i} = 1, 2, 3)$ represents the generalized coordinates, F is the generalized forces corresponding to generalized coordinates.

Substituting the Eqs. (4.31)–(4.36) into Eq. (4.37), yields

$$F = \frac{\partial}{\partial t}\left(\frac{\partial E_l}{\partial \dot{\theta}}\right) - \frac{\partial E_l}{\partial \theta} + \frac{\partial}{\partial t}\left(\frac{\partial \tilde{E}}{\partial \dot{\theta}}\right) - \frac{\partial \tilde{E}}{\partial \theta} + \frac{\partial \tilde{G}}{\partial \theta} \tag{4.38}$$

where

$$\begin{aligned} \frac{\partial}{\partial t}\left(\frac{\partial E_l}{\partial \dot{\theta}}\right) - \frac{\partial E_l}{\partial \theta} &= \begin{pmatrix} d_{i1} & d_{i2} \\ d_{i2} & d_{i3} \end{pmatrix} \begin{pmatrix} \ddot{\theta}_{i1} \\ \ddot{\theta}_{i4} \end{pmatrix} + \begin{pmatrix} \frac{1}{2}\frac{\partial d_{i1}}{\partial \theta_{i1}}\dot{\theta}_{i1} + \frac{\partial d_{i1}}{\partial \theta_{i4}}\dot{\theta}_{i4} & \left(\frac{\partial d_{i2}}{\partial \theta_{i4}} - \frac{1}{2}\frac{\partial d_{i3}}{\partial \theta_{i1}}\right)\dot{\theta}_{i4} \\ \left(\frac{\partial d_{i2}}{\partial \theta_{i1}} - \frac{1}{2}\frac{\partial d_{i1}}{\partial \theta_{i4}}\right)\dot{\theta}_{i1} & \frac{1}{2}\frac{\partial d_{i3}}{\partial \theta_{i4}}\dot{\theta}_{i4} + \frac{\partial d_{i3}}{\partial \theta_{i1}}\dot{\theta}_{i1} \end{pmatrix} \begin{pmatrix} \dot{\theta}_{i1} \\ \dot{\theta}_{i4} \end{pmatrix} \\ &= M_{li}\ddot{\theta} + C_{li}\dot{\theta} \quad (i = 1, 2, 3) \end{aligned} \tag{4.39}$$

$$\begin{aligned} \frac{\partial}{\partial t}\left(\frac{\partial \tilde{E}}{\partial \dot{\theta}}\right) - \frac{\partial \tilde{E}}{\partial \theta} &= \begin{pmatrix} m\tilde{d}_{i1} & m\tilde{d} \\ m\tilde{d} & m\tilde{d}_{i3} \end{pmatrix} \begin{pmatrix} \ddot{\theta}_{i1} \\ \ddot{\theta}_{i4} \end{pmatrix} + \begin{pmatrix} \frac{m}{2}\frac{\partial \tilde{d}_{i1}}{\partial \theta_{i1}}\dot{\theta}_{i1} + m\frac{\partial \tilde{d}_{i1}}{\partial \theta_{i4}}\dot{\theta}_{i4} & \left(m\frac{\partial \tilde{d}}{\partial \theta_{i4}} - \frac{m}{2}\frac{\partial \tilde{d}_{i3}}{\partial \theta_{i1}}\right)\dot{\theta}_{i4} \\ \left(m\frac{\partial \tilde{d}}{\partial \theta_{i1}} - \frac{m}{2}\frac{\partial \tilde{d}_{i1}}{\partial \theta_{i4}}\right)\dot{\theta}_{i1} & \frac{m}{2}\frac{\partial \tilde{d}_{i3}}{\partial \theta_{i4}}\dot{\theta}_{i4} + m\frac{\partial \tilde{d}_{i3}}{\partial \theta_{i1}}\dot{\theta}_{i1} \end{pmatrix} \begin{pmatrix} \dot{\theta}_{i1} \\ \dot{\theta}_{i4} \end{pmatrix} \\ &= \tilde{M}_i\ddot{\theta} + \tilde{C}_i\dot{\theta} \quad (\mathrm{i} = 1, 2, 3) \end{aligned} \tag{4.40}$$

$$\frac{\partial \tilde{G}}{\partial \theta} = G \tag{4.41}$$

In the case of unknown disturbances, the generalized forces can be written as

$$F = \tau - \tau' \tag{4.42}$$

where $\tau = (\,\tau_{i1} \quad \tau_{i4}\,)^{\mathrm{T}}$ $(\mathrm{i} = 1, 2, 3)$ represents the input torque of all driving links, $\tau' = (\,\tau'_{i1} \quad \tau'_{i4}\,)^{\mathrm{T}}(\mathrm{i} = 1, 2, 3)$ represents the unknown repeated disturbances.

According to Eqs. (4.38)–(4.42), dynamic equation of the HDCPR system can be derived as

$$M(\theta)\ddot{\theta} + C(\theta, \dot{\theta})\dot{\theta} + G(\theta) = \tau - \tau' \tag{4.43}$$

where $M = M_{li} + \tilde{M}_i$, $M(\theta)$ is the symmetric positive definite inertia matrix; $C = C_{li} + \tilde{C}_i$, $C(\theta, \dot{\theta})$ is the centrifugal force and Coriolis force matrix, and $\dot{M}(\theta) - 2C(\theta, \dot{\theta})$ is skew symmetric; $G(\theta)$ is the gravity matrix.

4.3.3 Adaptive ILC Design

An adaptive ILC method is proposed for control of the HDCPR system. Consider the dynamic model (4.43) of the HDCPR system in iteration domain, it can be described as

$$M(\theta^n(t))\ddot{\theta}^n(t) + C\left(\theta^n(t), \dot{\theta}^n(t)\right)\dot{\theta}^n(t) + G(\theta^n(t)) = \tau^n(t) - \tau'(t) \tag{4.44}$$

where $t \in [0, T]$ is the time, $n \in N$ denotes the operation or iteration number.

The system (4.44) meets the following conditions:

(B1) The desired input angular $\theta_d(t)$ is of the third-order continuity for $t \in [0, T]$.
(B2) For each iteration, the same initial conditions are satisfied:

$$\theta_d(0) - \theta^n(0) = 0, \dot{\theta}_d(0) - \dot{\theta}^n(0) = 0$$

In the iterative operational domain, a simple iterative learning control is applied as feedback PD control where information from previous operations is used. For system (16), consider the nth iterative operation with requirement B1 and B2 in the light of the following control law:

$$\tau^n(t) = K_p^n e^n(t) + K_d^n \dot{e}^n(t) + \tau^{n-1}(t) \tag{4.45}$$

where $K_p^n = 2nK_p^0$, $K_d^n = 2nK_d^0$, $K_p^0 = \Lambda K_d^0$, $K_d^0 > 0$ $(n = 1, 2, \cdots, N)$, K_p^0 represents the initial proportional control gain, K_d^0 represents the initial derivative control gain, K_p^0 and K_d^0 are the diagonal matrix. K_p^n and K_d^n are the control gains of the nth iteration $\tau^{-1}(t) = 0, e^n(t) = \theta_d(t) - \theta^n(t), \dot{e}^n(t) = \dot{\theta}_d(t) - \dot{\theta}^n(t)$.

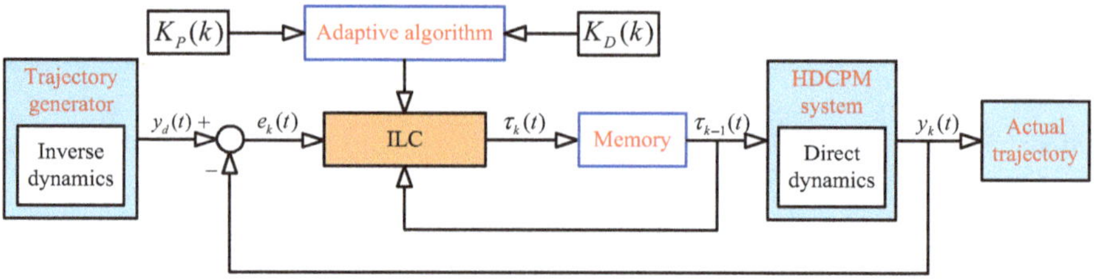

Fig. 4.30 Block diagram of adaptive iterative learning control system for the HDCPR

The gain computational method is used to adjust the PD gains from iteration to iteration. Such a control method acts not only in the time domain but also in the iterative domain, which is the main difference with the traditional control method. The block diagram of adaptive ILC system for the HDCPR is shown in Fig. 4.30.

Equation (4.44) can be linearized by Taylor formula along the desired value $\left(\theta_d(t), \dot{\theta}_d(t), \ddot{\theta}_d(t)\right)$ as follows

$$M(t)\ddot{e}^n(t) + [C(t) + C_1(t)]\dot{e}^n(t) + F(t)e^n(t) + \tau^n(t) - \tau'(t) = H(t) \tag{4.46}$$

where

$$M(t) = M(\theta_d(t))$$

$$C(t) = C\left(\theta_d(t), \dot{\theta}_d(t)\right)$$

$$C_1(t) = \left.\frac{\partial C}{\partial \dot{\theta}}\right|_{\theta_d(t), \dot{\theta}_d(t)} \dot{\theta}_d(t)$$

$$F(t) = \left.\frac{\partial M}{\partial \theta}\right|_{\theta_d(t)} \ddot{\theta}_d(t) + \left.\frac{\partial C}{\partial \theta}\right|_{\theta_d(t), \dot{\theta}_d(t)} \dot{\theta}_d(t) + \left.\frac{\partial G}{\partial \theta}\right|_{\theta_d(t)}$$

$$H(t) = M(\theta_d(t))\ddot{\theta}_d(t) + C\left(\theta_d(t), \dot{\theta}_d(t)\right)\dot{\theta}_d(t) + G(\theta_d(t))$$

For the n + 1th iteration, Eq. (4.46) can be rewritten as follows

$$M(t)\ddot{e}^{n+1}(t) + [C(t) + C_1(t)]\dot{e}^{n+1}(t) + F(t)e^{n+1}(t) + \tau^{n+1}(t) - \tau'(t) = H(t) \tag{4.47}$$

Now, one can state the following theorem.

Theorem Consider the HDCPR system (4.43) with properties B1 and B2, the adaptive iterative learning control law (4.45) is chosen such that the following conditions are satisfied:

$$\begin{cases} l_p = \lambda_{\min}\left(K_d^0 + 2C_1 - 2\Lambda M\right) > 0 \\ l_r = \lambda_{\min}\left(K_d^0 + 2C + 2F/\Lambda - 2\dot{C}_1/\Lambda\right) > 0 \\ l_p l_r \geq \|F/\Lambda - (C + C_1 - \Lambda M)\|_{\max}^2 \end{cases} \tag{4.48}$$

Then, for $t \in [0, T]$, the resulting HDCPR system guarantees $\theta^n(t) \rightarrow \theta_d(t)$ and $\dot{\theta}^n(t) \rightarrow \dot{\theta}_d(t)$ as $n \rightarrow \infty$, and it is asymptotically stable.

Proof To analyze the stability of the HDCPR system, a Lyapunov function candidate V is proposed as

$$V^n = \int_0^t \mathrm{e}^{-\rho\upsilon} y^{n^{\mathrm{T}}} K_d^0 y^n d\upsilon \geq 0 \tag{4.49}$$

where $y^n = \dot{e}^n + \Lambda e^n$, and ρ is a positive constant.

Define $\delta y^n = y^{n+1} - y^n$ and $\delta e^n = e^{n+1} - e^n$, then

$$\delta y^n = \delta \dot{e}^n + \Lambda \delta e^n \tag{4.50}$$

In view of Eqs. (4.45)–(4.47) and (4.50), one can get:

$$M\delta\dot{y}^n + \left(C + C_1 - \Lambda M + K_d^{n+1}\right)\delta y^n + [F - \Lambda(C + C_1 - \Lambda M)]\delta e^n = -K_d^{n+1} y^n \tag{4.51}$$

Define $\Delta V^n = V^{n+1} - V^n$, combining (4.49), (4.50), and (4.51), the following equation can be derived:

$$\begin{aligned} \Delta V^n &= \int_0^t \mathrm{e}^{-\rho\upsilon}\left(\delta y^{n^{\mathrm{T}}} K_d^0 \delta y^n + 2\delta y^{n^{\mathrm{T}}} K_d^0 y^n\right) d\upsilon \\ &= \frac{1}{2(n+1)} \int_0^t \mathrm{e}^{-\rho\upsilon}\left(\delta y^{n^{\mathrm{T}}} K_d^{n+1} \delta y^n + 2\delta y^{n^{\mathrm{T}}} K_d^{n+1} y^n\right) d\upsilon \\ &= \frac{1}{2(n+1)} \left\{ \int_0^t \mathrm{e}^{-\rho\upsilon} \delta y^{n^{\mathrm{T}}} K_d^{n+1} \delta y^n d\upsilon - 2\int_0^t \mathrm{e}^{-\rho\upsilon} \delta y^{n^{\mathrm{T}}} M \delta \dot{y}^n d\upsilon \right. \\ &\quad \left. - 2\int_0^t \mathrm{e}^{-\rho\upsilon} \delta y^{n^{\mathrm{T}}} \left[\left(C + C_1 - \Lambda M + K_d^{n+1}\right)\delta y^n + (F - \Lambda(C + C_1 - \Lambda M))\delta y^n\right] d\upsilon \right. \end{aligned} \tag{4.52}$$

Applying the partial integration, and considering $\delta y^n(0) = 0$, we have:

$$\begin{aligned} \int_0^t \mathrm{e}^{-\rho\upsilon} \delta y^{n^{\mathrm{T}}} M \delta \dot{y}^n d\upsilon &= \mathrm{e}^{-\rho\upsilon} \delta y^{n^{\mathrm{T}}} M \delta y^n + \rho \int_0^t \mathrm{e}^{-\rho\upsilon} \delta y^{n^{\mathrm{T}}} M \delta y^n d\upsilon \\ &\quad - \int_0^t \mathrm{e}^{-\rho\upsilon} \delta y^{n^{\mathrm{T}}} M \delta \dot{y}^n d\upsilon - \int_0^t \mathrm{e}^{-\rho\upsilon} \delta y^{n^{\mathrm{T}}} \dot{M} \delta y^n d\upsilon \end{aligned} \tag{4.53}$$

Substituting (4.53) into (4.52), one has:

$$\Delta V^n = \frac{1}{2(n+1)}\Big\{-\mathrm{e}^{-\rho v}\delta y^{n^{\mathrm{T}}} M\delta y^n - \rho\int_0^t \mathrm{e}^{-\rho v}\delta y^{n^{\mathrm{T}}} M\delta y^n dv$$
$$- 2\int_0^t \mathrm{e}^{-\rho v}\delta y^{n^{\mathrm{T}}}[(F - \Lambda(C + C_1 - \Lambda M))\delta y^n]dv \tag{4.54}$$
$$- \int_0^t \mathrm{e}^{-\rho v}\delta y^{n^{\mathrm{T}}}\big(2C_1 - 2\Lambda M + K_d^{n+1}\big)\delta y^n dv\Big\}$$

Consider the following equation:

$$\int_0^t \mathrm{e}^{-\rho v}\delta y^{n^{\mathrm{T}}} K_d^{n+1}\delta y^n dv = 2(n+1)\int_0^t \mathrm{e}^{-\rho v}\delta y^{n^{\mathrm{T}}} K_d^0\delta y^n dv \geq \int_0^t \mathrm{e}^{-\rho v}\delta y^{n^{\mathrm{T}}} K_d^0\delta y^n dv \tag{4.55}$$

Combing (4.50) and (4.54), and consider (4.55), one can obtain the following equation:

$$\Delta V^n \leq \frac{1}{2(n+1)}\Big\{-\mathrm{e}^{-\rho v}\delta y^{n^{\mathrm{T}}} M\delta y^n - \rho\int_0^t \mathrm{e}^{-\rho v}\delta y^{n^{\mathrm{T}}} M\delta y^n dv - \int_0^t \mathrm{e}^{-\rho v}\delta \dot{e}^{n^{\mathrm{T}}}\big(2C_1 - 2\Lambda M + K_d^0\big)\delta\dot{e}^n dv$$
$$- 2\Lambda\int_0^t \mathrm{e}^{-\rho v}\delta e^{n^{\mathrm{T}}}\big(2C_1 - 2\Lambda M + K_d^0\big)\delta\dot{e}^n dv - 2\int_0^t \mathrm{e}^{-\rho v}\delta\dot{e}^{n^{\mathrm{T}}}[F - \Lambda(C + C_1 - \Lambda M)]\delta e^n dv$$
$$- \Lambda^2\int_0^t \mathrm{e}^{-\rho v}\delta e^{n^{\mathrm{T}}}\big(2C_1 - 2\Lambda M + K_d^0\big)\delta e^n dv - 2\Lambda\int_0^t \mathrm{e}^{-\rho v}\delta e^{n^{\mathrm{T}}}[F - \Lambda(C + C_1 - \Lambda M)]\delta e^n dv\Big\}$$

For the sake of simplicity, we further have:

$$\Delta V^n \leq \frac{1}{2(n+1)}\Big\{-\mathrm{e}^{-\rho v}\delta y^{n^{\mathrm{T}}} M\delta y^n - \rho\int_0^t \mathrm{e}^{-\rho v}\delta y^{n^{\mathrm{T}}} M\delta y^n dv$$
$$-\rho\Lambda\int_0^t \mathrm{e}^{-\rho v}\delta e^{n^{\mathrm{T}}} l_p\delta e^n dv - \Lambda \mathrm{e}^{-\rho v}\delta e^{n^{\mathrm{T}}} l_p\delta e^n - \int_0^t \omega \mathrm{e}^{-\rho v}dv\Big\} \tag{4.56}$$

where

$$\omega = \delta\dot{e}^{n^{\mathrm{T}}}\big(2C_1 - 2\Lambda M + K_d^0\big)\delta\dot{e}^n$$
$$+ 2\Lambda\delta\dot{e}^{n^{\mathrm{T}}}[F/\Lambda - (C + C_1 - \Lambda M)]\delta e^n$$
$$+ \Lambda^2\delta e^{n^{\mathrm{T}}}\big(K_d^0 + 2C + 2F/\Lambda - 2\dot{C}_1/\Lambda\big)\delta e^n$$

Let $Q = F/\Lambda - (C + C_1 - \Lambda M)$, then from (4.48), we can get:

$$\omega \geq l_p\|\delta\dot{e}\|^2 + 2\Lambda\delta\dot{e}^{\mathrm{T}} Q\delta e + \Lambda^2 l_r\|\delta e\|^2 \tag{4.57}$$

Using the Cauchy-Schwartz inequality equation:

$$\delta \dot{e}^{\mathrm{T}} Q \delta e \geq \|\delta \dot{e}\| \|Q\|_{\max} \|\delta e\|$$

According to (4.48), (4.57) can be rewritten as

$$\omega \geq l_p \left(\|\delta \dot{e}\| - \frac{\Lambda}{l_p} \|Q\|_{\max} \|\delta e\| \right)^2 + \Lambda^2 \left(l_p - \frac{1}{l_r} \|Q\|_{\max}^2 \right) \|\delta e\|^2 \geq 0$$

Due to the $M(\theta)$ is the symmetric positive definite inertia matrix, and from (4.48), (4.56), we can conclude that $\Delta V^n \leq 0$. Therefore, $V^{n+1} - V^n \leq 0$.

Since K_d^0 is a positive definite matrix, $V^n > 0$ and V^n is bounded. As a result, $y^n(t) \to 0$ when $n \to \infty$. Then, $\dot{e}^n(t) \to 0$ and $e^n(t) \to 0$.

From above analysis it can be seen that for $t \in [0, T]$ under the action of controller (4.45) the tracking errors converge arbitrarily close to zero. Thus, $\theta^n(t) \to \theta_d(t)$ and $\dot{\theta}^n(t) \to \dot{\theta}_d(t)$ as $n \to \infty$. This completes the proof of Theorem.

4.3.4 Numerical Results and Discussion

In this section, simulation study is carried out to illustrate the performance of the HDCPR control system. All the simulations are performed with the software named MATLAB 2010, in which a three-dimensional simulation model of the HDCPR whose schematic sketch is shown in Fig. 4.28 has been established. Three groups of the HDPMs have the same structure parameters and the mechanisms have the symmetrical design in the three-dimensional space. The parameters of the HDCPR are listed in Table 4.2.

The desired path is selected as

Table 4.2 Parameters of the HDCPR

System parameters	Value
m	20 kg
g	9.81 m/s^2
h	1 m
a	1 m
l_1	0.2 m
l_2	0.5 m
l_3	0.5 m
l_4	0.28 m
l_5	0.51 m
Density of the links	7.85×10^{-3} kg/m^3
Cross sectional area of the Links	2×10^{-4} m^2

$$\begin{cases} X = 0.25\cos(0.5\pi t) + 0.5 \\ Y = 0.25\sin(0.5\pi t) + \sqrt{3}/6 \quad t \in [0, 4\text{s}] \\ Z = 0.5 \end{cases} \tag{4.58}$$

The initial control gains are set as follows

$$K_p^0 = K_d^0 = diag\,(200, 200)$$

Figure 4.31 shows the sport demo interface of the HDCPR control system when the end-effector tracks the desired circle trajectory in different time. From the spot demo interface of Fig. 4.31, it can be noted that the HDCPR system runs smoothly, and the movement of all joints is fluency, so by which tracking the desired trajectory movement is accompanied well.

Figure 4.32 shows the desired trajectory and actual trajectory of the end-effector along the given circle motion path in the different iterations. From it one can see

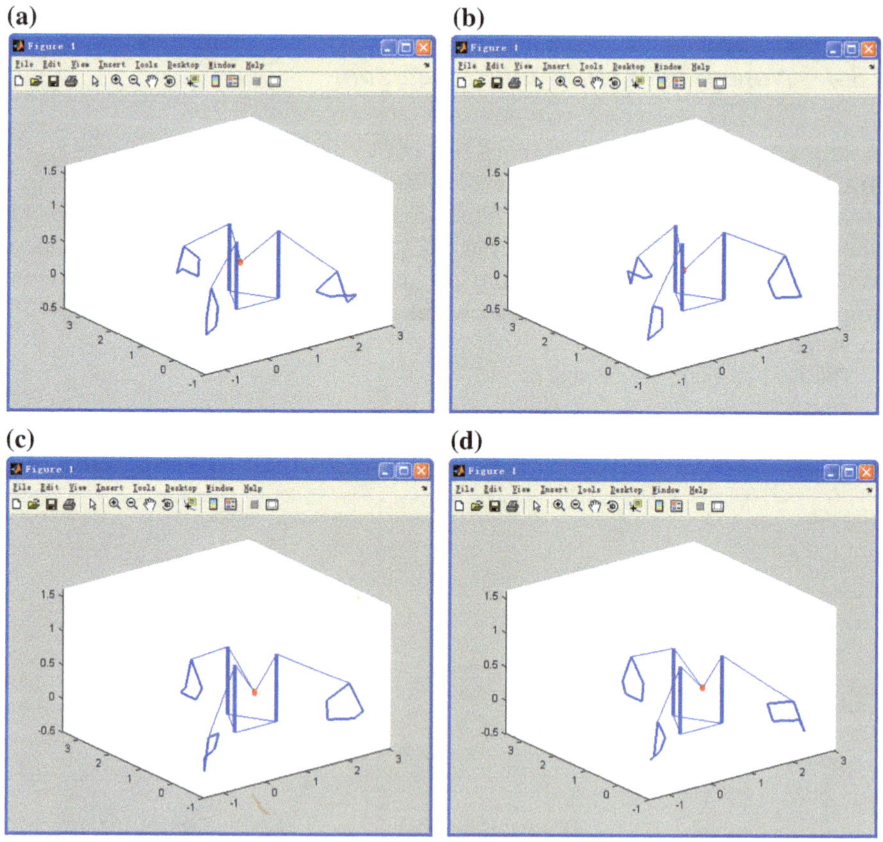

Fig. 4.31 Spot demo interface of the HDCPR: **a** t = 1 s, **b** t = 2 s, **c** t = 3 s, **d** t = 4 s

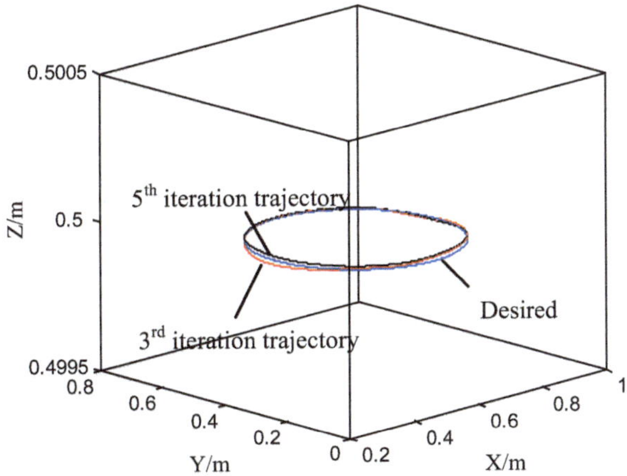

Fig. 4.32 Trajectory tracking of the end-effector

that, the actual trajectory tracks the desired trajectory with high accuracy, which also illustrate the correctness of the dynamic model of the HDCPR.

The tracking errors of the end-effector in X, Y and Z directions under the controller can be seen in Figs. 4.33, 4.34 and 4.35. respectively. It is clearly that the tracking errors are small and the corresponding tracking performance improvement from iteration to iteration, and eventually could converge close to zero. From the simulations, we can see that the performance of the proposed control strategy is effective for the HDCPR system.

Figure 4.36 displays the position tracking error curves of the six driving links of three groups of the HDPMs in different iterations. From it one can see that, at the initial iteration, the maximum position errors of the which driven by CV motors are 0.0008, 0.0017, and 0.0007 rad, respectively; the maximum position errors of the

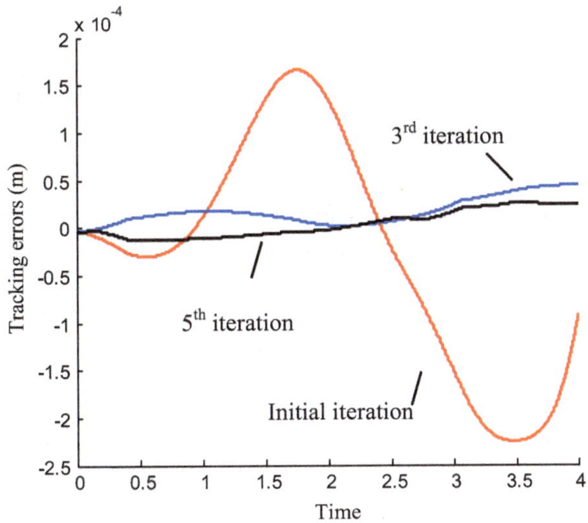

Fig. 4.33 Tracking errors in *X*-direction

Fig. 4.34 Tracking errors in *Y*-direction

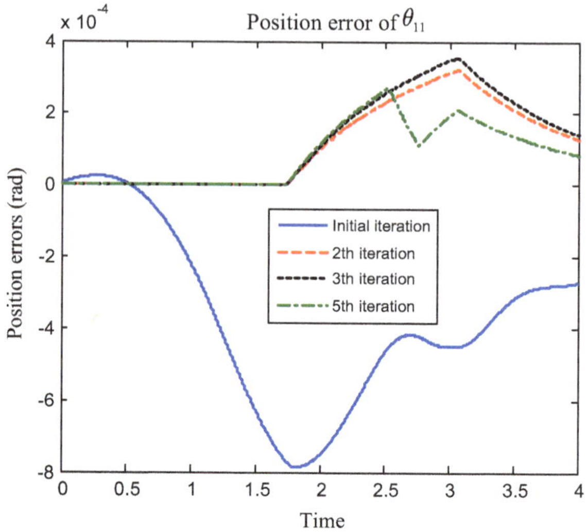

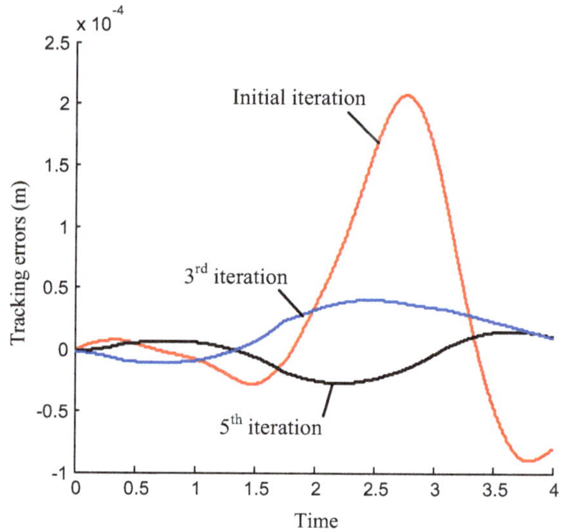

which driven by servomotors are 0.0024, 0.0026, and 0.0019 rad, respectively; finally, after five iterations, are reduced to 0.0003, 0.0001, and 0.0002 rad, respectively; are reduced to 0.0001, 0.00005, and 0.0001 rad, respectively.

Figure 4.37 shows the desired curves and tracking curves of cable length in different iterations. It is illustrated in Fig. 4.37, which shows the cable length vary reasonably and is consistent with the desired motion law.

The driving force of ith cable exerted on the end-effector is defined as a tension cable force Ti. Given the total and inertial forces together through every virtual displacement in X, Y, and Z directions, and using the assumptions along with

Fig. 4.35 Tracking errors in *Z*-direction

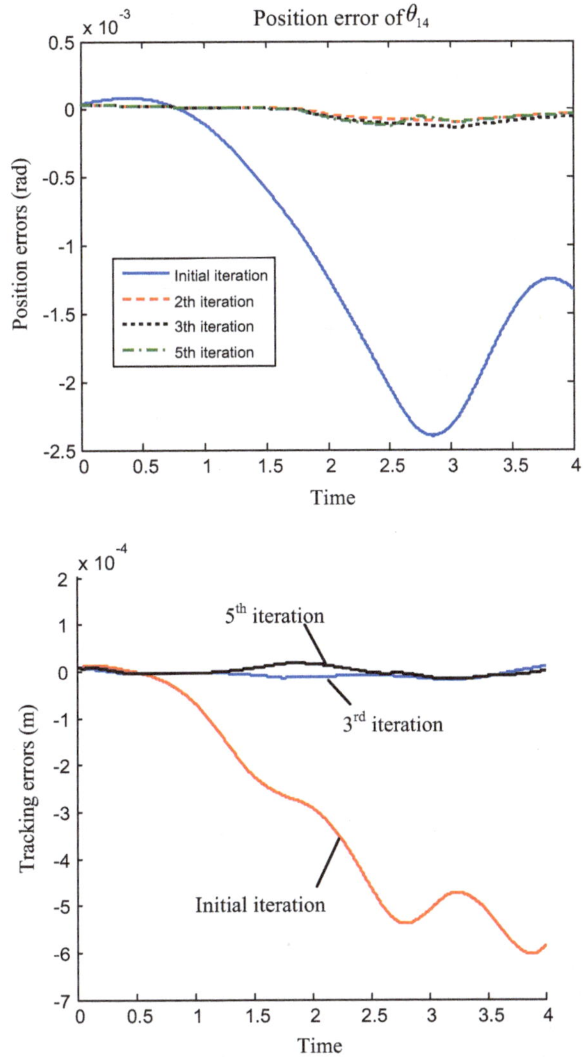

various substitutions and algebraic manipulations of the CPR are derived, the dynamic model is expressed as

$$N_{3\times3}T_{3\times1} = M_{3\times1} \tag{4.59}$$

$$N_{3\times3} = \begin{bmatrix} \frac{X_1 - X}{l'_1} & \frac{X_2 - X}{l'_2} & \frac{X_3 - X}{l'_3} \\ \frac{Y_1 - Y}{l'_1} & \frac{Y_2 - Y}{l'_2} & \frac{Y_3 - Y}{l'_3} \\ \frac{Z_1 - Z}{l'_1} & \frac{Z_2 - Z}{l'_2} & \frac{Z_3 - Z}{l'_3} \end{bmatrix} \tag{4.60}$$

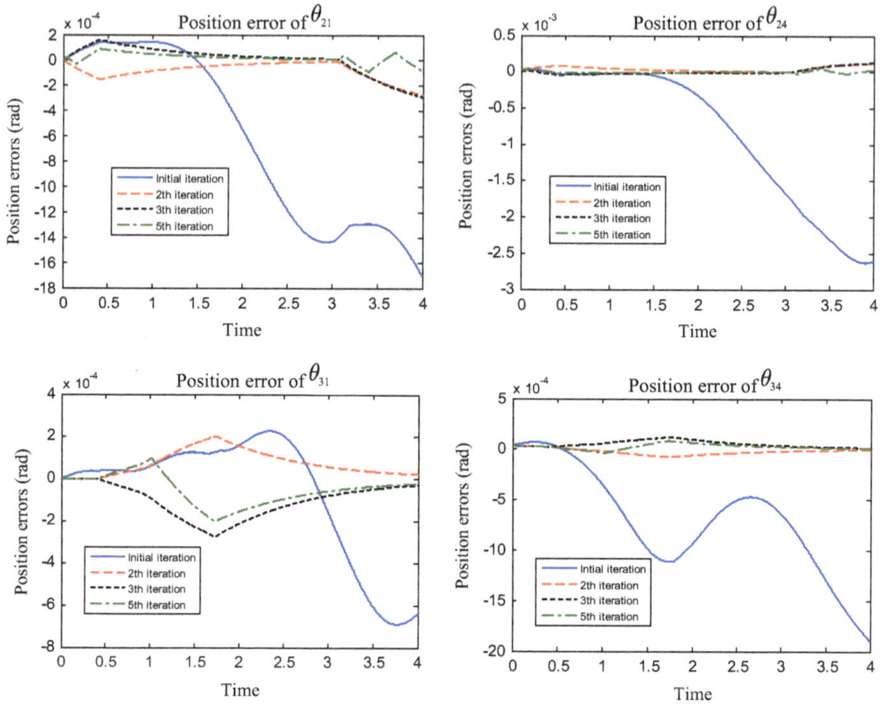

Fig. 4.36 Position tracking error curves of the six driving links in different iterations

where $N_{3\times 3}$ is the matrix of coordinate transformer, $T_{3\times 1} = [T_1 \quad T_2 \quad T_3]^{\mathrm{T}}$ is the vector of cable tension, $M_{3\times 1} = [m\ddot{x} \quad m\ddot{y} \quad mg + m\ddot{z}]^{\mathrm{T}}$ is the vector of inertia force, m is mass of the end-effector, and g the gravitational acceleration.

Figure 4.38 depicts the curves of cable tension exerted on the end-effector related to the circle trajectory. Since the end-effector moves on a horizontal circle trajectory, the changes of cable tensions in Fig. 4.38 are reasonable and the transition is smooth.

Referring to Figs. 4.37 and 4.38, it should be noted that, while the cable length tracking performance are improved from iteration to iteration, the cable forces to drive the end-effector are nearly the same from iteration to iteration after a few iterations. From the above simulation results, it may be concluded that the proposed control system can achieve a relatively favorable control performance and has high robustness.

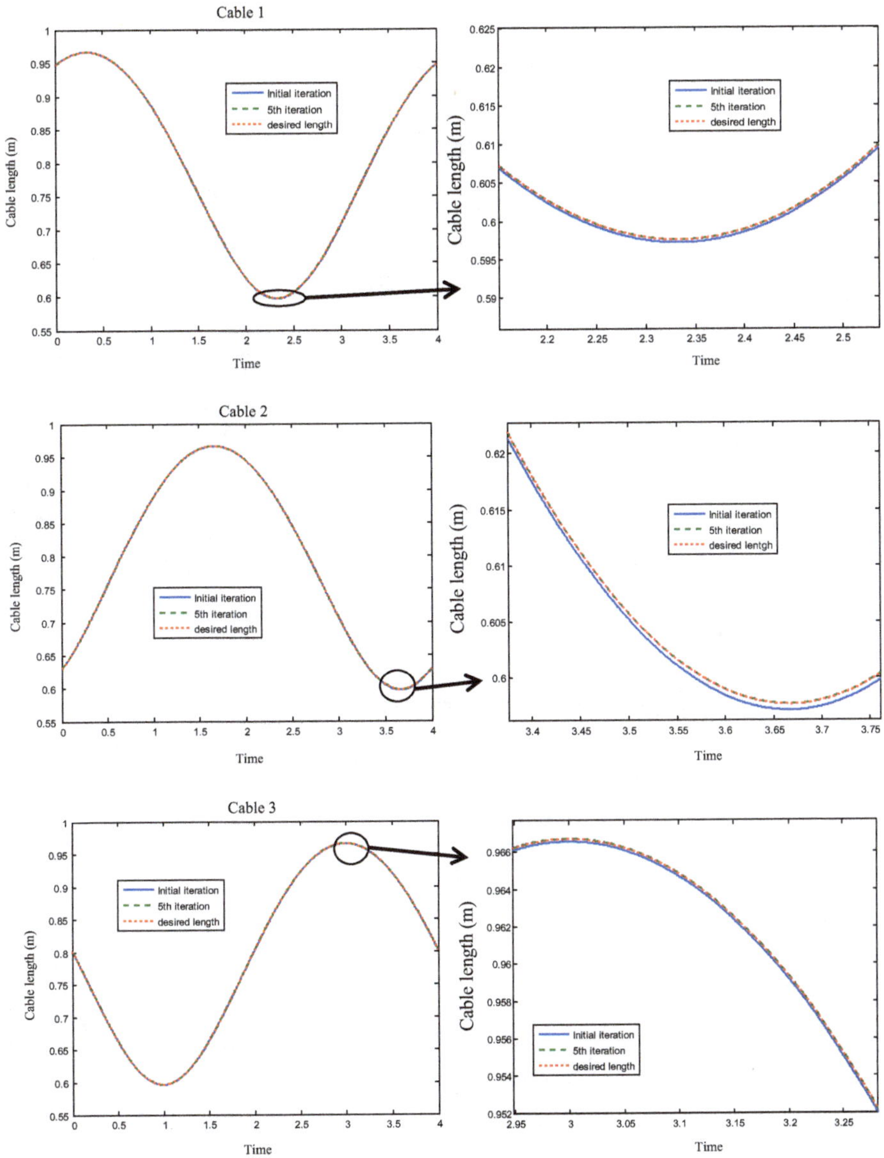

Fig. 4.37 Cable length tracking curves in different iterations

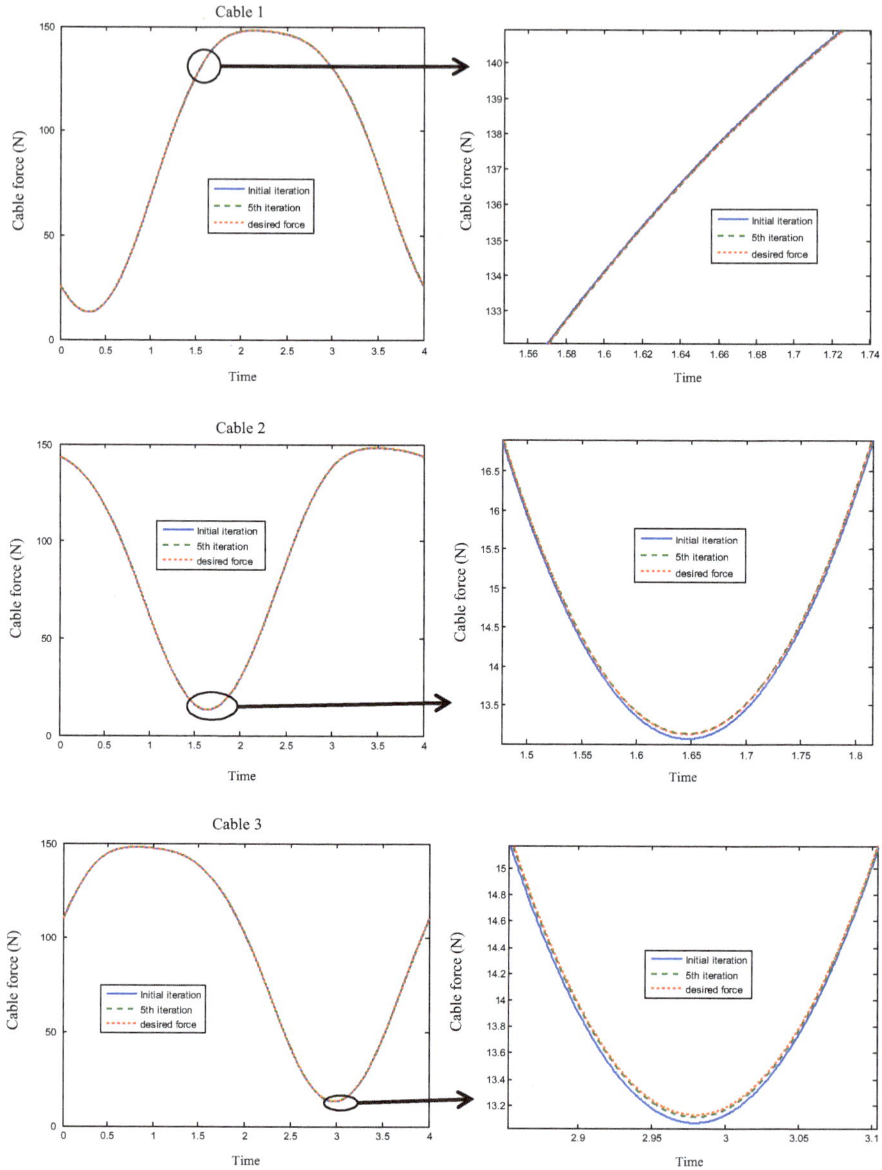

Fig. 4.38 Cable force tracking curves in different iterations

4.4 Integrated Mechanism Design and Control for Completely Restrained Hybrid-Driven Based Cable Parallel Manipulators

4.4.1 Integrated Optimization Design

In the following, our interest is focused on the integrated optimization design of mechanism and control for the HDCPR. In this work, we will take into account simultaneously workspace, kinematic and dynamic performances, and control performance indices that have a direct impact on the optimum performance of the whole HDCPR systems by using the concept of multi-objective optimization by the GAs.

The classical proportional-integral-derivative (PID) controller is still widely used in industrial applications due to its design simplicity and its excellent performance. Specifically, most of the manipulators employed in industrial operations are controlled by PID algorithms. In this investigation, a PID method is proposed for control of the HDCPR system. It is possible to find a set of suitable values for the PID controller to obtain the trade-off control performance of the HDCPR system by using optimization.

Consider the dynamic model of the HDCPR system, it can be expressed as

$$\boldsymbol{M}(\boldsymbol{\theta}(t))\ddot{\boldsymbol{\theta}}(t)+\boldsymbol{C}\big(\boldsymbol{\theta}(t),\dot{\boldsymbol{\theta}}(t)\big)\dot{\boldsymbol{\theta}}(t)+\boldsymbol{G}\big(\boldsymbol{\theta}(t),\dot{\boldsymbol{\theta}}(t)\big)=\boldsymbol{\tau}(t)-\boldsymbol{\tau}'(t) \tag{4.61}$$

In accordance with the dynamic model (4.61) of the HDCPR system, the PID controller can be described as

$$\boldsymbol{\tau}(t)=k_p\boldsymbol{e}(t)+k_i\int_0^t\boldsymbol{e}(t)dt+k_d\frac{d\boldsymbol{e}(t)}{dt} \tag{4.62}$$

where k_p, k_i, and k_d represent the proportional control gain, derivative control gain, and integral control gain, respectively. $\boldsymbol{e}(t)=\boldsymbol{\theta}_d(t)-\boldsymbol{\theta}(t)$ is the tracking error, $\boldsymbol{\theta}_d(t)$ is desired input angular, $\boldsymbol{\theta}(t)$ actual angular displacement.

Equation (4.62) can be linearized by Taylor formula along the desired value $\big(\boldsymbol{\theta}_d(t),\dot{\boldsymbol{\theta}}_d(t),\ddot{\boldsymbol{\theta}}_d(t)\big)$ as follows:

$$\boldsymbol{M}(t)\ddot{\boldsymbol{e}}(t)+[\boldsymbol{C}(t)+\boldsymbol{C}_1(t)]\dot{\boldsymbol{e}}(t)+\boldsymbol{F}(t)\boldsymbol{e}(t)+\boldsymbol{\tau}(t)-\boldsymbol{\tau}'(t)=\boldsymbol{H}(t) \tag{4.63}$$

where

$$\boldsymbol{M}(t)=\boldsymbol{M}(\boldsymbol{\theta}_d(t))$$

$$\boldsymbol{C}(t)=\boldsymbol{C}\big(\boldsymbol{\theta}_d(t),\dot{\boldsymbol{\theta}}_d(t)\big)$$

$$C_1(t) = \left.\frac{\partial C}{\partial \dot{\theta}}\right|_{\theta_d(t),\dot{\theta}_d(t)} \dot{\theta}_d(t) + \left.\frac{\partial G}{\partial \dot{\theta}}\right|_{\theta_d(t),\dot{\theta}_d(t)}$$

$$F(t) = \left.\frac{\partial M}{\partial \theta}\right|_{\theta_d(t)} \ddot{\theta}_d(t) + \left.\frac{\partial C}{\partial \theta}\right|_{\theta_d(t),\dot{\theta}_d(t)} \dot{\theta}_d(t) + \left.\frac{\partial G}{\partial \theta}\right|_{\theta_d(t)}$$

$$H(t) = M(\theta_d(t))\ddot{\theta}_d(t) + C\big(\theta_d(t), \dot{\theta}_d(t)\big)\dot{\theta}_d(t) + G(\theta_d(t))$$

Mathematically, a general optimization problem can be defined as follows [50]

find $\Re$
minimize $f(\Re)$
subject to $g(\Re) \leq 0$

where $\Re$ is the design variable vector, $f(\Re)$ is the objective function, $g(\Re) \leq 0$ is the set of inequality constraints.

In order that the integrated optimization design of mechanism and control for the HDCPR can meet the requirement for better dynamic performance and higher tracking precision, and all kinds of the constraint conditions, the following work needed to be performed. Firstly, we should determine definitely that parameters can be chosen as design variables. As shown in the Fig. 4.28 and consider the proportional, integral, and derivative gain of the PID controller, the design variables can be expressed as follows

$$\Re = \left(l_j, p, s, h, a, b, k_p, k_i, k_d\right)^{\mathrm{T}} \tag{4.64}$$

4.4.2 *Constraints*

In integrated optimization problem, it is essential to take into account the workspace where the robot will move, the kinematical capacities of the manipulator and specificities of the task to be realized.

The workspace of the HDCPR is characterized as the set of points where the end-effector can be positioned while all cables are in tension ($T_i > 0$). However, it is not sufficient to obtain the actual workspace of the HDCPR if it only depends on the tension condition; we should also consider the following constraints:

1. Geometrical constraints. Considering the geometrical dimensions of the HDCPR and the manipulator emplacement in a chain of machining (e.g. assembly), the geometrical constraints can be written as

$$\chi_{\min} < \chi < \chi_{\max} \tag{4.65}$$

where $\chi = (l_j, p, s, h, a, b)^{\mathrm{T}}$, $\chi_{\min}$ and $\chi_{\max}$ are the minimum and the maximum limit of the permitted χ, respectively.

2. The workspace of the HDPM. the set of points (X, Y, Z) where the end-effector can be positioned are influenced by cable length L_i. Cable length Li is calculated by the coordinates joint C (x'_C, z'). Thus, the workspace of the joint C in x'-z' plane is one of the constraints of the workspace of the HDCPR, and it can be expressed as

$$\begin{cases} |l_1 - l_2| < L_{AC} < l_1 + l_2 \\ |l_3 - l_4| < L_{CE} < l_3 + l_4 \\ L_i = s - \sqrt{(p + l_5 - x'_C)^2 + (h - z'_C)^2} \end{cases} \tag{4.66}$$

where

$$L_{AC} = \sqrt{x'^2_C + z'^2_C},\ L_{CE} = \sqrt{(x'_C - l_5)^2 + z'^2_C}.$$

3. The motion range of the end-effector can be expressed as

$$\begin{cases} 0 < X = \frac{L_1^2 - L_2^2 + a^2}{2a} < a \\ 0 < Y = \frac{L_1^2 - L_4^2 + b^2}{2b} < b \\ 0 < Z = h - \sqrt{L_1^2 - X^2 - Y^2} < h \end{cases} \tag{4.67}$$

A general numerical workspace generation approach is employed here. The possible motion range of joint C in the HDPM is first discretized into a number of points. According to the Eq. (4.66), a series of the cable length L_i can be calculated. Then, from Eq. (4.67), various combinations of L_i can calculate the end-effector coordinates (X, Y, Z). Thus, a series of points in Cartesian coordinate system are determined. At the end, the points are checked and those satisfying the tension condition ($T_i > 0$) will generate the workspace.

4. Mobility condition constraints. Due to the uncontrollability of the CV motor, the full rotation of the link associated with the CV motor in the HDPM is a necessary condition. Therefore, the mobility condition of the HDPM is very important in order to fulfill its function. Mobility of the HDPM is defined such that the two input links operate as an independent driving unit and are capable of revolving around their rotating shafts, respectively. According to the theorem above, the constraints of the double crank planar five-bar linkage can be written as follows

$$l_{\max} + l_{\min 2} + l_{\min 1} - l_m - l_n \le 0 \tag{4.68}$$

where $l_{\max}$, $l_{\min 1}$ and $l_{\min 2}$ are the lengths of the longest link and the two shortest links of the HDPM, and $l_{\min 1}$ and $l_{\min 2}$ are the same as l_1 and l_4; l_m and l_n are the lengths of the other two links.

5. Actuating power constraints. The CV motor actuating power and servomotor actuating power of the HDPMs of the HDCPR can be calculated by

$$\begin{cases} P_{i1} = \tau_{i1} \times \dot{\theta}_{i1} \\ P_{i4} = \tau_{i4} \times \dot{\theta}_{i4} \end{cases} \tag{4.69}$$

where P_{i1} is the CV motor power of the ith group of the HDPM, P_{i4} is the servomotor power of the ith group of the HDPM. The CV motor provides main power and motion required, and the servomotor acts as a motion regulation device which suffers from the limited power capacity and high cost. Thus, the actuating power constraints of the HDPMs of the HDCPR can be written as

$$P_{i1} > P_{i4} \tag{4.70}$$

6. Control performance constraints. The inequality and equality constraints of the control system are used to restrict the overshoot, rising time, steady-state error, and the drive torque within the limitation. In addition, control parameters have to define range due to the need of optimization design. (k_p^*, k_i^*, k_d^*) can be obtained. And then, we could expand control parameters range with the center of (k_p^*, k_i^*, k_d^*). The inequality and equality constraints can be written in the following form

$$\begin{cases} t_s \le t_s^+ \\ \varsigma \le \varsigma^+ \\ \delta \le \delta^+ \\ \tau \le \tau^+ \\ (1-\beta)k_p^* \le k_p \le (1+\beta)k_p^* \\ (1-\beta)k_i^* \le k_i \le (1+\beta)k_i^* \\ (1-\beta)k_d^* \le k_d \le (1+\beta)k_d^* \end{cases} \tag{4.71}$$

where t_s, δ, ς are the rising time, overshoot and steady-state error, τ is the drive torques of the HDPM $\beta \in [0, 1]$.

The optimal objective is to obtain the optimum performance of mechanism and control for the whole HDCPR system. A straightforward way to define the objective function is to combine performance measures of mechanism and control in a weighted sum manner. The HDCPR comprises the CPR and four groups of HDPMs whose drive system consists of four groups of CV motors and servomotors. Referring to Fig. 4.28, the driving torque of link DE can be expressed as

$$\tau_{14} = J_1 \ddot{\theta}_{14} \tag{4.72}$$

where τ_{14}, J_1 and $\ddot{\theta}_{14}$ are the driving torque, moment of inertia and angular acceleration of the link DE, and the angular acceleration is proportional to the driving torque. In addition, the angular acceleration is bound up with the comprehensive performance of the system. Based on the above discussions, the objective function of the mechanism can therefore be chosen as

$$f_1(\Re) = \min\left(\max\sum_{i=1}^{4}\left|\ddot{\theta}_{i4}\right|\right) \tag{4.73}$$

The design objective of the control system can be set to minimize the integral square error of the trajectory tracking of the end-effector of the HDCPR control system. Hence, the objective function of the control system can be described as

$$f_2(\Re) = \min\left(\int_0^{\mathrm{T}} e^2(t)\mathrm{d}t\right) \tag{4.74}$$

where T and e(t) are motion time and tracking error of the end-effector of the HDCPR, respectively.

Combining Eqs. (4.73) and (4.74), the objective function of mechanism and control for the whole HDCPR system can be derived

$$f(\Re) = \alpha_1 f_1(\Re) + \alpha_2 f_2(\Re) \tag{4.75}$$

where α_1 and α_2 $(0 < \alpha_1, \alpha_2 < 1)$ are the weighting coefficients, and $\alpha_1 + \alpha_2 = 1$.

The GAs can be viewed as a class of optimization methods based on the genetic processes of biological organisms. The development of populations of living organisms takes place according to the principles of natural selection and survival of the fittest. The GA process involves the creation of an initial population, aptitude evaluation, selection, and genetic operations required for reproduction. There is no mathematical proof of the fact that the GAs, once converged, will reach a global minimum. In this investigation, we opted for the GAs due to their good adequacy with the complicated multiobjective optimization problem, which offer several advantages such as overcoming the obstacle of the local minima (or maxima) and enabling the obtention of the optimal Pareto front of a multi-objective optimization problem.

4.4.3 Results and Discussion

In this section, separated optimization design and integrated optimization design for the completely restrained HDCPR with 3 Degrees of Freedom are comparatively implemented with MATLAB Version 7.10 software. The flowchart of integrated optimization procedure for the HDCPR can be described in Fig. 4.39. The parameters of the HDCPR are given by: the cross sectional area of the links of five-bar linkage = 2×10^{-4} m^2, the density of the links of five-bar linkage = 7.85×10^3 $\mathrm{kg/m}^3$, the mass of the end-effector = 30 kg, the acceleration due

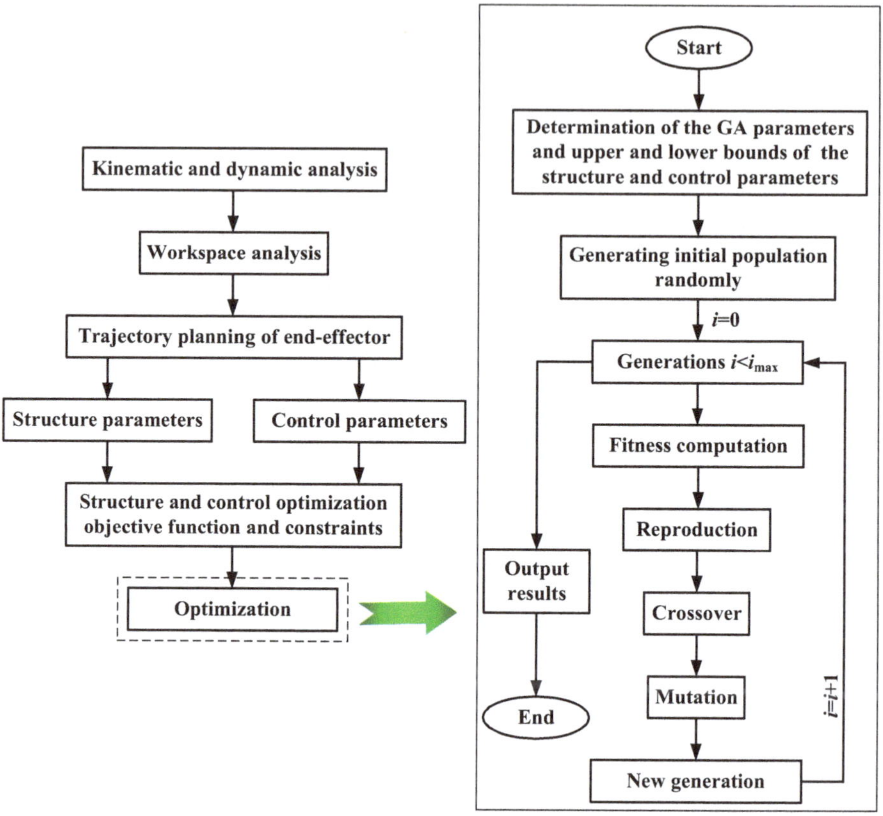

Fig. 4.39 Flow diagram of integrated optimization design of the HDCPR

to gravity = 9.81 m/s^2. The parameters of the GAs are listed in Table 4.3. The weighting coefficients α_1 and α_2 are $\alpha_1 = 0.5, \alpha_2 = 0.5$. During simulation, the end-effector of the HDCPR is asked to move along the given spatial trajectory. This section presents two motion cases of the end-effector.

The hardware control system of the HDCPR consists of the HDCPR device, the man-machine interface, an i5-4430 3.0 GHz PC with Windows 7, an IMC2600P PCI motion control card, a PCI 8735 data acquisition card, six

Table 4.3 Parameters of the GAs

GA parameters	Value
Population size	100
Generations	200
Chromosome length	20
Generation gap	0.85
Crossover rate	0.8
Mutation rate	0.05

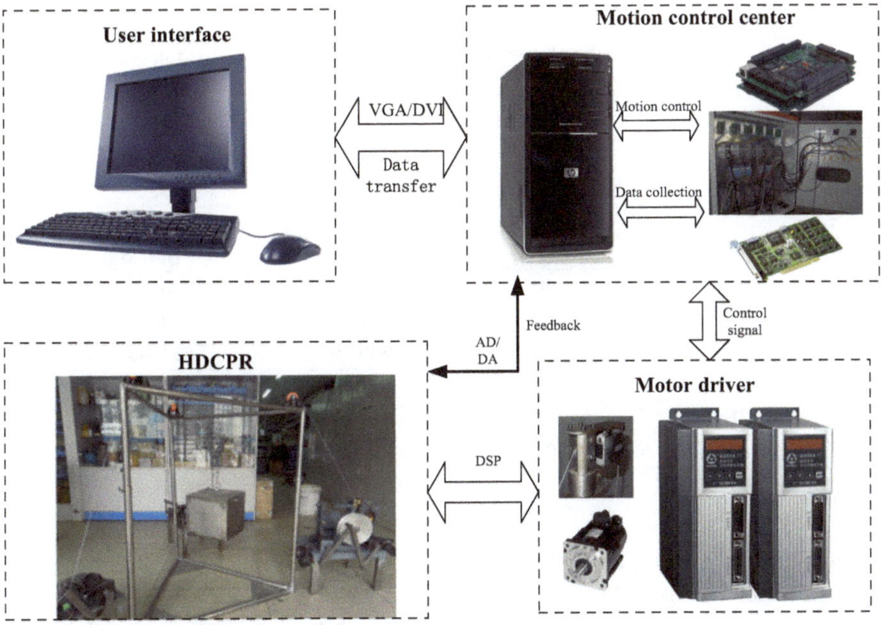

Fig. 4.40 Hardware control system of the HDCPR

YZ-ACSD608 AC motors, and three MPS-S-1000 displacement sensors with the 0.2% FS measurement accuracy, three ZNLBM-100 tension sensors with the 0.1% FS measurement accuracy, a MPU6050 three-axis accelerometer with the 16384 LSB/g measurement accuracy (see Fig. 4.40).

4.4.3.1 Case 1: Ellipse Trajectory

The ellipse trajectory is given by the following expression

$$\begin{cases} x = 0.2\cos(0.5\pi t) + 0.5 \\ y = 0.25\sin(0.5\pi t) + 0.5 \quad 0 \le t \le 4(s) \\ z = 0.6 \end{cases} \tag{4.76}$$

In this example, the random disturbances are

$$\tau'_{i4} = rand(1)\cos(0.5\pi t) \; \tau'_{i4} = rand(1)\sin(0.5\pi t)$$

Table 4.4 shows the comparison of optimization results for the separated optimization design and integrated optimization design. Figure 4.41a–c shows the convergence plots for separated optimization design and integrated optimization design, respectively. Figure 4.42 shows the spot demo interface of the HDCPR

Table 4.4 Comparison of optimization results for case 1

Variables	Lower bound	Upper bound	Separated design	Integrated design
l_1/mm	50	350	181.2	162.2
l_2/mm	150	700	409.3	495.3
l_3/mm	150	700	491.4	487.6
l_4/mm	50	350	299.7	288.5
l_5/mm	150	700	590.9	504.8
a/mm	500	1500	987.4	1001.6
b/mm	500	1500	995.3	1011.9
p/mm	500	2000	921.6	1065.7
s/mm	1500	3500	2497.9	2259.4
h/mm	700	1500	998.4	1006.1
k_p	0	10	4.7	7.3
k_i	0	10	2.0	2.8
kd	0	10	0.3	0.6

during following the trajectory of the end-effector in different time. From the spot demo interface of Fig. 4.42, it can be noted that the end-effector runs smoothly, and the movement of all joints is fluency, so by which tracking the desired trajectory movement is accompanied well. Figure 4.43 shows the curves of cable tension exerted on the end-effector of the HDCPR related to the ellipse trajectory. Since the end-effector moves on a horizontal ellipse trajectory, the changes of cable tensions in Fig. 4.43 are reasonable and the transition is smooth.

Figure 4.44 shows comparative curves of angular acceleration of the servomotor between the separated design and integrated design of the HDCPR. In accordance with Fig. 4.44, the maximum angular acceleration of the integrated design is smaller than that of the separated design. Since the angular acceleration is bound up with the comprehensive performance of the system, the performance of the integrated design system is superior to that of the separated design system.

For comparison, Fig. 4.45 illustrates the desired trajectory and actual trajectory of the end-effector along the given ellipse motion path under the separated design and integrated design in X, Y and Z directions. From it one can see that, the actual trajectory tracks the desired trajectory with high accuracy under the integrated design. The corresponding tracking errors in X, Y and Z directions are illustrated by Fig. 4.46. As can be seen from Fig. 4.46, the simulation results have less error for the integrated design compared to the separated design. Ellipse trajectory tracking errors are listed in Table 4.5.

The cables are driven only by coils and servomotors which is a classical approach for the CPRs. However, for the range and speed of obtainable output motions, it may be limited by certain servomotor power capacity and high cost. This

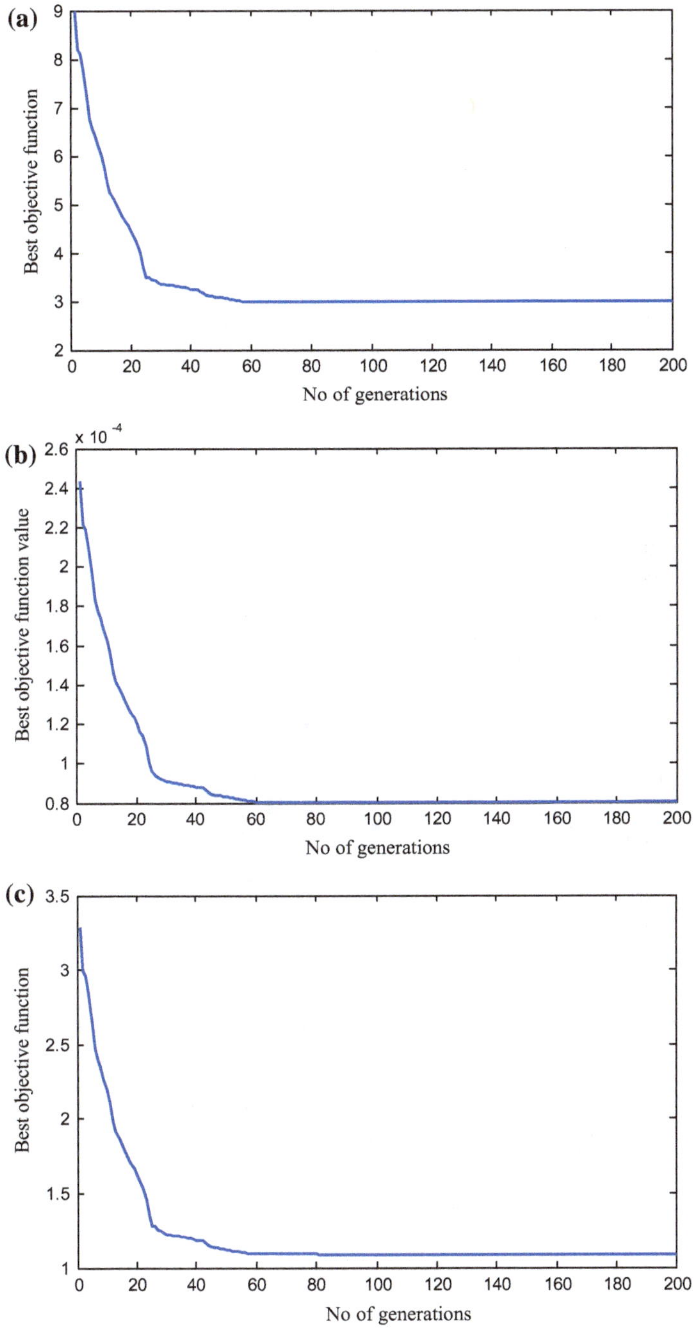

Fig. 4.41 Convergence plots for case 1, **a** f_1, **b** f_2 under separated design, **c** f under integrated design

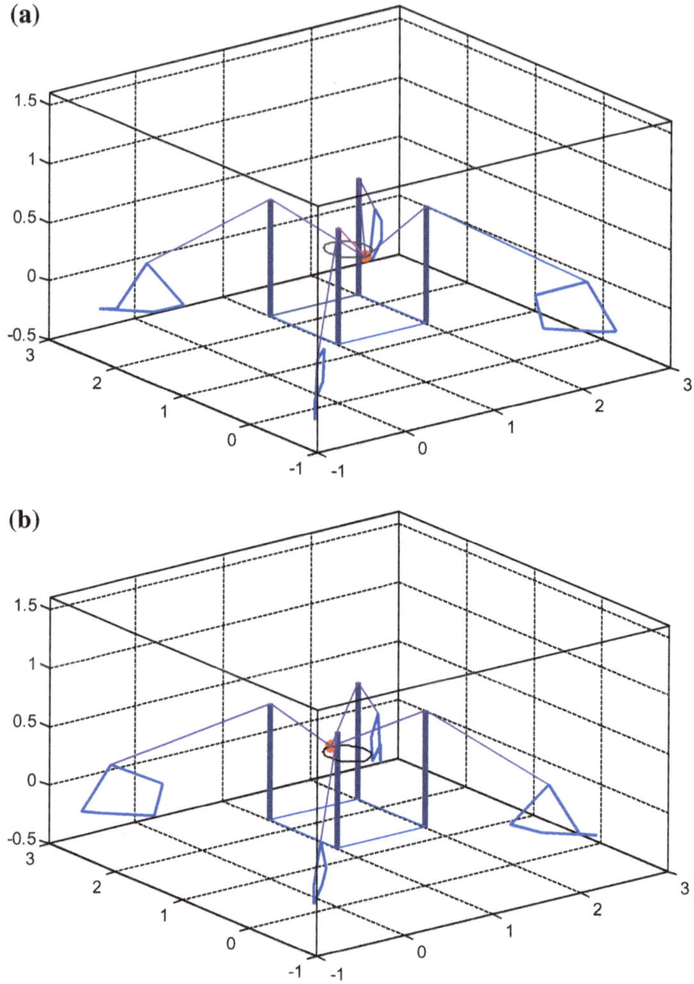

Fig. 4.42 Spot demo interface of the HDCPR for case 1: **a** t = 1 s, **b** t = 3 s

problem can be solved by using hybrid-driven machines instead of coils and servomotors. For the hybrid-driven machines, the CV motor provided the majority of power supply to drive the mechanism's motion, and the servomotor is real-time controllable and off-line programmable. Therefore, the hybrid-driven mechanisms

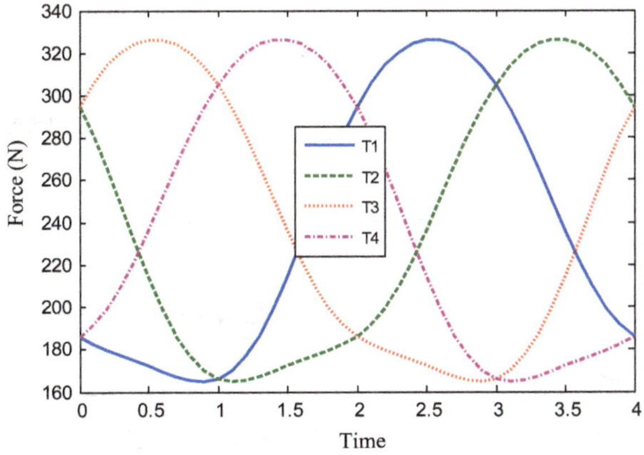

Fig. 4.43 Curves of cable tension for Case 1

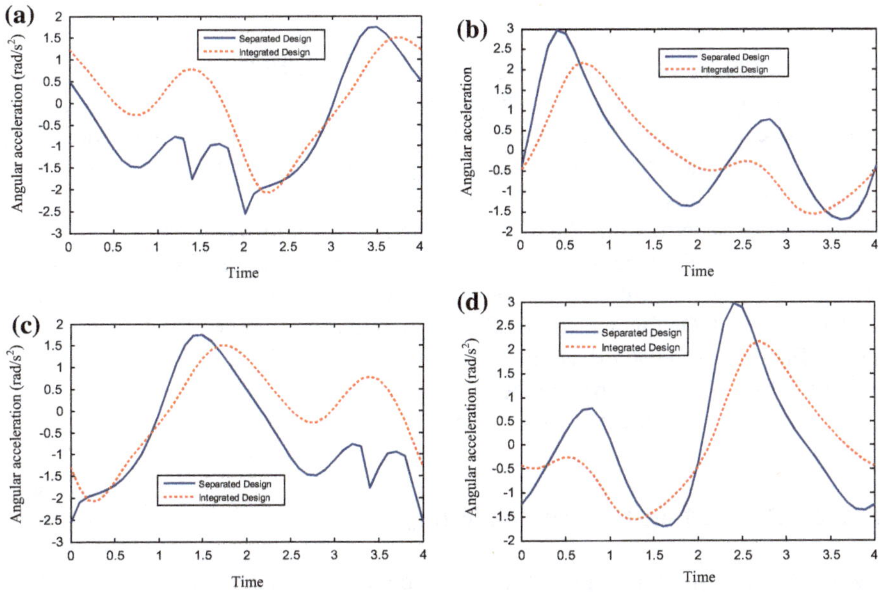

Fig. 4.44 Comparison of angular acceleration curves between the separated design and integrated design of the HDCPR for Case 1, **a** the first servomotor, **b** the second servomotor, **c** the third servomotor, **d** the forth servomotor

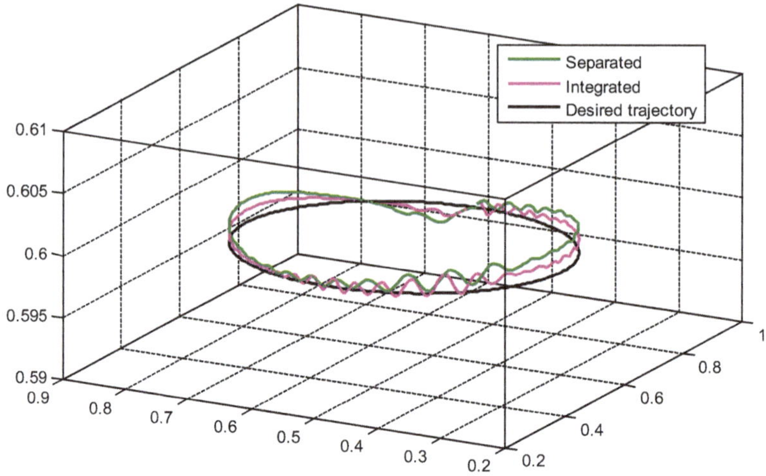

Fig. 4.45 Trajectory tracking of the end-effector during the separated and integrated design for Case 1

take the advantage of the complementary characteristics of two types of motors to generate programmable output motions with high power capacities in low cost.

A classical CPR is simulated with the same mechanical parameters and ellipse trajectory. τ_i is the motor torque of the CPR. Figures 4.47 and 4.48 display the actuating torque curves of the CPR and HDCPR, respectively. From Figs. 4.47 and 4.48, it can be seen that the changes of the actuating torques are reasonable and the transition is smooth since the end-effector moves on a horizontal ellipse trajectory. Figure 4.49 illustrates comparison of torque curves between the four groups of the HDPMs of the HDCPR and the CPR. The maximum torque of the four groups of the HDPMs of the HDCPR is smaller than that of the CPR, and according to the HDCPR, the CV motors provide the majority of torque supply and the servomotors mainly act as a motion regulation device as depicted in Fig. 4.49.

4.4.3.2 Case 2: Line Segment Trajectory

The equation of the line segment in global coordinate system is:

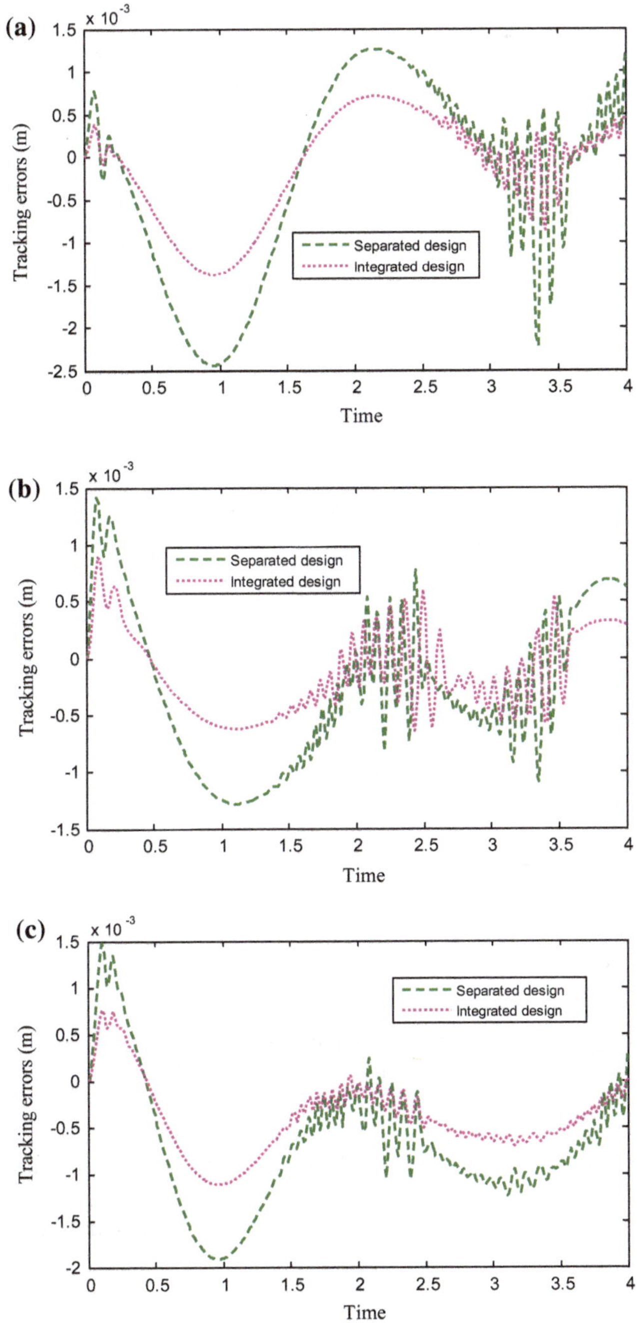

Fig. 4.46 Comparison of tracking errors curves between the separated design and integrated design in **a** *X*-direction, **b** *Y*-direction and **c** *Z*-direction for Case 1

Table 4.5 Comparison of optimization results for Case 2

t/s	Separated design			Integrated design		
	x/mm	*y/mm*	*z/mm*	*x/mm*	*y/mm*	*z/mm*
0	0	0	0	0	0	0
0.2	0.18	1.23	1.19	0.11	0.63	0.72
0.4	−0.62	0.26	0.13	−0.34	0.12	0.11
0.6	−1.56	−0.45	−0.95	−0.88	−0.24	−0.53
0.8	−2.27	−0.99	−1.74	−1.26	−0.49	−0.99
1.0	−2.31	−1.25	−1.91	−1.34	−0.57	−1.11
1.2	−1.95	−1.24	−1.56	−1.12	−0.61	−0.92
1.4	−1.07	−1.15	−1.02	−0.62	−0.54	−0.61
1.6	−0.08	−0.92	−0.63	−0.06	−0.48	−0.32
1.8	0.75	−0.35	−0.18	0.40	−0.12	−0.14
2.0	1.72	0.02	−0.09	0.68	−0.18	−0.01
2.2	1.26	−0.72	−0.98	0.72	−0.22	−0.31
2.4	1.08	−0.55	−0.92	0.61	−0.24	−0.39
2.6	0.73	−0.28	−0.78	0.47	0.12	−0.48
2.8	0.45	−0.43	−1.01	0.30	−0.26	−0.61
3.0	0.11	−0.49	−1.11	0.11	−0.25	−0.65
3.2	0.26	−0.15	−1.19	−0.45	−0.21	−0.6
3.4	1.05	0.40	−1.05	−0.42	−0.56	−0.58
3.6	−0.09	0.24	−0.84	−0.12	0.45	−0.49
3.8	0.43	0.67	−0.38	0.25	0.32	−0.19
4.0	1.26	0.62	0.32	0.44	0.31	−0.01

$$\begin{cases} X = \frac{t}{6} + \frac{1}{3} \\ Y = \frac{t}{4\sqrt{3}} + \frac{\sqrt{3}}{6} \\ Z = \frac{t}{6} + \frac{1}{3} \end{cases} \quad 0 \le t \le 2(s) \tag{4.77}$$

In this example, the random disturbances are

$$\tau'_{i4} = rand(1)\cos t \; \tau'_{i4} = rand(1)\sin t$$

The comparison of optimization results for the separated optimization design and integrated optimization design are given in Table 4.6. Figure 4.50 shows the line segment trajectory tracking under the separated design and integrated design.

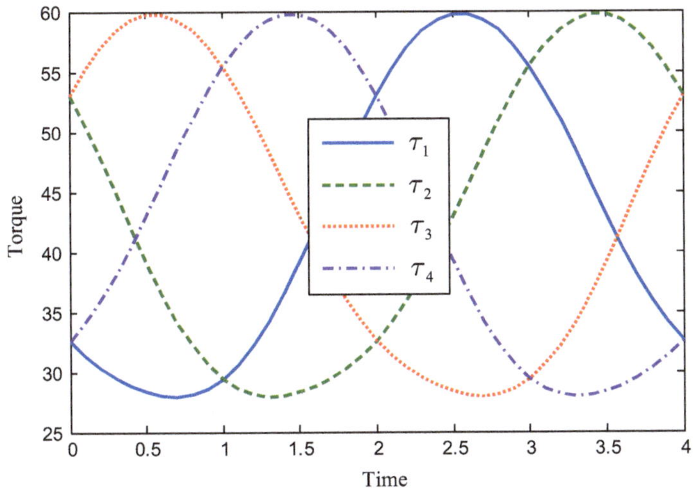

Fig. 4.47 Torque curves of the CPM for Case 1

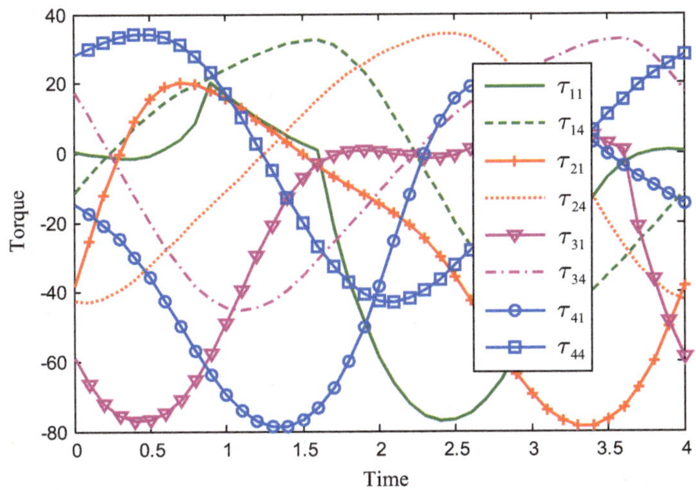

Fig. 4.48 Torque curves of the four groups of HDPMs of the HDCPM for Case 1

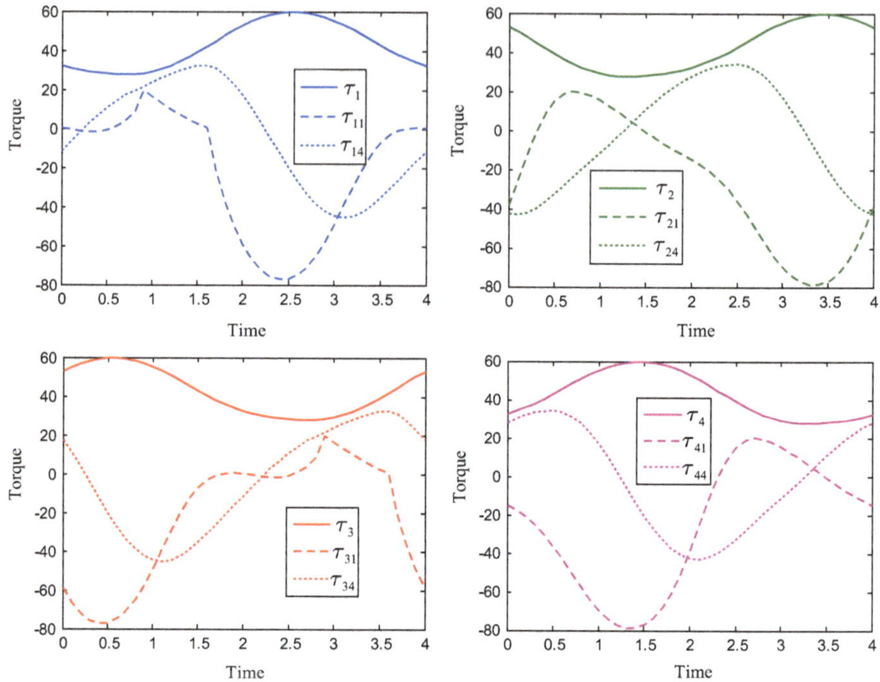

Fig. 4.49 Comparison of torque curves between four groups of the HDPMs of HDCPR and the CPR for Case 1

Table 4.6 Ellipse trajectory tracking errors

Variables	Lower bound	Upper bound	Separate design	Integrated design
l_1/mm	50	350	192.5	201.3
$l2/mm$	150	700	483.1	499.5
$l3/mm$	150	700	489.5	500.7
$l4/mm$	50	350	285.6	267.2
l_5/mm	150	700	573.8	510.6
a/mm	500	1500	1002.9	996.4
b/mm	500	1500	987.4	1002.3
p/mm	500	2000	964.5	1038.9
s/mm	1500	3500	2278.3	2315.7
h/mm	700	1500	994.8	1002.4
k_p	0	10	4.2	6.9
ki	0	10	1.8	2.6
k_d	0	10	0.3	0.5

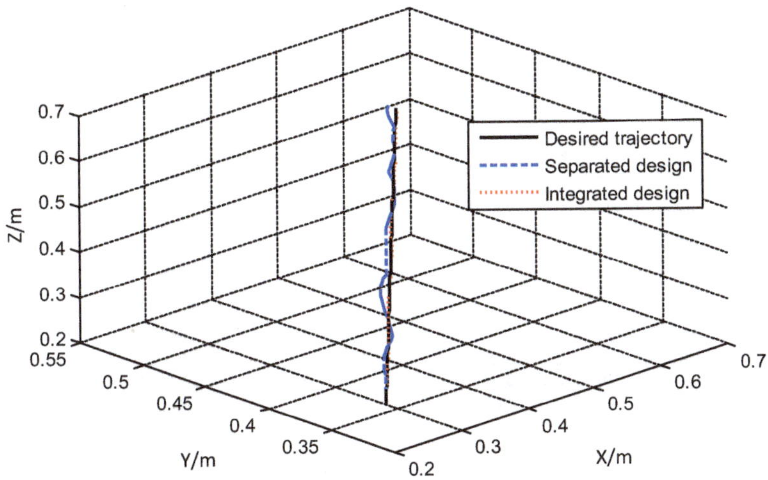

Fig. 4.50 Trajectory tracking of the end-effector during the separated and integrated design for Case 2

The tracking errors of the end-effector in X, Y and Z directions under the separated optimization design and integrated optimization design can be seen in Fig. 4.51. It is clearly that the tracking errors are small in the integrated optimization design and the corresponding tracking performance is better than that of separated optimization design. Table 4.3 shows the line segment trajectory tracking errors.

From the above two simulation examples, integrated optimization design can get less trajectory tracking errors to compare with separated optimization design under the influence of random disturbances and achieve a relatively favourable dynamic performance. Table 4.7

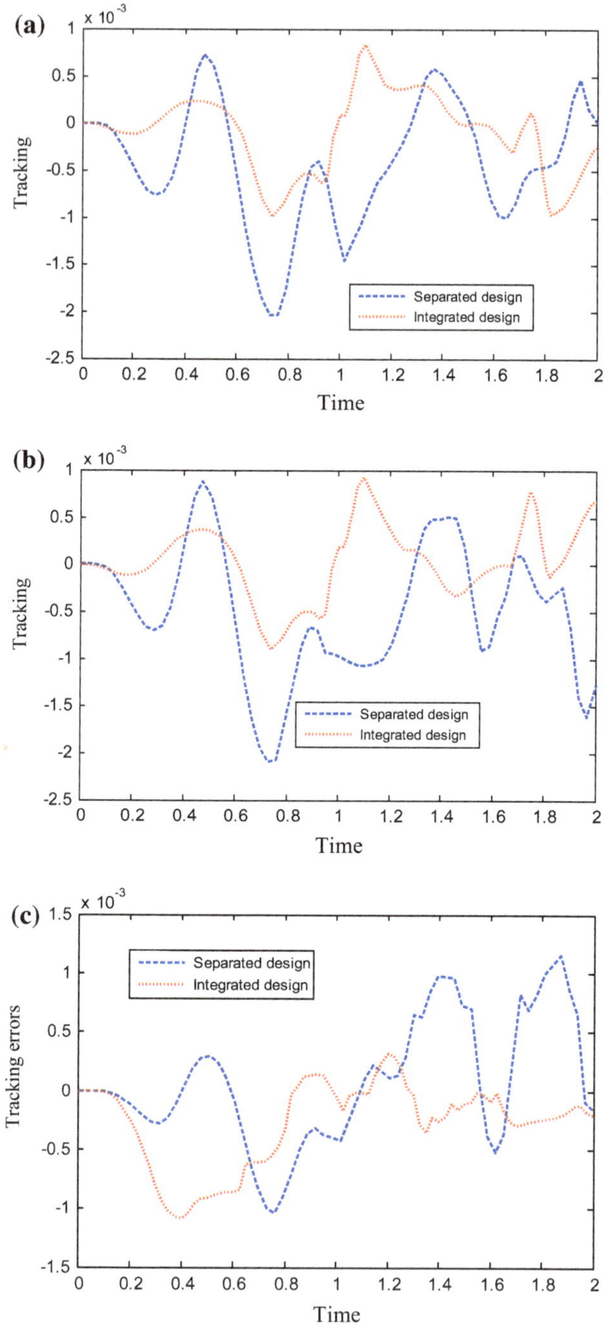

Fig. 4.51 Comparison of tracking errors curves between the separated design and integrated design in **a** *X*-direction, **b** *Y*-direction and **c** *Z*-direction for Case 2

Table 4.7 Line segment trajectory tracking errors

t/s	Separate design			Integrated design		
	x/mm	y/mm	z/mm	x/mm	y/mm	z/mm
0	0	0	0	0	0	0
0.2	−0.45	−0.42	−0.13	−0.14	−0.15	−0.25
0.4	0.02	0.23	−0.02	0.26	0.31	−1.09
0.6	−0.53	−0.59	−0.08	−0.03	0.08	−0.86
0.8	−1.67	−1.57	−0.83	−0.83	−0.62	−0.32
1.0	−1.32	−0.98	−0.91	0.11	0.15	−0.05
1.2	−0.41	−0.87	0.15	0.37	0.39	−0.28
1.4	0.52	0.49	0.97	0.22	−0.20	−0.25
1.6	−0.85	−0.75	−0.94	0.09	−0.05	−0.07
1.8	−0.48	−0.38	0.93	−0.73	0.05	−0.24
2.0	0	−1.26	−0.17	−0.25	0.68	−0.21

4.5 Conclusions

This chapter describes the theoretical investigations of a class of the completely restrained hybrid-driven based cable parallel manipulator (HDCPR) with three translational motions. The HDCPR combines the advantages of four groups of HDPMs and a 4-cable parallel manipulator in a way to provide a solution for moving heavy objects with high efficiency and high performance. Detailed design and implementation of a virtual prototype model of the HDCPR is presented as well. Dynamic modeling of the full system including both the mechanical system and the driving system is addressed. The inverse kinematic and dynamic problems of the HDCPR system are resolved on condition that an operation path of the end-effector has been planned. Then, the main factors, which influence the workspace of the HDCPR such as workspace of the HDPM and the length of cables, are analyzed based on the structure constraints of the HDCPR. Finally, illustrative examples for trajectory tracking and workspace demonstrate that the manipulator design is reasonable and the associated models are correct. The prototype verification of the designed HDCPR is currently under investigation. Furthermore, more work is in progress to implement control system design, design modification and optimum synthesis for the HDCPR. In addition, the methodology presented could be used as a guide for the design and analysis of other manipulators.

Dynamic modeling and adaptive iterative learning control of a incompletely restrained HDCPR with three translational motions, which combines the HDPM with the CPR in a way to provide a solution for moving heavy objects with high efficiency and high-performance, was presented. The dynamic equation of the whole HDCPR system is addressed by Lagrange formulation. In view of nonlinear, time-varying characteristics and repetitive unknown disturbances of the HDCPR system, a controller incorporates adaptive traditional feedback PD control with iterative learning control is designed for the high precision trajectory tracking of the HDCPR system. In addition, the proposed control method is asymptotically stable based on Lyapunov stability analysis. Finally, simulation studies for trajectory tracking control of the HDCPR system illustrate that the dynamic equation is correct and the proposed control method is effective.

Constraints from the mechanism and control disciplines are imposed on the integrated optimization process with the aim of obtaining the optimum mechanics parameters and control performance using the GAs. As an example, separated optimization design and integrated optimization design for the completely restrained HDCPR with 3 Degrees of Freedom are comparatively investigated on the basis of trajectory tracking control of the HDCPR system. The sport demos of the HDCPR control system can be displayed using a 3D modeler developed in MATLAB for the purpose of visualization when the end-effector tracks the desired trajectory in different time. The results presented verify the validity of the developed integrated optimization design methodology.

References

1. Zi B, Cao J, Zhu Z et al (2013) Design, dynamics, and workspace of a hybrid-driven-based cable parallel manipulator. Math Probl Eng
2. Chien CJ (2008) A combined adaptive law for fuzzy iterative learning control of nonlinear systems with varying control tasks. IEEE Trans Fuzzy Syst 16(1):40–51
3. Yousef H, Hamdy M, El-Madbouly E et al (2009) Adaptive fuzzy decentralized control for interconnected MIMO nonlinear subsystems. Automatica 45(2):456–462
4. Neupert J, Arnold E, Schneider K et al (2010) Tracking and anti-sway control for boom cranes. Control Eng Practice 18(1):31–44
5. Tayebi A (2004) Adaptive iterative learning control for robot manipulators. Automatica 40 (7):1195–1203
6. Ouyang PR, Zhang WJ, Gupta MM (2006) An adaptive switching learning control method for trajectory tracking of robot manipulators. Mechatronics 16(1):51–61
7. Tayebi A, Abdul S, Zaremba MB et al (2008) Robust iterative learning control design: application to a robot manipulator. IEEE/ASME Trans Mechatron 13(5):608–613
8. Lee JH, Lee KS (2007) Iterative learning control applied to batch processes: An overview. Control Eng Practice 15(10):1306–1318
9. Ouyang PR, Petz BA, Xi FF (2011) Iterative learning control with switching gain feedback for nonlinear systems. J Comput Nonlinear Dyn 6(1):011020
10. Zi B, Cao J, Qian S (2014) Iterative learning tracking control of a hybrid-driven based three-cable parallel manipulator. Adv Mech Eng 6:248385
11. You B, Kim M, Lee D et al (2011) Iterative learning control of molten steel level in a continuous casting process. Control Eng Practice 19(3):234–242
12. Fonseca IM, Bainum PM, Lourenção PTM (2002) Structural and control optimization of a space structure subjected to the gravity-gradient torque. Acta Astronaut 51(10):673–681
13. Zi B, Ding H, Cao J et al (2014) Integrated mechanism design and control for completely restrained hybrid-driven based cable parallel manipulators. J Intell Rob Syst 74(3–4):643–661
14. Gamage LB, de Silva CW, Campos R (2012) Design evolution of mechatronic systems through modeling, on-line monitoring, and evolutionary optimization. Mechatronics 22 (1):83–94

15. Zha XF (2002) Optimal pose trajectory planning for robot manipulators. Mech Mach Theory 37(10):1063–1086
16. Porfirio CR, Odloak D (2011) Optimizing model predictive control of an industrial distillation column. Control Engineering Practice 19(10):1137–1146
17. Canfield RA, Meirovitch L (1994) Integrated structural design and vibration suppression using independent modal space control. AIAA journal 32(10):2053–2060
18. Zhang WJ, Li Q, Guo LS (1999) Integrated design of mechanical structure and control algorithm for a programmable four-bar linkage. IEEE/ASME Trans Mechatron 4(4):354–362
19. Zhu Y, Qiu J, Tani J et al (1999) Simultaneous optimization of structure and control for vibration suppression. J Vib Acoust 121(2):237–243
20. Zhang X (2004) Integrated optimal design of flexible mechanism and vibration control. Int J Mech Sci 46(11):1607–1620
21. Yan HS, Yan GJ (2009) Integrated control and mechanism design for the variable input-speed servo four-bar linkages. Mechatronics 19(2):274–285
22. Gogate GR, Matekar SB (2012) Optimum synthesis of motion generating four-bar mechanisms using alternate error functions. Mech Mach Theory 54:41–61
23. Kaveh A, Rahami H (2006) Nonlinear analysis and optimal design of structures via force method and genetic algorithm. Comput Struct 84(12):770–778
24. Fogel DB (1994) An introduction to simulated evolutionary optimization. IEEE Trans Neural Networks 5(1):3–14

Chapter 5
Error Analysis of the Hybrid-Driven Based Cable-Suspended Parallel Robots

5.1 Preamble

The geometric error is an utmost important consideration factor when designing a machine tool or robot manipulator [1]. The actual kinematic parameters of the manipulator deviate from their nominal values, which are referred to as kinematic errors, and the kinematic errors would result in the end-effector errors if the nominal kinematics were used to estimate the position of the manipulator [2]. Due to the above problems, kinematic parameter estimation technique, and kinematic modeling and calibration is an effective way to improve the absolute accuracy of the manipulator [3–5]. On the basis of the error modeling, it is expected that the model enables the designer to clarify those errors affecting the uncompensable position errors and they thereby should be eliminated or at least minimized with the aid of sensitivity analysis [6, 7], and sensitivity analysis plays an important role in dynamic analysis of the system [8].

In this chapter aiming at a class of HDCPR, which combines the HDPM with the CPR in a way to provide a solution for moving heavy objects with high efficiency and high-performance, explicit expressions of error mapping functions in terms of position have been formulated thanks to closed loop vector chain and direct differential modeling strategy. In this investigation, a comprehensive and multifunctional error model is systematically defined to incorporate properly all possible geometric errors into a mathematic model so that it can meet the requirements for the tolerance design, assembly and kinematic calibration. Additionally, sensitivity analysis of the HDCPR is also carried out.

The remainder of this chapter is organized as follows: Sect. 5.2 describes the design model of the HDCPR. Then, mechanics analysis and error modelling of the HDCPR is described based on the differentiation of kinematic equations in Sect. 5.3. Sensitivity analysis of the HDCPR is provided in Sect. 5.4. In Sect. 5.5, illustrative simulation studies highlight its performances. Finally, conclusions and future studies are summarized in Sect. 5.6.

B. Zi and S. Qian, *Design, Analysis and Control of Cable-Suspended Parallel Robots and Its Applications*, DOI 10.1007/978-981-10-1753-7_5

5.2 Mechanism Description

Figure 5.1 presents the three-dimensional design model of the HDCPR with three translational motions. For each cable, one end is connected to the end-effector, the other one rolls through a pulley fixed on the top of the relative cable tower and then is fed into the HDPM. The HDCPR comprises of two modules: (1) the CPR consisting of three-cable tower, cable tower frame, three cables, pulley struts, pulleys, girder, cargo (i.e. end-effector); (2) three groups of HDPMs containing three-phase asynchronous motors, servomotors, reducers and double crank five-bar linkage; the asynchronous motors are connected using the pulley transmission mechanisms, while the servomotors and the reducers are linked by couplings. At the same time, the pulley transmission mechanisms and reducers are joined to the double crank planar five-bar linkage. The double crank planar five-bar linkage is made up of a large crank disk, a long connecting rod, a small crank disk, and a short connecting rod. Moreover, each cable tower is equipped with a cable guide pulley, and the radius of each pulley is assumed to be zero.

A simple kinematic sketch of the HDCPR structure model with the associated coordinate systems are depicted in Fig. 5.2. At the bottom of one cable tower, a global coordinate system (OXYZ) is established. The end-effector has location coordinates G(x, y, z). The distance between each cable tower top Pi(xi, yi, zi) and the end-effector (load) is Li (i = 1, 2, 3). The three cable towers have the same height and are arrayed in a triangle on the cable tower frame. Cable tower frame can

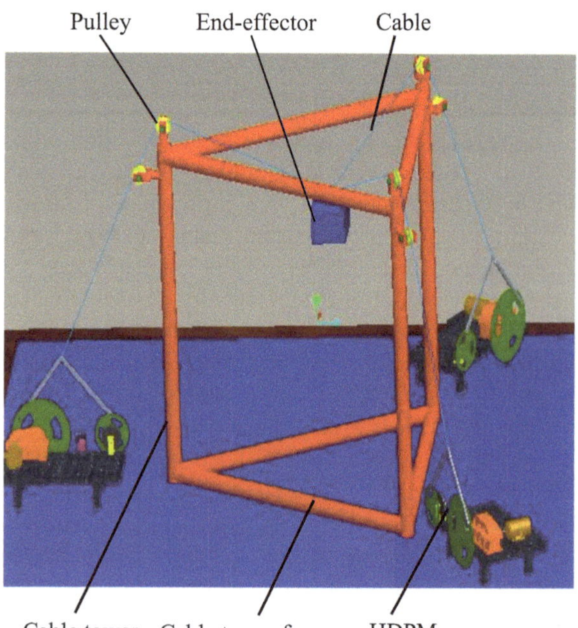

Fig. 5.1 Three-dimensional model of the HDCPM

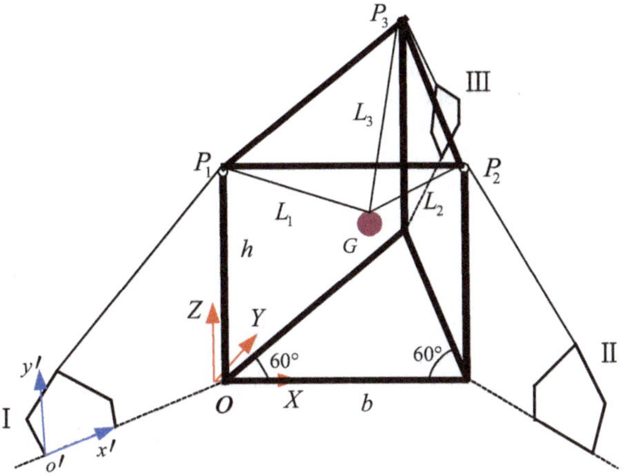

Fig. 5.2 Kinematic sketch of the HDCPM

be used to enhance the stiffness of three-cable tower. The tension force exerted by the cables on the heavy loads (end-effector) has a great influence on each cable tower top position Pi(xi, yi, zi) and each total cable length L0i. In addition, the deformations from cable towers and cable guide pulley and assembly clearance also have a bad influence for the movement of the HDCPR. For simplicity, we assume that the deformation and assembly clearance from each cable tower and each cable guide pulley could be seen as position variance Δxi, Δyi, Δzi of the ith cable tower top Pi(xi, yi, zi). In addition, the cable is treated as a massless body compared to heavy loads, and we assume that the deformation and assembly error from the each cable could be seen as cable length variance ΔL0i of the ith total cable L0i.

5.3 Mechanics Analysis and Error Modelling

Figure 5.3 displays the kinematic sketch of the ith group of the HDPM and ith cable tower. Link A_iE_i is fixed on the ground. The links A_iB_i and D_iE_i are driven by a CV motor and a servomotor, respectively. In addition, these links drive B_iC_i and C_iD_i, so that the joint Ci is the output. This section assumes that the links have the same density, and the centroid is in the center of links. In addition, the deformations of the links A_iB_i, B_iC_i, C_iD_i, D_iE_i are also neglected. For more details on the deformation analysis of the links of the HDPM, refer to Sect. 5.5. The length of the links are l_{i1}, l_{i2}, l_{i3}, l_{i4}, l_{i5}, the angel between links and x′-axis are $\theta_{i1}, \theta_{i2}, \theta_{i3}, \theta_{i4}, 0°$, respectively. ($x_{ic}$, y_{ic}) is the position coordinates of the joint C_i in the local coordinate system.

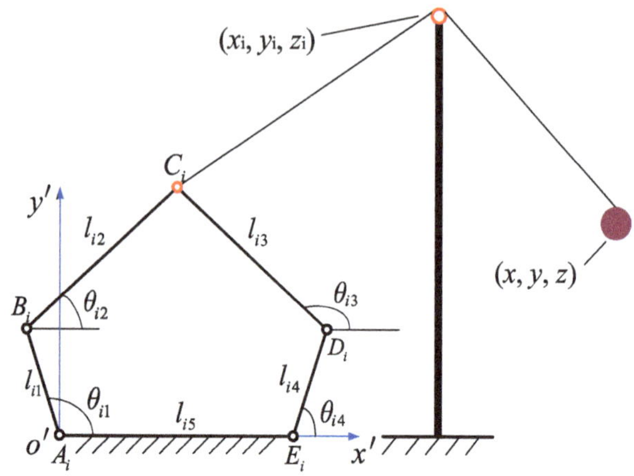

Fig. 5.3 Kinematic sketch of the *i*th group of the HDPM and *i*th cable tower the HDPM and *i*th cable tower

Referring to Fig. 5.3, based on the method of vector chain, the closed loop polygon vector equation of the first group of the HDPM can be expressed as

$$\overrightarrow{A_1B_1} + \overrightarrow{B_1C_1} = \overrightarrow{A_1E_1} + \overrightarrow{E_1D_1} + \overrightarrow{D_1C_1} \tag{5.1}$$

Equation (5.1) can also be written in the complex exponential form

$$l_{11}e^{i\theta_{11}} + l_{12}e^{i\theta_{12}} = l_{13}e^{i\theta_{13}} + l_{14}e^{i\theta_{14}} + l_{15} = x_{1c} + y_{1c}i \tag{5.2}$$

Developing Eq. (5.2) into real part and imaginary part, respectively, we have

$$\begin{cases} x_{1c} = l_{11}c_{11} + l_{12}c_{12} \\ y_{1c} = l_{11}s_{11} + l_{12}s_{12} \\ x_{1c} = l_{13}c_{13} + l_{14}c_{14} + l_{15} \\ y_{1c} = l_{13}s_{13} + l_{14}s_{14} \end{cases} \tag{5.3}$$

where the expressions of the elements of the above functions can be easily expressed in the following notations: $s_{ij} = \sin\theta_{ij}$ and $c_{ij} = \cos\theta_{ij}$.

According to Eq. (5.3), we eliminate the driven angular displacement θ_{12}, θ_{13}, then simplify and order it. So the constraint equations of the HDPM can be obtained as

$$u_1\colon 2l_{11}y_{1c}s_{11} + 2l_{11}x_{1c}c_{11} + l_{12}^2 - l_{11}^2 - x_{1c}^2 - y_{1c}^2 = 0 \tag{5.4}$$

$$u_2\colon 2l_{14}y_{1c}s_{14} + 2l_{14}(x_{1c} - l_{15})c_{14} + l_{13}^2 + 2l_{15}x_{1c} - l_{14}^2 - l_{15}^2 - x_{1c}^2 - y_{1c}^2 = 0 \tag{5.5}$$

Clearly, the constraint equations of the first group of the HDPMs have been derived. Since the three groups of the HDPMs are identical, the constraint equations of the second and third groups of the HDPMs can also be derived in the same way.

Combining Eqs. (5.4) and (5.5), the whole constraint equations of the three groups of the HDPMs can be written as follows

$$u_{ik}(l_{i1}, l_{i2}, l_{i3}, l_{i4}, l_{i5}, \theta_{i1}, \theta_{i4}, x_{ic}, y_{ic}) = 0. \quad i \in \{1, 2, 3\} \ k \in \{1, 2\} \tag{5.6}$$

By differentiating Eq. (5.6) with respect to all the variables

$$\begin{aligned} \frac{\partial u_{ik}}{\partial x_{ic}} dx_{ic} + \frac{\partial u_{ik}}{\partial y_{ic}} dy_{ic} = & -\frac{\partial u_{ik}}{\partial l_{i1}} dl_{i1} - \frac{\partial u_{ik}}{\partial l_{i2}} dl_{i2} - \frac{\partial u_{ik}}{\partial l_{i3}} dl_{i3} - \frac{\partial u_{ik}}{\partial l_{i4}} dl_{i4} - \frac{\partial u_{ik}}{\partial l_{i5}} dl_{i5} \\ & - \frac{\partial u_{ik}}{\partial \theta_{i1}} d\theta_{i1} - \frac{\partial u_{ik}}{\partial \theta_{i4}} d\theta_{i4} \end{aligned} \tag{5.7}$$

Substituting Eq. (5.6) into Eq. (5.7), the error model of the three groups of the HDPMs can be expressed in the matrix form

$$\begin{bmatrix} dx_{ic} \\ dy_{ic} \end{bmatrix} = \begin{bmatrix} l_{i1}c_{i1} - x_{ic} & l_{i1}s_{i1} - y_{ic} \\ l_{i4}c_{i4} + l_{i5} - x_{ic} & l_{i4}s_{i4} - y_{ic} \end{bmatrix}^{-1} \begin{bmatrix} y_{ic}s_{i1} - x_{ic}c_{i1} + l_{i1} & 0 \\ -l_{i2} & 0 \\ 0 & -l_{i3} \\ 0 & -y_{ic}s_{i4} - (x_{ic} - l_{i5})c_{i4} + l_{i4} \\ 0 & l_{i4}c_{i4} - x_{ic} + l_{i5} \\ -l_{i1}y_{ic}c_{i1} + l_{i1}x_{ic}s_{i1} & 0 \\ 0 & -l_{i4}y_{ic}c_{i4} + l_{i4}(x_{ic} - l_{i5})s_{i4} \end{bmatrix}^T \begin{bmatrix} dl_{i1} \\ dl_{i2} \\ dl_{i3} \\ dl_{i4} \\ dl_{i5} \\ d\theta_{i1} \\ d\theta_{i4} \end{bmatrix} \tag{5.8}$$

In a compact form, Eq. (5.8) can be rewritten as

$$de_{ic} = J_{ic} dq_{ic} \tag{5.9}$$

where $de_{ic} = [dx_{ic} \ dy_{ic}]^T$

$$dq_{ic} = [dl_{i1} \ dl_{i2} \ dl_{i3} \ dl_{i4} \ dl_{i5} \ d\theta_{i1} \ d\theta_{i4}]^T$$

de_{ic} and dq_{ic} represent the position error of the joint C_i and the error sources for kinematics' parameters of the three groups of the HDPMs, respectively. $J_{ic}(i \in \{1, 2, 3\})$ is the Jacobian matrix of the three groups of the HDPMs, and we have

$$J_{ic} = \begin{bmatrix} l_{i1}c_{i1} - x_{ic} & l_{i1}s_{i1} - y_{ic} \\ l_{i4}c_{i4} + l_{i5} - x_{ic} & l_{i4}s_{i4} - y_{ic} \end{bmatrix}^{-1} \begin{bmatrix} y_{ic}s_{i1} - x_{ic}c_{i1} + l_{i1} & 0 \\ -l_{i2} & 0 \\ 0 & -l_{i3} \\ 0 & -y_{ic}s_{i4} - (x_{ic} - l_{i5})c_{i4} + l_{i4} \\ 0 & l_{i4}c_{i4} - x_{ic} + l_{i5} \\ -l_{i1}y_{ic}c_{i1} + l_{i1}x_{ic}s_{i1} & 0 \\ 0 & -l_{i4}y_{ic}c_{i4} + l_{i4}(x_{ic} - l_{i5})s_{i4} \end{bmatrix}^T \tag{5.10}$$

Equation (5.9) describes the mapping relation from the errors of the bar length and the drive angular displacement to the position error of the joint C_i. For each group of the HDPMs, the position error of the joint C_i relates to 7 error sources. Consider the three groups of the HDPMs, there are 21 error sources to the position error of the joint C_i for the HDPMs.

According to the geometric relationship of the CPR, as shown in Figs. 5.2 and 5.3, the constraint equations of the CPR are given by the following expressions

$$f_i : L_i^2 = (x - x_i)^2 + (y - y_i)^2 + (z - z_i)^2, \quad i \in \{1, 2, 3\} \tag{5.11}$$

To facilitate the establishment of the error model of the CPR, Eq. (11) can be rewritten as

$$f_i(x, y, z, x_i, y_i, z_i, L_i) = 0 \tag{5.12}$$

By differentiating Eq. (5.12) with respect to all the variables

$$\begin{aligned} \frac{\partial f_i}{\partial x}dx + \frac{\partial f_i}{\partial y}dy + \frac{\partial f_i}{\partial z}dz = & -\frac{\partial f_i}{\partial x_1}dx_1 - \frac{\partial f_i}{\partial y_1}dy_1 - \frac{\partial f_i}{\partial z_1}dz_1 - \frac{\partial f_i}{\partial x_2}dx_2 - \frac{\partial f_i}{\partial y_2}dy_2 - \frac{\partial f_i}{\partial z_2}dz_2 \\ & - \frac{\partial f_i}{\partial x_3}dx_3 - \frac{\partial f_i}{\partial y_3}dy_3 - \frac{\partial f_i}{\partial z_3}dz_3 - \frac{\partial f_i}{\partial L_1}dL_1 - \frac{\partial f_i}{\partial L_2}dL_2 - \frac{\partial f_i}{\partial L_3}dL_3 \end{aligned} \tag{5.13}$$

Substituting Eq. (5.11) into Eq. (5.13), the error model of the CPR is shown as follows

$$\begin{bmatrix} dx \\ dy \\ dz \end{bmatrix} = \begin{bmatrix} x - x_1 & y - y_1 & z - z_1 \\ x - x_2 & y - y_2 & z - z_2 \\ x - x_3 & y - y_3 & z - z_3 \end{bmatrix}^{-1} \begin{bmatrix} x - x_1 & 0 & 0 \\ y - y_1 & 0 & 0 \\ z - z_1 & 0 & 0 \\ 0 & x - x_2 & 0 \\ 0 & y - y_2 & 0 \\ 0 & z - z_2 & 0 \\ 0 & 0 & x - x_3 \\ 0 & 0 & y - y_3 \\ 0 & 0 & z - z_3 \\ L_1 & 0 & 0 \\ 0 & L_2 & 0 \\ 0 & 0 & L_3 \end{bmatrix}^T \begin{bmatrix} dx_1 \\ dy_1 \\ dz_1 \\ dx_2 \\ dy_2 \\ dz_2 \\ dx_3 \\ dy_3 \\ dz_3 \\ dL_1 \\ dL_2 \\ dL_3 \end{bmatrix} \tag{5.14}$$

In a compact form, Eq. (5.14) can be rewritten as

$$de_0 = J_0 dq_0 \tag{5.15}$$

where $de_0 = [dx\ dy\ dz]^T$

$$dq_0 = [dx_1\ dy_1\ dz_1\ dx_2\ dy_2\ dz_2\ dx_3\ dy_3\ dz_3\ dL_1\ dL_2\ dL_3]^T$$

de_0 represents the position error of the end-effector of the CPR. dq_0 is the error sources for kinematics parameters of the CPR, which including the position error of the pulley center and the length error of the cable between the pulley center and the end-effector. J0 is the Jacobian matrix of the CPR, and we can write J0 in matrix form as

$$J_0 = \begin{bmatrix} x-x_1 & y-y_1 & z-z_1 \\ x-x_2 & y-y_2 & z-z_2 \\ x-x_3 & y-y_3 & z-z_3 \end{bmatrix}^{-1} \begin{bmatrix} x-x_1 & 0 & 0 \\ y-y_1 & 0 & 0 \\ z-z_1 & 0 & 0 \\ 0 & x-x_2 & 0 \\ 0 & y-y_2 & 0 \\ 0 & z-z_2 & 0 \\ 0 & 0 & x-x_3 \\ 0 & 0 & y-y_3 \\ 0 & 0 & z-z_3 \\ L_1 & 0 & 0 \\ 0 & L_2 & 0 \\ 0 & 0 & L_3 \end{bmatrix}^T \tag{5.16}$$

Equation (5.15) shows the mapping relationship from the errors of the position of the pulley center and the cable length to the position error of the end-effector of the CPR, and there are 12 error sources.

According to Fig. 5.3, the relationship between the three groups of the HDPMs and the CPR can be expressed as

$$p_i : (a_i - x_{ic})^2 + (h - y_{ic})^2 = (L_{0i} - L_i)^2, \quad i \in \{1, 2, 3\} \tag{5.17}$$

where a_i is the distance between the joint A_i and the bottom of the cable pillar. L_{0i} represents the length of the ith total cable.

Similarly, in order to establish the error model of the HDCPR, Eq. (5.17) can be rewritten as

$$p_i(L_i, a_i, x_{ic}, y_{ic}, L_{0i}) = 0 \tag{5.18}$$

By differentiating Eq. (5.18) with respect to all the variables

$$\begin{aligned} \frac{\partial p_i}{\partial L_1}dL_1 + \frac{\partial p_i}{\partial L_2}dL_2 + \frac{\partial p_i}{\partial L_3}dL_3 = -\frac{\partial p_i}{\partial a_1}da_1 - \frac{\partial p_i}{\partial a_2}da_2 - \frac{\partial p_i}{\partial a_3}da_3 - \frac{\partial p_i}{\partial x_{1c}}dx_{1c} - \frac{\partial p_i}{\partial y_{1c}}dy_{1c} \\ - \frac{\partial p_i}{\partial x_{2c}}dx_{2c} - \frac{\partial p_{ii}}{\partial y_{2c}}dy_{2c} - \frac{\partial p_i}{\partial x_{3c}}dx_{3c} - \frac{\partial p_i}{\partial y_{3c}}dy_{3c} - \frac{\partial p_i}{\partial L_{01}}dL_{01} - \frac{\partial p_i}{\partial L_{02}}dL_{02} - \frac{\partial p_i}{\partial L_{03}}dL_{03} \end{aligned} \tag{5.19}$$

Substituting Eq. (5.17) into Eq. (5.19), we obtain

$$\begin{bmatrix} dL_1 \\ dL_2 \\ dL_3 \end{bmatrix} = \begin{bmatrix} L_1 - L_0 & 0 & 0 \\ 0 & L_2 - L_0 & 0 \\ 0 & 0 & L_3 - L_0 \end{bmatrix}^{-1} \begin{bmatrix} a_1 - x_{1c} & 0 & 0 \\ 0 & a_1 - x_{2c} & 0 \\ 0 & 0 & a_1 - x_{3c} \\ x_{1c} - a_1 & 0 & 0 \\ y_{1c} - h & 0 & 0 \\ 0 & x_{2c} - a_2 & 0 \\ 0 & y_{2c} - h & 0 \\ 0 & 0 & x_{3c} - a_3 \\ 0 & 0 & y_{3c} - h \\ L_1 - L_{01} & 0 & 0 \\ 0 & L_2 - L_{02} & 0 \\ 0 & 0 & L_3 - L_{03} \end{bmatrix}^T \begin{bmatrix} da_1 \\ da_2 \\ da_3 \\ dx_{1c} \\ dy_{1c} \\ dx_{2c} \\ dy_{2c} \\ dx_{3c} \\ dy_{3c} \\ dL_{01} \\ dL_{02} \\ dL_{03} \end{bmatrix} \tag{5.20}$$

Equation (5.20) can be rewritten in a compact form

$$de_L = J_L dq_L \tag{5.21}$$

where $de_L = [dL_1 \; dL_2 \; dL_3]^T$ represents the length error of the cable between the pulley center and the end-effector. dq_L is the 12 error sources, which including the position error between the joint A_i and the bottom of the cable pillar, the position error of the joint C_i and the length error of the total cable. J_L is the Jacobian matrix of the driven cable, which associates the CPR with the HDPMs, and we get

$$J_L = \begin{bmatrix} L_1 - L_0 & 0 & 0 \\ 0 & L_2 - L_0 & 0 \\ 0 & 0 & L_3 - L_0 \end{bmatrix}^{-1} \begin{bmatrix} a_1 - x_{1c} & 0 & 0 \\ 0 & a_1 - x_{2c} & 0 \\ 0 & 0 & a_1 - x_{3c} \\ x_{1c} - a_1 & 0 & 0 \\ y_{1c} - h & 0 & 0 \\ 0 & x_{2c} - a_2 & 0 \\ 0 & y_{2c} - h & 0 \\ 0 & 0 & x_{3c} - a_3 \\ 0 & 0 & y_{3c} - h \\ L_1 - L_{01} & 0 & 0 \\ 0 & L_2 - L_{02} & 0 \\ 0 & 0 & L_3 - L_{03} \end{bmatrix}^T \tag{5.22}$$

On the basis of the foregoing kinematics of the three groups of the HDPMs and the CPR, the constraint equations of the HDCPR can be expressed as

$$\begin{cases} 2l_{i1}y_{ic}s_{i1} + 2l_{i1}x_{ic}c_{i1} + l_{i2}^2 - l_{i1}^2 - x_{ic}^2 - y_{ic}^2 = 0 \\ 2l_{i4}y_{ic}s_{i4} + 2l_{i4}(x_{ic} - l_{i5})c_{i4} + l_{i3}^2 + 2l_{i5}x_{ic} \\ -l_{i4}^2 - l_{i5}^2 - x_{ic}^2 - y_{ic}^2 = 0 \\ L_i^2 = (x - x_i)^2 + (y - y_i)^2 + (z - z_i)^2 \\ (a_i - x_{ic})^2 + (h - y_{ic})^2 = (L_{0i} - L_i)^2 \end{cases} \quad i \in \{1, 2, 3\} \tag{23}$$

By combining Eqs. (5.9), (5.15) and (5.21), the error model of the HDCPR can be obtained as

$$de = J_O \begin{bmatrix} dq_1 \\ J_L \begin{bmatrix} da \\ J_{1c}dq_{1c} \\ J_{2c}dq_{2c} \\ J_{3c}dq_{3c} \\ dL_O \end{bmatrix} \end{bmatrix} \tag{5.24}$$

where

$$da = [da_1\ da_2\ da_3]^T, \quad dL_O = [dL_{01}\ dL_{02}\ dL_{03}]^T,$$

$$dq_1 = [dx_1\ dy_1\ dz_1\ dx_2\ dy_2\ dz_2\ dx_3\ dy_3\ dz_3]^T.$$

In a compact form, Eq. (5.24) can be rewritten as

$$de = Jdq \tag{5.25}$$

where

$$de = de_O = [dx\ dy\ dz]^T$$

$$dq = [\ dx_i\ dy_i\ dz_i\ da_i\ dl_{i1}\ dl_{i2}\ dl_{i3}\ dl_{i4}\ dl_{i5}\ d\theta_{i1}\ d\theta_{i4}\ dL_{Oi}]^T$$

de is the position error of the end-effector, *dq* represents the 36 error sources for kinematic parameters of the CPR and the three groups of the HDPMs, J is the Jacobian matrix of the HDCPR.

5.4 Sensitivity Analysis

In error analysis, error sensitivity is represented by the Jacobian matrix. Based on the foregoing error model, the position error mapping function is described by the following formulation

$$\Delta e = J\Delta q \tag{5.26}$$

Taking norm on both sides of Eq. (5.26) yields

$$\Delta e = \sqrt{\Delta e^T \Delta e} = \Delta q^T J^T J \Delta q = \sum_{i=1}^{36}\sum_{j=1}^{36}(\sum_{k=1}^{3} J_{ki}J_{kj})\Delta q_i \Delta q_j \tag{5.27}$$

where Δq_i stands for the ith element in Δq, J_{ki} is the element of row k and column i in J.

In order to investigate the influences of the error sources on the position accuracy of the end-effector of the HDCPR, a probability model is needed. Assume that all elements in Δq are statistically independent, allowing the mean value of Δq to be zero, i.e. $E(\Delta q) = 0$. Thus the variation of Δe can be expressed as

$$D(\Delta e) = \sum_{i=1}^{36}(\sum_{k=1}^{3} J_{ki}^2)D(\Delta q_i) \tag{5.28}$$

Hence, the standard deviation of can be represented as

$$\sigma(\Delta e) = \sqrt{\sum_{i=1}^{36}(\sum_{k=1}^{3} J_{ki}^2)\sigma^2(\Delta q_i)} \tag{5.29}$$

With a view to describing the standard deviation of Δe caused by the unit standard deviation of Δq_i, the sensitivity of Δe with respect to Δq_i is defined as

$$S_i = \sqrt{\sum_{k=1}^{3} J_{ki}^2} \tag{5.30}$$

Note that S_i varies with the configuration of the HDCPR, thus its mean value $\overline{S_i}$ can be employed as an index to evaluate the influence of Δq_i on the position accuracy in the overall workspace. To distinguish the sensitivity S_i and $\overline{S_i}$, we named S_i the local sensitivity and $\overline{S_i}$ the global sensitivity, respectively. So $\overline{S_i}$ is defined as [6]

$$\overline{S_i} = (\int_V S_i dV)/V \tag{5.31}$$

where V is the volume of the overall workspace.

For a better identifying and analyzing the error sources, the sensitivity coefficient is defined by using normalization method

$$s_{ki} = \overline{S_i} \Big/ \sum\nolimits_{i=1}^{36} \overline{S_i} \tag{5.32}$$

5.5 Results and Discussion

In this study, simulations were carried out to highlight the performances of the HDCPR. All the simulations are performed with MATLAB and Abaqus 6.10 software for three groups of the HDPMs having the same structure parameters and the mechanisms having the symmetrical design in the three-dimensional space. The kinematics parameters of the HDCPR are listed in Table 5.1.

As the power source of the HDCPR, numerical results of the HDPM are presented firstly. Since the three groups of the HDPMs are identical, error results of the first group of the HDPM is only given.

Figure 5.4 displays the bar charts of the sensitivity coefficients of the kinematics parameters of the first group of the HDPM. Table 5.2 shows the numerical values of the sensitivity coefficients. The notations 1, 2, 3, 4, 5, 6 and 7 denote Δl_{11}, Δl_{12}, Δl_{13}, Δl_{14}, Δl_{15}, $\Delta \theta_{11}$ and $\Delta \theta_{14}$, respectively, as shown in the horizontal axis (shown in Fig. 5.4). Referring to Fig. 5.4, on one hand, it can be noted that the bar length and Δl_{12}, Δl_{13} and Δl_{15} are the most sensitive to the position error of the joint C_1, On the other hand, the drive angular displacements $\Delta \theta_{11}$ and $\Delta \theta_{14}$ are much more insensitive to the geometric errors than the bar length, and $\Delta \theta_{14}$ is not as much sensitive as $\Delta \theta_{11}$. Therefore, the sensitivity coefficients can be used as the weights for the optimal tolerance design to improve the position accuracy of the HDPM.

In the following discussion, we will conduct the study of sensitivity analysis of the whole HDCPR.

Figure 5.5 displays the bar charts of the sensitivity coefficients of the kinematics parameters of the HDCPR, and numerical values of the sensitivity coefficients are given in Table 5.3. The notations 1–36 denote Δx_1, Δy_1, Δz_1, Δx_2, Δy_2, Δz_2, Δx_3,

Table 5.1 Kinematics parameters of the HDCPR

System parameters	Value (m)
h	1
b	1
a	1.51
L_0	2
l_1	0.2
l_2	0.5
l_3	0.5
l_4	0.28
l_5	0.51

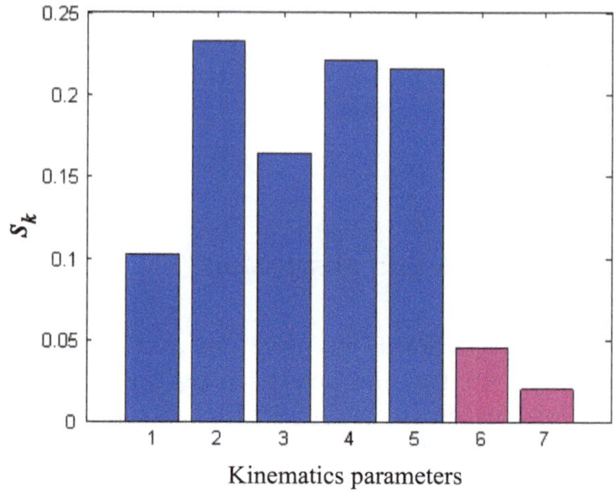

Fig. 5.4 Bar charts of the sensitivity coefficients

Table 5.2 Sensitivity coefficients of the first group of the HDPM

Sensitivity coefficient $s_k(\Delta q_{1c})$	Value
Δl_{11}	0.1022
Δl_{12}	0.2321
Δl_{13}	0.1639
Δl_{14}	0.2203
Δl_{15}	0.2154
$\Delta \theta_{11}$	0.0456
$\Delta \theta_{14}$	0.0205

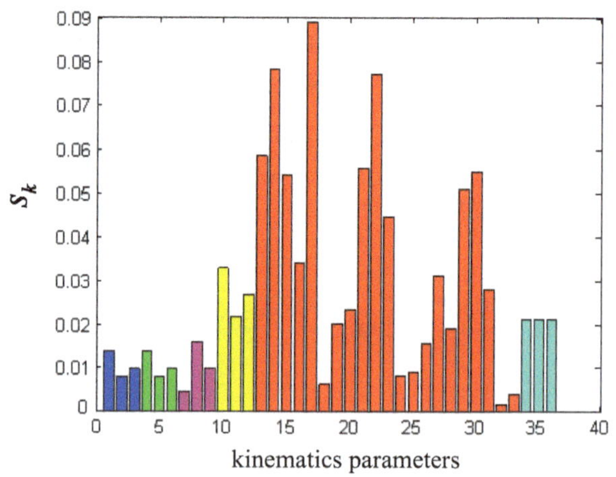

Fig. 5.5 Bar charts of the sensitivity coefficients

Table 5.3 Sensitivity coefficients of the HDCPR

Sk(Δq)	Group 1	Group 2	Group 3
Δx	0.0137	0.0137	0.0046
Δy	0.0080	0.0080	0.0159
Δz	0.0098	0.0098	0.0098
Δa	0.0328	0.0216	0.0266
Δl_1	0.0583	0.0233	0.0312
Δl_2	0.0781	0.0555	0.0190
Δl_3	0.0539	0.0771	0.0508
Δl_4	0.0340	0.0446	0.0547
$\Delta \mathrm{l}_5$	0.0889	0.0082	0.0280
$\Delta\theta_1$	0.0063	0.0090	0.0017
$\Delta\theta_4$	0.0202	0.0155	0.0042
ΔL_0	0.0210	0.0210	0.0211

Δy_3, Δz_3, Δa_1, Δa_2, Δa_3, Δl_{11}, Δl_{12}, Δl_{13}, Δl_{14}, Δl_{15}, $\Delta\theta_{11}$, $\Delta\theta_{14}$, Δl_{21}, Δl_{22}, Δl_{23}, Δl_{24}, Δl_{25}, $\Delta\theta_{21}$, $\Delta\theta_{24}$, Δl_{31}, Δl_{32}, Δl_{33}, Δl_{34}, Δl_{35}, $\Delta\theta_{31}$, $\Delta\theta_{34}$, ΔL_{01}, ΔL_{02} and ΔL_{03}, respectively, as shown in the horizontal axis (see Fig. 5.5).

From Fig. 5.5 and Table 5.3, it can be seen that the parameters of the HDPMs is the main error sources, while the parameters of the CPR is the secondary error sources. Assuming the employed criterion for the coefficients comparison is the value of sk above 0.05. Among these error sources, the sensitivity coefficients of Δl_{11}, Δl_{12}, Δl_{13}, Δl_{15}, Δl_{22}, Δl_{23}, Δl_{33} and Δl_{34} are higher than the others, So the 8 error elements are regarded as the sensitive error sources to the position error of the end-effector of the HDCPR.

In order to investigate the kinematic sensitivity property further, the following kinematic sensitivity simulation is performed. During simulation, the end-effector of the HDCPR is asked to move along the given spatial trajectory. The spatial trajectory is expressed as

$$\begin{cases} x = 0.5 + 0.25\cos t \\ y = \sqrt{3}/6 + 0.25\sin t \\ z = 0.5 \end{cases} \tag{5.33}$$

The cables are driven only by coils and servomotors which is a classical approach for the CPRs. However, for the range and speed of obtainable output motions, it may be limited by certain servomotor power capacity and high cost. This problem can be solved by using HDPMs instead of coils and servomotors. For the HDPM, the CV motor provided the majority of power supply to drive the mechanism's motion, and the servomotor is real-time controllable and off-line programmable. Therefore, the HDPM takes the advantage of the complementary characteristics of two types of motors to generate programmable output motions with high power capacities in low cost.

Let T_i denote the driving force of the ith cable exerted on the end-effector, m denotes mass of the end-effector, and g denotes acceleration of gravity. In terms of dynamic equations Eq. (5.34) based on Euler equation of Newtonian mechanics, it is valid only for $T_i > 0$, i.e., the cables are in tension.

$$\begin{pmatrix} \frac{x_1-x}{L_1} & \frac{x_2-x}{L_2} & \frac{x_3-x}{L_3} \\ \frac{y_1-y}{L_1} & \frac{y_2-y}{L_2} & \frac{y_3-y}{L_3} \\ \frac{z_1-z}{L_1} & \frac{z_2-z}{L_2} & \frac{z_3-z}{L_3} \end{pmatrix} \begin{pmatrix} T_1 \\ T_2 \\ T_3 \end{pmatrix} = \begin{pmatrix} m\ddot{x} \\ m\ddot{y} \\ mg + m\ddot{z} \end{pmatrix} \quad (5.34)$$

The actuating power of the CV motor and servomotor of the HDPMs of the HDCPR can be calculated by

$$\begin{cases} P_{CVi} = \tau_{i1} \times \dot{\theta}_{i1} \\ P_{SVi} = \tau_{i4} \times \dot{\theta}_{i4} \end{cases} \quad (5.35)$$

where P_{CVi} is the CV motor power of the ith group of the HDPMs, P_{SVi} is the servomotor power of the ith group of the HDPMs, τ_{i1} and τ_{i4} is the torque input by the CV motor and by the servomotor, respectively.

As the motion trajectory of the end-effector is the same, and as the structural parameters of the classical approach for the CPR are the same, the servomotor actuating power of the CPR can be obtained by

$$P_i = T_i \times \dot{L}_i \quad (5.36)$$

where P_i is the ith servomotor power, T_i represents the ith cable tension, $\dot{L}_i$ denotes the ith cable velocity.

Assuming density of the links is 7.85×103 kg/m^3 and the cross sectional area of the links is 2×10^{-4} m^2. Figure 5.6 shows absolute value power curves of three groups of the HDPMs of the HDCPR. Figure 5.7 shows the servomotor absolute value power curves of the classical approach for the CPR. For comparison, absolute value power curves of the first group of HDPM of the HDCPR and the first servomotor of the CPR are given in Fig. 5.8. From the simulations, one can see that the CV motors provide the majority of power supply and the servomotors mainly act as a motion regulation device as depicted in Figs. 5.6, 5.7 and 5.8. Figure 5.9 shows the curves of cable tension exerted on the end-effector related to the circle trajectory. Since the end-effector moves on a horizontal circle trajectory, the changes of cable tensions in Fig. 5.9 are reasonable and the transition is smooth.

Figures 5.10 and 5.11 show the strain contours of the HDPM under the cable tension exerted on the joint C, as shown in Fig. 5.9, by using Abaqus 6.10 Finite Element Analysis software. The strain of each link of the HDPM is microscopic, and the maximum strain is below 10^{-4}. Figure 5.12 shows the maximum deformation curve of the HDPM. The maximum deformation appears in the start-up process, and the deformation is less than 10^{-4} mm, which is negligible relative to integral structure of the HDPM.

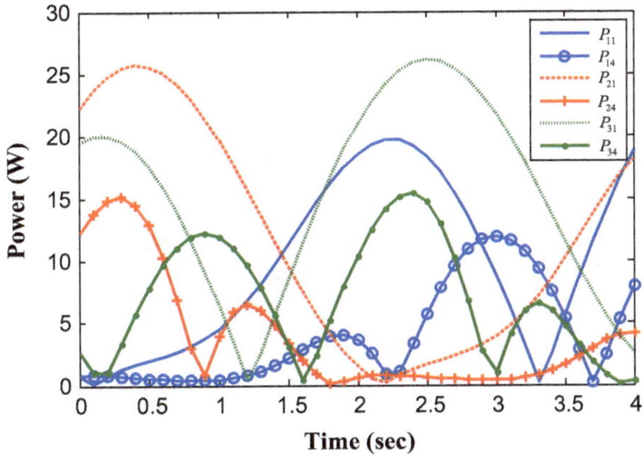

Fig. 5.6 Motor absolute value power curves

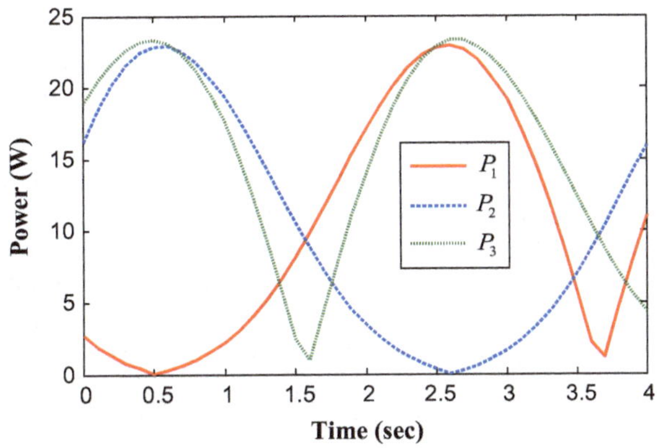

Fig. 5.7 Servomotor absolute value power curves

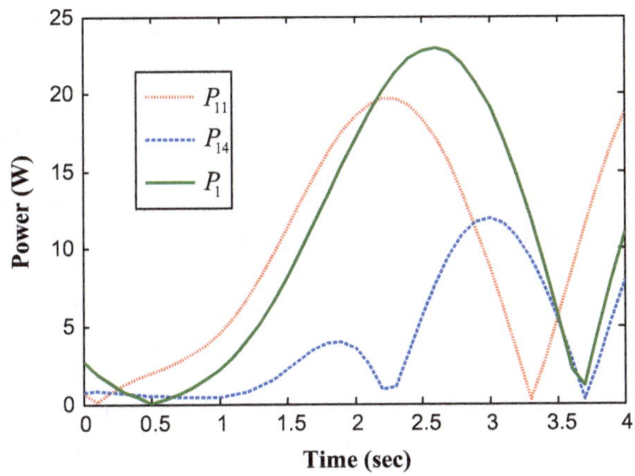

Fig. 5.8 Comparison of absolute value power curves

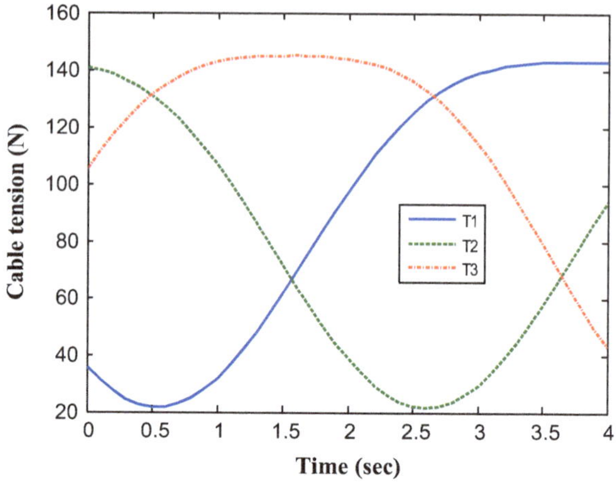

Fig. 5.9 Curves of cable tension

Figure 5.13 displays theoretical trajectory and deformation trajectory of the joint C of the HDPM related to the circle trajectory, and the tracking errors of the joint C of the HDPM in x′ and y′ directions can be seen in Figs. 5.14 and 5.15. respectively. Referring to Figs. 5.13, 5.14 and 5.15, it should be noted that, the theoretical trajectory and deformation trajectory of the joint C of the HDPM are almost coincident, and the maximum errors in x′ and y′ directions are less than 0.003 mm and 0.015 mm, respectively. From the above simulation results, it may be concluded that the deformations of the links of the HDPM can be neglected.

Figures 5.16, 5.17, 5.18, 5.19, 5.20, 5.21, 5.22, 5.23, 5.24 and 5.25 show the local sensitivity curves of with respect to. Referring to Figs. 5.16, 5.17, 5.18, 5.19, 5.20, 5.21, 5.22, 5.23, 5.24 and 5.25, the parameters of five-bar length, and are more sensitive than the others. Obviously, the changes of local error sensitivity caused by each kinematics parameters of the HDCPR are similar to the position accuracy in the overall workspace. In addition, it can be noted that some error sources have a characteristic of periodic variation. It results from the symmetrical position and identical structure of the three groups of the HDPMs. From the simulations, we can see that the sensitivity method is proved to be effective on the basis of the structural characteristic of the HDCPR.

Assume that all elements in are statistically independent and allowing the mean value of to be zero, i.e.. The normal distribution variance values of 36 elements in are displayed in Table 5.4.

Figure 5.26 illustrates the standard deviation plots of the position error of the end-effector of the HDCPR at various X-Y plane positions about different Z-axis coordinate values, where z is 0.3, 0.45, 0.6, 0.75, 0.9 m, respectively. At each elevation, the standard deviation of the planar position error caused by the unit standard deviation of each kinematics parameter error is calculated and displayed using a combination surface plot in Fig. 5.26a, b. It demonstrates that when the end-effector moves at the upper levels in the workspace, the standard deviation is generally more than that is at the middle and lower levels, as depicted in Fig. 5.26.

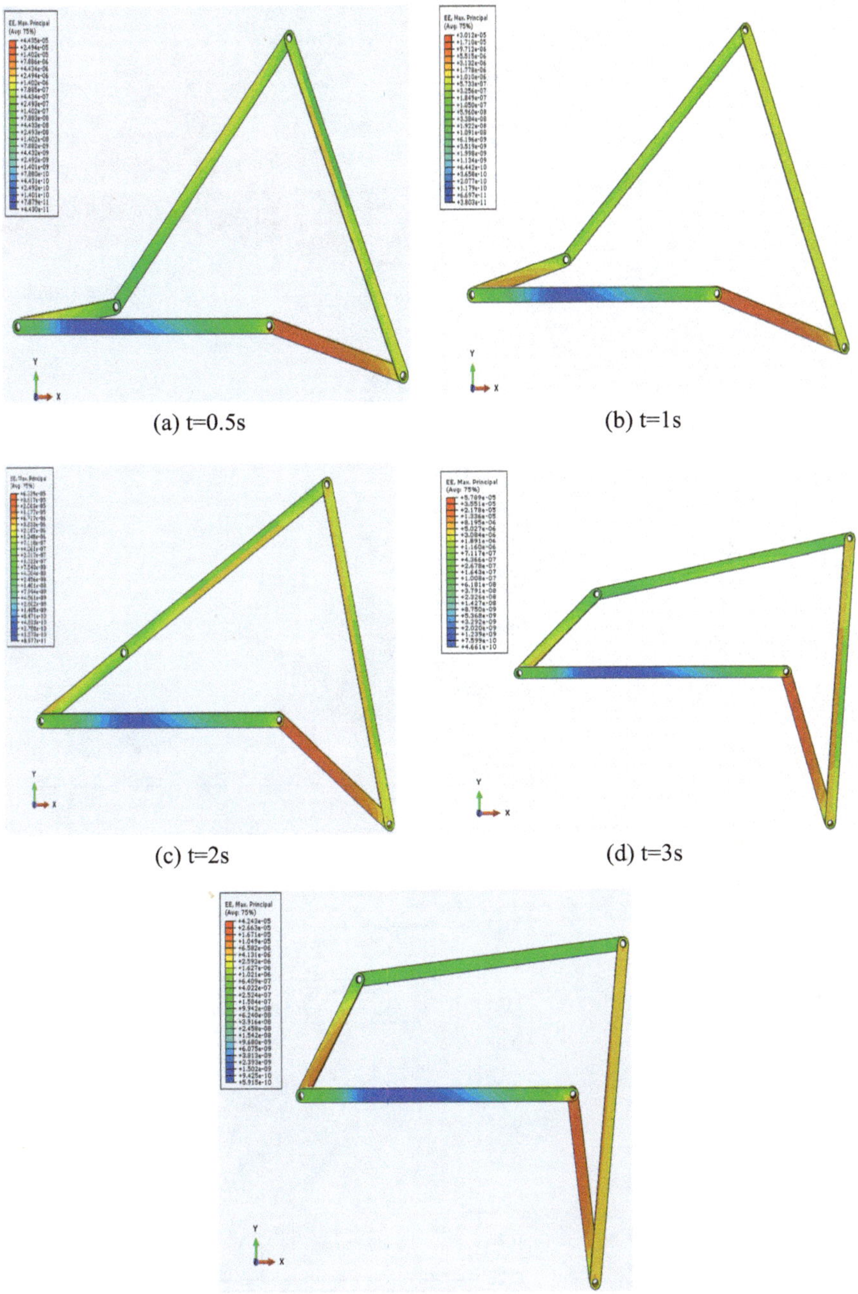

(a) t=0.5s

(b) t=1s

(c) t=2s

(d) t=3s

(e) t=4s

Fig. 5.10 Strain contours of the HDPM

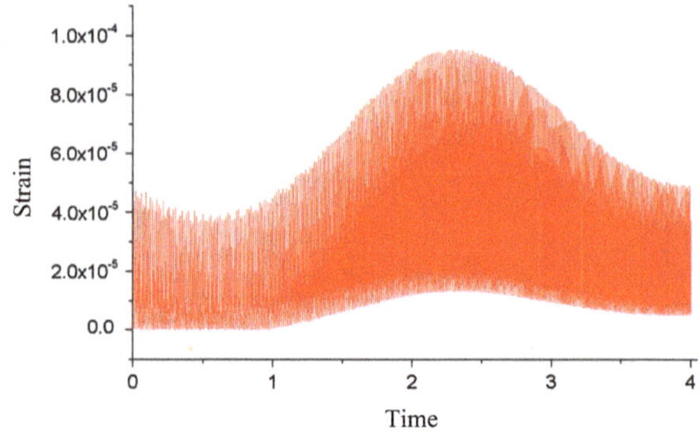

Fig. 5.11 Maximum strain curve of the HDPM

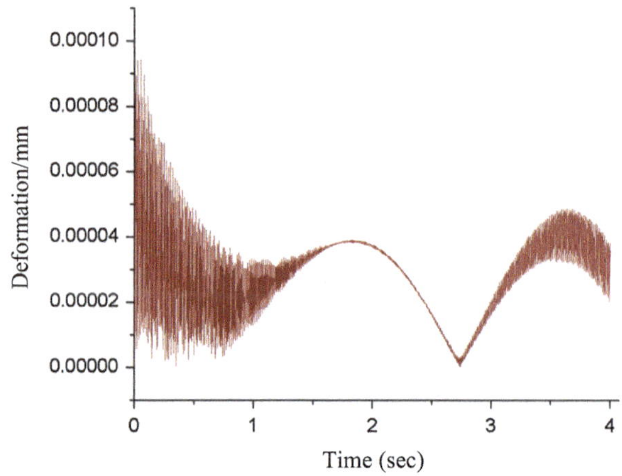

Fig. 5.12 Maximum deformation curve of the HDPM

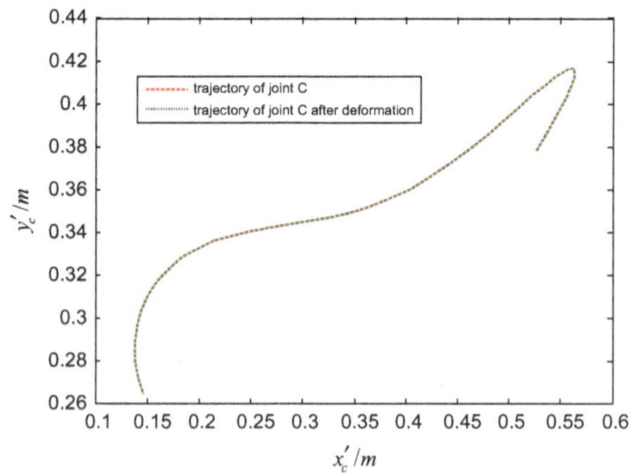

Fig. 5.13 Theoretical trajectory and deformation trajectory of joint C of the HDPM

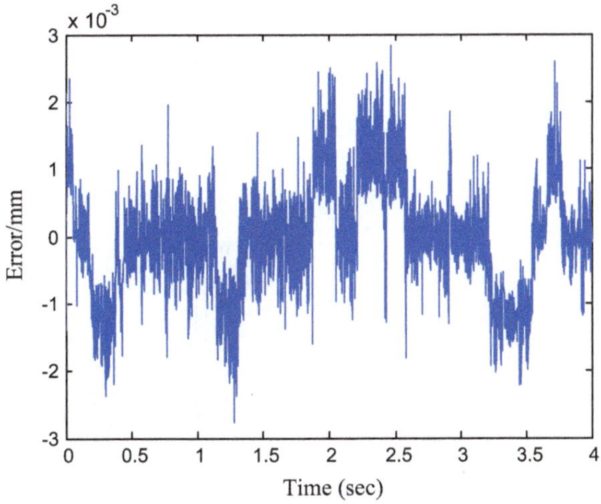

Fig. 5.14 Error curve of joint C in x'-direction

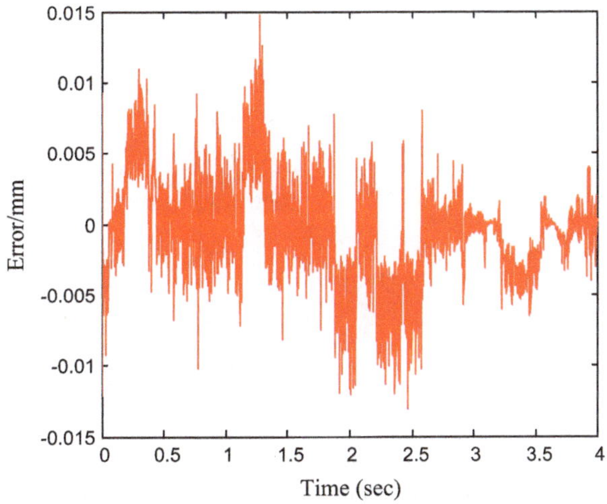

Fig. 5.15 Error curve of joint C in y'-direction

In addition, it attains maximum value at the edge of the levels. Therefore, for the sake of the position precision of the end-effector, we should make the end-effector of the HDCPR move at the middle and lower levels in the workspace.

According to the motion trajectory in Eq. (5.33) and the variance values as shown in Table 5.4, one can further validate that whether or not the sensitivity coefficients of , , , , , , and are the sensitive error sources to the position error of the end-effector of the HDCPR. Figures 5.27, 5.28, 5.29, 5.30, 5.31, 5.32, 5.33, 5.34, 5.35 and 5.36 display the desired planning trajectory and the actual motion trajectory under the influences of different error sources. Note that the blue trajectory represents desired planning trajectory and red trajectory represents actual motion trajectory.

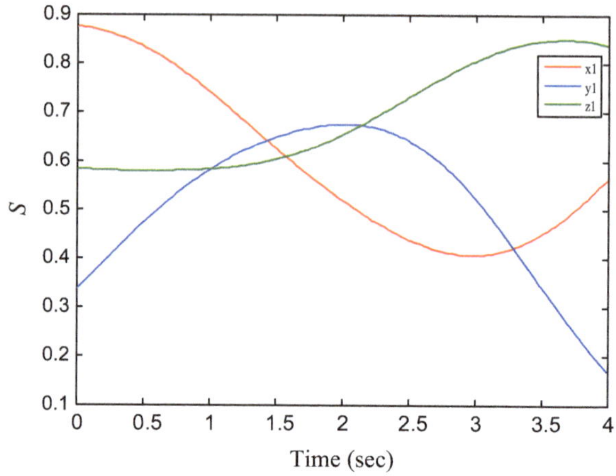

Fig. 5.16 Sensitivity with respect to x_1, y_1 and z_1

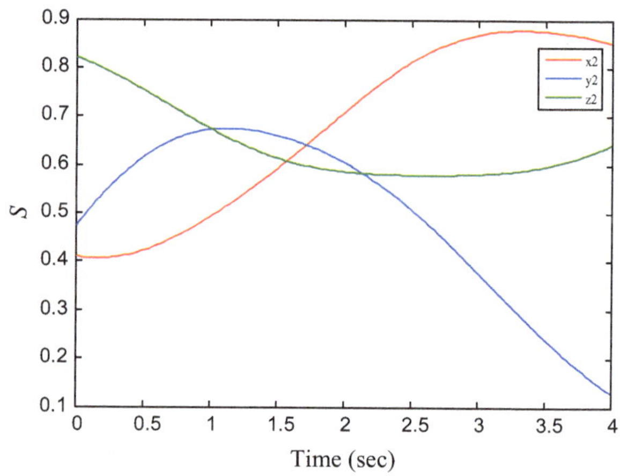

Fig. 5.17 Sensitivity with respect to x_2, y_2 and z_2

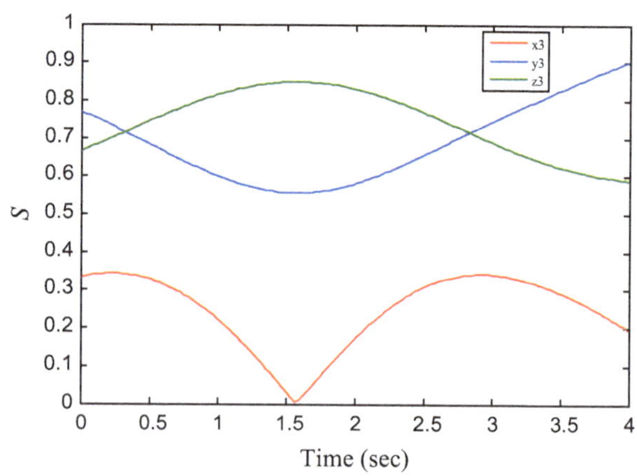

Fig. 5.18 Sensitivity with respect to x_3, y_3 and z_3

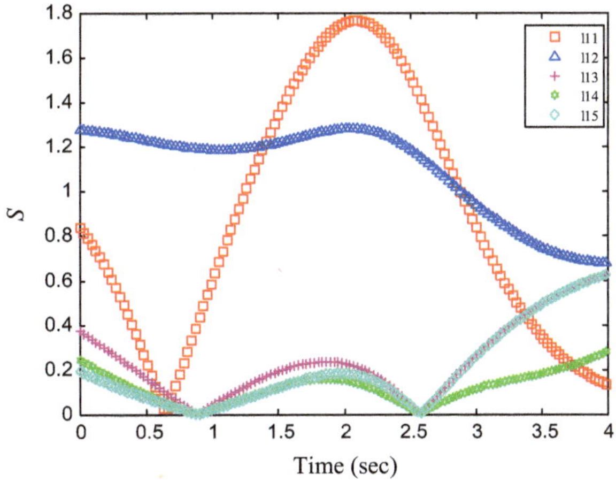

Fig. 5.19 Sensitivity with respect to $l_{11}, l_{12}, l_{13}, l_{14}$ and l_{15}

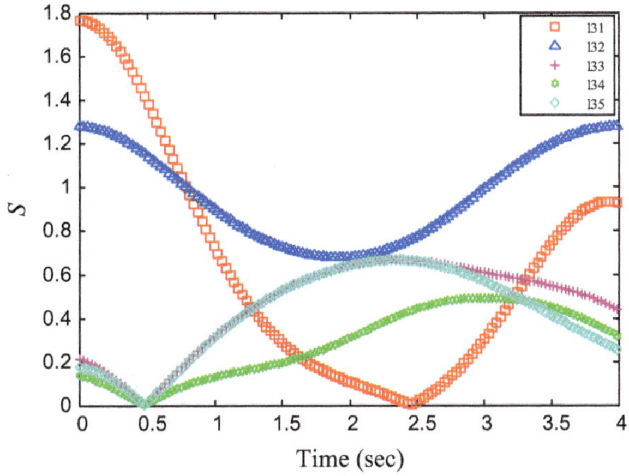

Fig. 5.20 Sensitivity with respect to $l_{31}, l_{32}, l_{33}, l_{34}$ and l_{35}

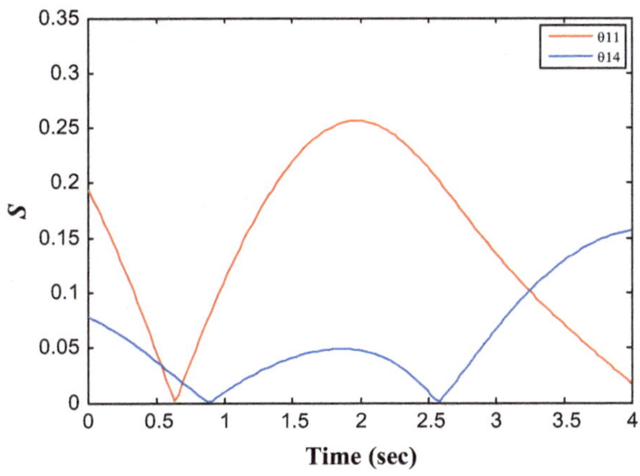

Fig. 5.21 Sensitivity with respect to θ_{11} and θ_{14}

Fig. 5.22 Sensitivity with respect to θ_{21} and θ_{24}

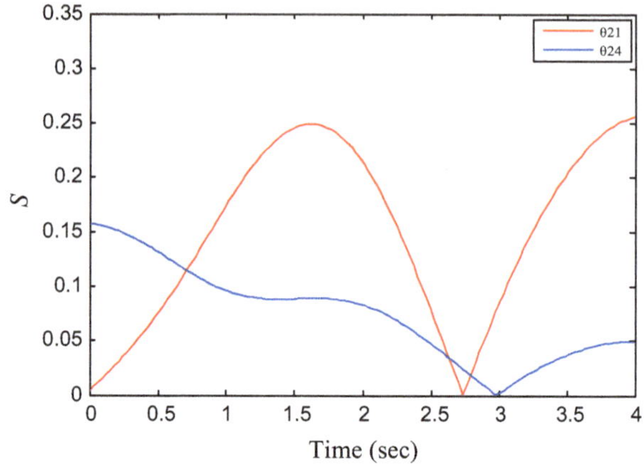

Fig. 5.23 Sensitivity with respect to θ_{31} and θ_{34}

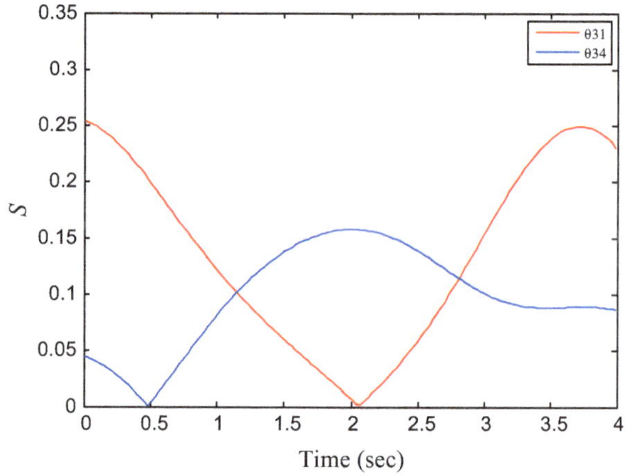

Fig. 5.24 Sensitivity with respect to a_1, a_2 and a_3

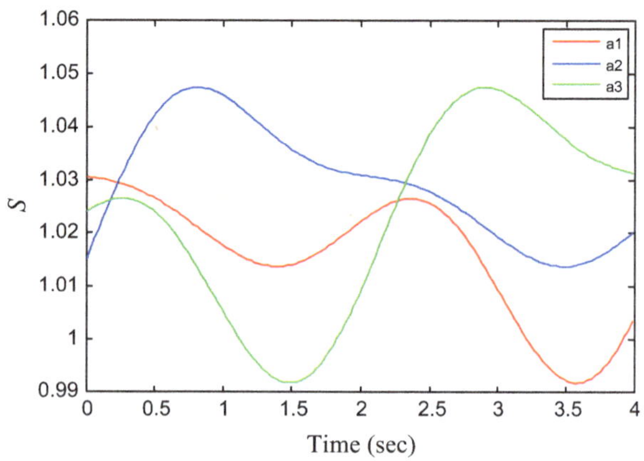

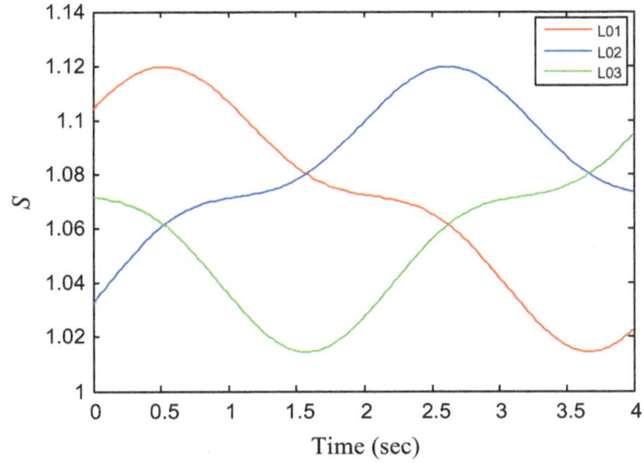

Fig. 5.25 Sensitivity with respect to L_{01}, L_{02} and L_{03}

Table 5.4 The variance values of the HDCPM

$D(\Delta q)$	Group 1	Group 2	Group 3
Δx	0.0001	0.0001	0.0001
Δy	0.0001	0.0001	0.0001
Δz	0.0003	0.0003	0.0003
Δa	0.0001	0.0001	0.0001
Δl_1	0.00012	0.00012	0.00012
Δl_2	0.00012	0.00012	0.00012
Δl_3	0.00012	0.00012	0.00012
Δl_4	0.00012	0.00012	0.00012
Δl_5	0.00012	0.00012	0.00012
$\Delta \theta_1$	0.000001	0.000001	0.000001
$\Delta \theta_4$	0.000001	0.000001	0.000001
ΔL_0	0.0001	0.0001	0.0001

Referring to Figs. 5.27, 5.28, 5.29, 5.30, 5.31, 5.32, 5.33, 5.34, 5.35 and 5.36 and analyzing the graphs of the desired and actual trajectory, it should be noted that the 8 sensitive error elements have larger effect on the motion trajectory of the end-effector than those of Δa_i and $\Delta L_{0i} (i \in \{1, 2, 3\})$. Δx_i, Δy_i, Δz_i, $\Delta \theta_{i1}$ and $\Delta \theta_{i4}$ $(i \in \{1, 2, 3\})$ have less influence on the motion trajectory of the end-effector. On the basis of the results of simulations, one can see that Δl_{11}, Δl_{12}, Δl_{13}, Δl_{15}, Δl_{22}, Δl_{23}, Δl_{33} and Δl_{34} are the most sensitive error sources, in comparison to the other error sources. In real systems, beside the main geometric errors of the HDCPR the influences on motion trajectory errors have: temperature changes which cause thermal deformations, deformations of manipulator parts caused by operating loads, the properties of the axis controls.

According to the results listed Table 5.3 and Figs. 5.5, 5.27, 5.28, 5.29, 5.30, 5.31, 5.32, 5.33, 5.34, 5.35 and 5.36, it has been also proved that the desired planning trajectory and the actual motion trajectory under the influences of different

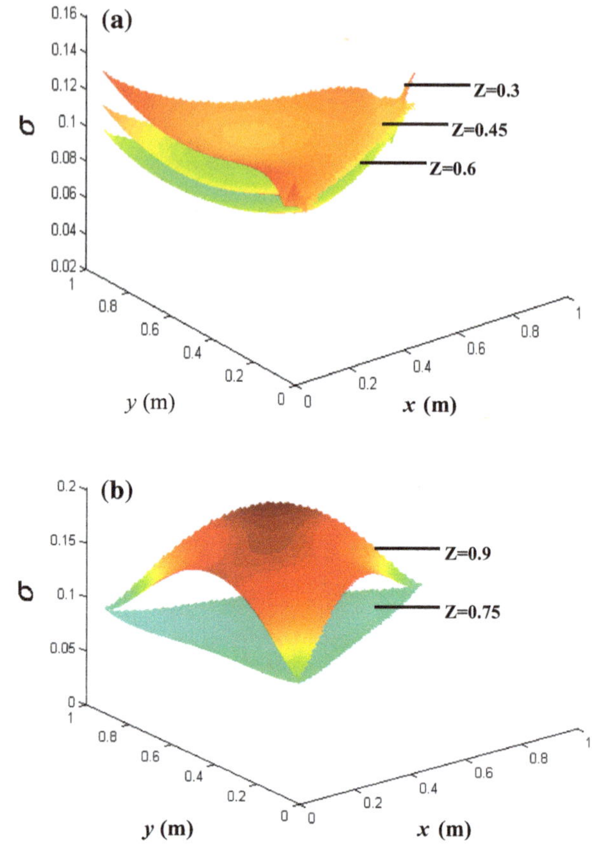

Fig. 5.26 Standard deviation of position error with different z values

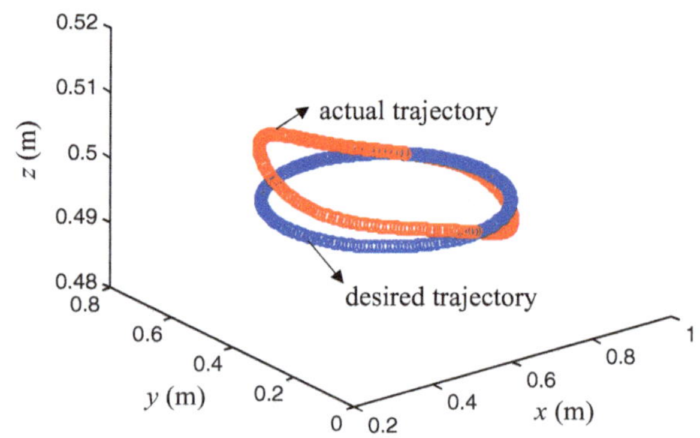

Fig. 5.27 Motion trajectory with respect to Δl_{11}, Δl_{12}, Δl_{13} and Δl_{15}

error sources were in agreement with the aforementioned sensitivity analysis results, i.e., the parameters of the HDPMs is the main error sources, while the parameters of the CPR is the secondary error sources.

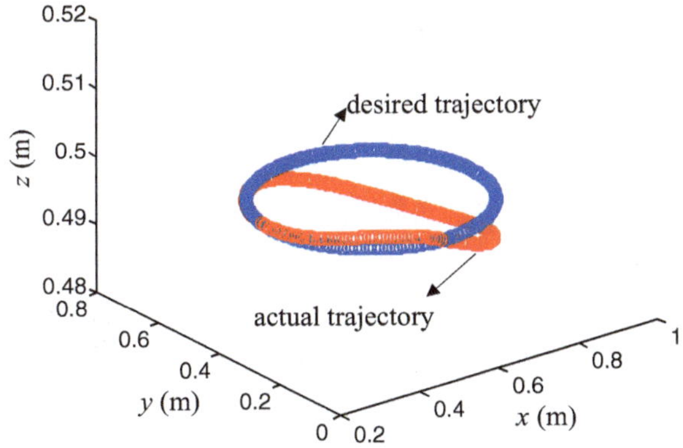

Fig. 5.28 Motion trajectory with respect to Δl_{22} and Δl_{23}

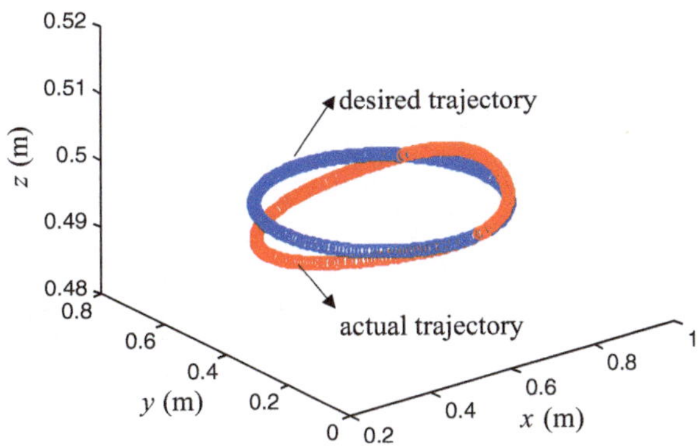

Fig. 5.29 Motion trajectory with respect to Δl_{33} and Δl_{34}

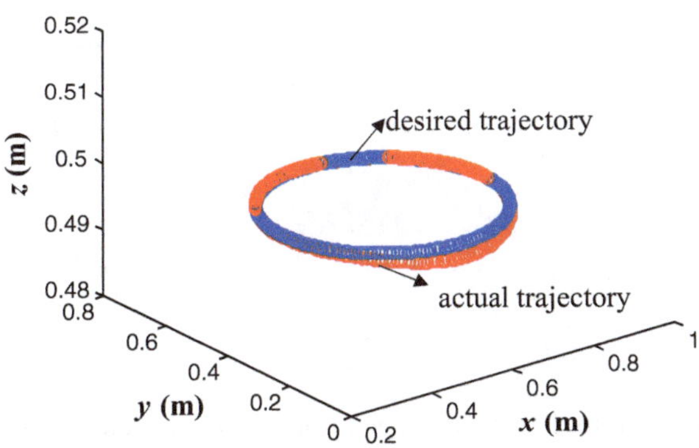

Fig. 5.30 Motion trajectory with respect to Δa_1, Δa_2 and Δa_3

Software compensation can be used to further improve position accuracy to some extent, but with the current computer technology, even the best tuned system will still produce parts with considerable systematic errors [9]. In addition, it has

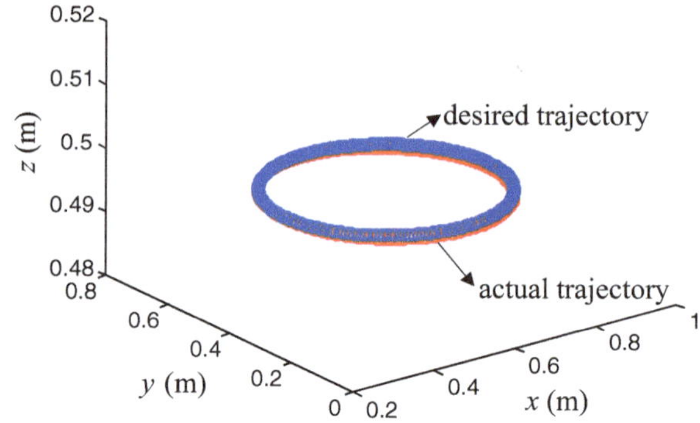

Fig. 5.31 Motion trajectory with respect to ΔL_{01}, ΔL_{02} and ΔL_{03}

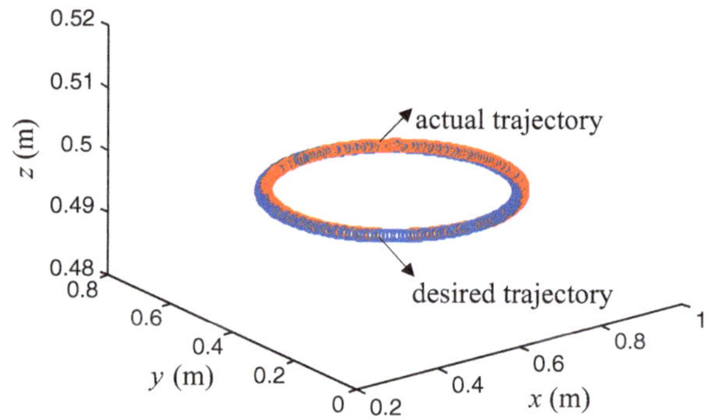

Fig. 5.32 Motion trajectory with respect to Δx_1, Δx_2 and Δx_3

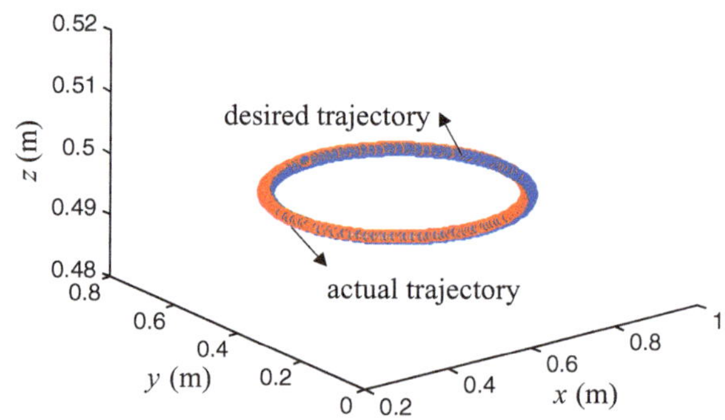

Fig. 5.33 Motion trajectory with respect to Δy_1, Δy_2 and Δy_3

been found that some position errors are uncompensable by software, and these errors are associated with the relative geometric source errors. As a result, it is possible to take the kinematic calibration, manufacturing and machining technology, and assembly process to improve the position accuracy.

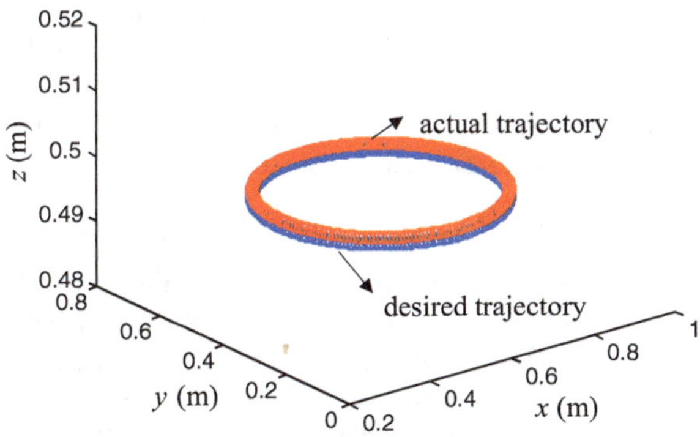

Fig. 5.34 Motion trajectory with respect to Δz_1, Δz_2 and Δz_3

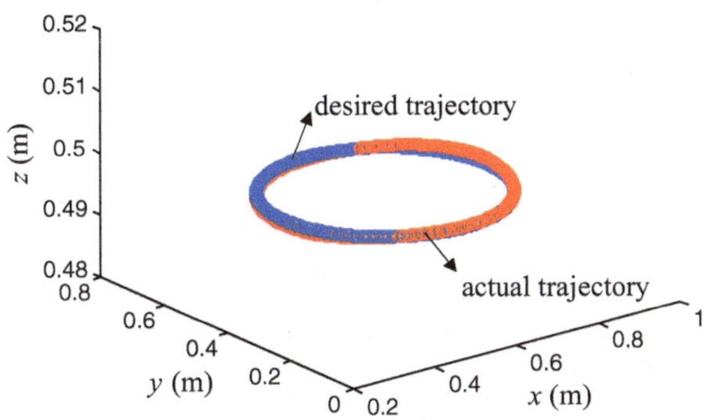

Fig. 5.35 Motion trajectory with respect to $\Delta\theta_{11}$, $\Delta\theta_{21}$ and $\Delta\theta_{31}$

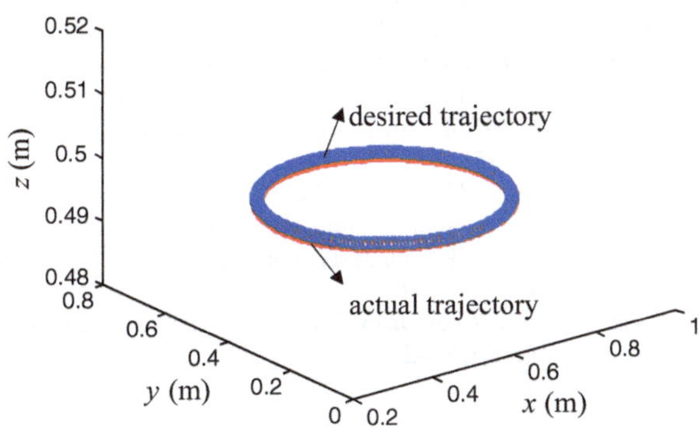

Fig. 5.36 Motion trajectory with respect to $\Delta\theta_{14}$, $\Delta\theta_{24}$ and $\Delta\theta_{34}$

5.6 Conclusions

This chapter presents a class of hybrid-driven based three-cable parallel manipulator (HDCPR) with three translational motions, which combines the HDPM with the CPR in a way to provide a solution for moving heavy objects with high efficiency and high-performance. In this investigation, on the basis of the method of vector chain and direct differential, the mapping relationship between 36 error sources of kinematics parameters and position error of the end-effector for the HDCPR is established, which is also suitable for other parallel mechanisms. In addition, the effects of the error sources on the end-effector of the HDCPR are discussed by means of the sensitivity analysis. Illustrative simulation studies are carried out for the validation of the proposed methodology. It is also revealed that the position error of the end-effector of the HDCPR depends not only on the kinematic parameters of the mechanism, but also the geometrical dimensions. Error model and sensitivity analysis provide theoretical foundation for the further research on optimum design, kinematic calibration and error compensation of the HDCPR. The prototype verification of the designed HDCPR is a topic for further research and is currently under investigation. In addition, the methodology presented in this work could be used as a guide for the design and accuracy analysis of other manipulators.

References

1. Zi B, Ding H, Wu X et al (2014) Error modeling and sensitivity analysis of a hybrid-driven based cable parallel manipulator. Precis Eng 38(1):197–211
2. He R, Zhao Y, Yang S et al (2010) Kinematic-parameter identification for serial-robot calibration based on POE formula. IEEE Trans Rob 26(3):411–423
3. Yongbo W, Pessi P, Huapeng WU et al (2009) Accuracy analysis of hybrid parallel robot for the assembling of ITER. Fusion Eng Des 84(7–11):1964–1968
4. Santolaria J, Aguilar JJ, Yagüe JA et al (2008) Kinematic parameter estimation technique for calibration and repeatability improvement of articulated arm coordinate measuring machines. Precis Eng 32(4):251–268
5. Shi H, Su HJ, Dagalakis N et al (2013) Kinematic modeling and calibration of a flexure based hexapod nanopositioner. Precis Eng 37(1):117–128
6. Huang T, Li Y, Tang G et al (2002) Error modeling, sensitivity analysis and assembly process of a class of 3-DOF parallel kinematic machines with parallelogram struts. Sci China Ser E Technol Sci 45(5):467–476
7. Liu H, Huang T, Chetwynd DG (2011) A general approach for geometric error modeling of lower mobility parallel manipulators. J Mech Robot 3(2):021013
8. Zi Z (2011) Sensitivity analysis approaches applied to systems biology models. IET Syst Biol 5 (6):336–346
9. Tong K, Lehtihet EA, Joshi S (2004) Software compensation of rapid prototyping machines. Precis Eng 28(3):280–292

Chapter 6
Performance Indices of the Hybrid-Driven Based Cable-Suspended Parallel Robots

6.1 Preamble

Performance evaluation plays an important role in the design process of the manipulator [1–4]. It deals with problems in terms of light weight, large reachable workspace, fast motion and heavy payload capability, and high reconfigurability [5–9]. The purpose of performance analysis is aiming at enhancing the performance indices such as workspace, singularity, conditioning index [10, 11], stiffness index [12], and payload index [13, 14], etc.

In this chapter, to illustrate the advantage of the performance indices, firstly, hybrid-driven based cable-suspended parallel manipulators (HDCPR) is presented, then Kinematics and dynamics of the cable parallel manipulators with and without HDPM are studied based on Lagrange method separately. Finally, some performance indices, such as conditioning index, stiffness index, and payload index are investigated and compared based on above analysis. In Sect. 6.2, The cable parallel manipulator with the HDPM has the advantages of both the cable parallel manipulator and the HDPM. Kinematics and dynamics of the cable parallel manipulators with and without HDPM are studied based on Lagrange method separately, and drive torques and drive powers are also given to compare the load carrying capacity of the two manipulators. Following that workspace, stiffness performance, singularity analysis of the two manipulators is carried out. As an example, a completely restrained cable parallel manipulator with 3 Degrees of Freedom is investigated on the basis of the above design objectives. In Sect. 6.3, conditioning index, stiffness index, and payload index are investigated and compared. Kinematics and Jacobian analysis of the proposed HDCPR are performed. Additionally, the conditioning index, stiffness index, and payload index are given to reveal performance variation of the HDCPR on the basis of Jacobian matrix. To illustrate the advantage of the performance indices, the 3-DOF HDCPR, improved HDCPR, and cable parallel manipulator are compared. The results demonstrate the validity of the performance evaluation developed of the HDCPR.

B. Zi and S. Qian, *Design, Analysis and Control of Cable-Suspended Parallel Robots and Its Applications*, DOI 10.1007/978-981-10-1753-7_6

6.2 Comparative Study of Cable Parallel Manipulators with and Without Hybrid-Driven Planar Five-Bar Mechanism

6.2.1 System Description and Kinematics

The HDCPR is that the HDPM is introduced into a CPR in place of the conventional servomechanism as a power source of the CPR system. Figure 6.1 presents the 3D design model of the completely restrained CPR and HDCPR with three translational motions. The HDCPR suspends an end-effector and restrains all motion degrees of freedom for the object by using four cables when the end-effector moves within the workspace. For each cable, one end is connected to the end-effector, and the other one rolls through a pulley fixed on the top of the relative cable tower rack and then is fed into the HDPM. The HDCPR comprises of two modules: (1) the CPR consisting of four-cable tower racks, four cables, pulley struts, pulleys, girder, cargo (i.e. end-effector); (2) four groups of HDPMs containing three-phase asynchronous motors, servomotors, reducers and double crank five-bar linkage; the asynchronous motors are connected by the pulley transmission mechanisms, while the servomotors and the reducers are linked by couplings. Moreover, each tower rack of the CPR is equipped with a cable guide pulley, and the pulley transmission mechanisms and reducers are joined to the double crank planar five-bar linkage.

6.2.1.1 Kinematics of the CPR

A simple schematic sketch of the CPR structure model with the associated coordinate systems are depicted in Fig. 6.2. At the bottom of one cable tower rack, a global coordinate system O(XYZ) is established. The end-effector has location coordinates G(X, Y, Z). The distance between each cable tower rack top $P_i(X_i, Y_i, Z_i)$ and the end-effector is L_i (i = 1, 2, 3, 4). The four cable tower racks have same height and are arrayed in a triangle on the ground, when deformation is ignored. In order to simplify the model, the cables are treated as a massless body with no deformation.

The relationship between the cable length L_i and the end-effector location G(X, Y, Z) of the CPR, can be easily obtained as follows

$$L_i = \sqrt{(X - X_i)^2 + (Y - Y_i)^2 + (Z - Z_i)^2} \quad (\mathrm{i} = 1, 2, 3, 4) \tag{6.1}$$

According to Fig. 6.2 and Eq. (6.1), we can obtain

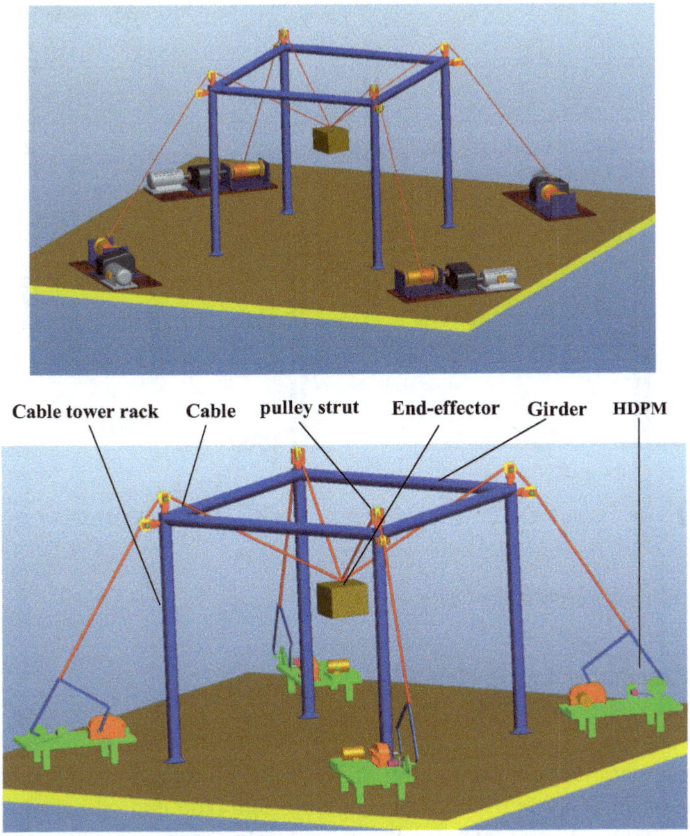

Fig. 6.1 3D model of the CPR and HDCPR tower

$$\begin{cases} L_1^2 = X^2 + Y^2 + (Z-h)^2 \\ L_2^2 = (X-a)^2 + Y^2 + (Z-h)^2 \\ L_3^2 = (X-a)^2 + (Y-b)^2 + (Z-h)^2 \\ L_4^2 = X^2 + (Y-b)^2 + (Z-h)^2 \end{cases} \tag{6.2}$$

Hence, the end-effector coordinates (X, Y, Z) can be obtained by

$$\begin{cases} X = \frac{L_1^2 - L_2^2 + a^2}{2a} \\ Y = \frac{L_1^2 - L_4^2 + b^2}{2b} \\ Z = h - \sqrt{L_1^2 - X^2 - Y^2} \end{cases} \tag{6.3}$$

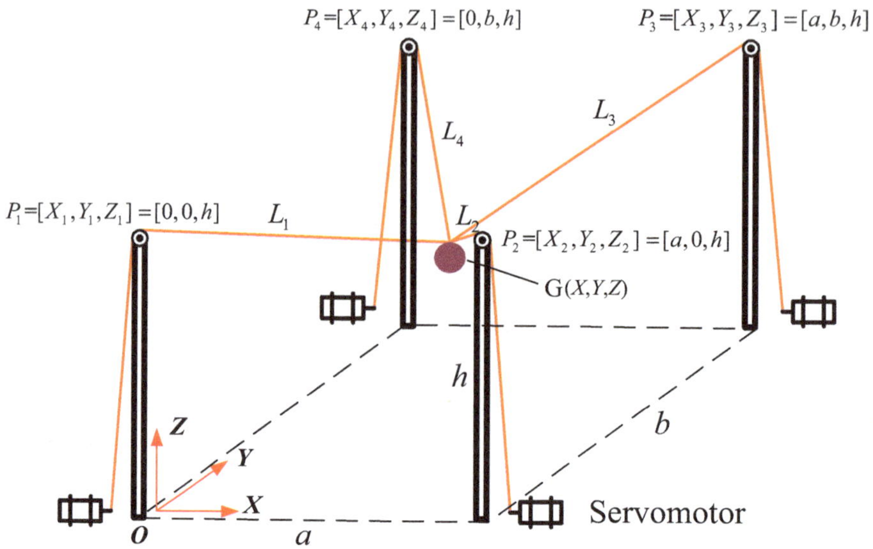

Fig. 6.2 Schematic sketch of the CPR

6.2.1.2 Kinematics of the HDCPR

A simple schematic sketch of the HDCPR structure model with the associated coordinate systems are depicted in Fig. 6.3b. At the bottom of one cable tower rack, a global coordinate system O(XYZ) is established. The end-effector has location coordinates G(X, Y, Z). The distance between each cable tower rack top P_i (X_i, Y_i, Z_i) and the end-effector is L_i (i = 1, 2, 3, 4). The four cable tower racks have same height and are arrayed in a triangle on the ground, when deformation is ignored. In order to simplify the model, the cables are treated as a massless body with no deformation. At the bottom of the HDPM, a local coordinate system $(x'o'z')$ is established. This section assumes that the links have the same density, and the centroid is in the center of links. The length of the links of the HDPM are l_1, l_2, l_3, l_4, l_5, the angel between links and x'-axis are $\theta_{i1}, \theta_{i2}, \theta_{i3}, \theta_{i4}, 0°$, the mass of links are m_1, m_2, m_3, m_4, the rotational inertia of links are J_1, J_2, J_3, J_4, the mass of the end-effector is m, respectively.

Link AB driven by a CV motor and link ED driven by a servomotor are the two driving links (see Fig. 6.3a), and the joint C is the output. There are two independent constraints in the HDPM, which can be used to derive the kinematic equation of the HDPM

$$\begin{cases} l_1 \cos\theta_{i1} + l_2 \cos\theta_{i2} - l_3 \cos\theta_{i3} - l_4 \cos\theta_{i4} - l_5 = 0 \\ l_1 \sin\theta_{i1} + l_2 \sin\theta_{i2} - l_3 \sin\theta_{i3} - l_4 \sin\theta_{i4} = 0 \end{cases} \tag{6.4}$$

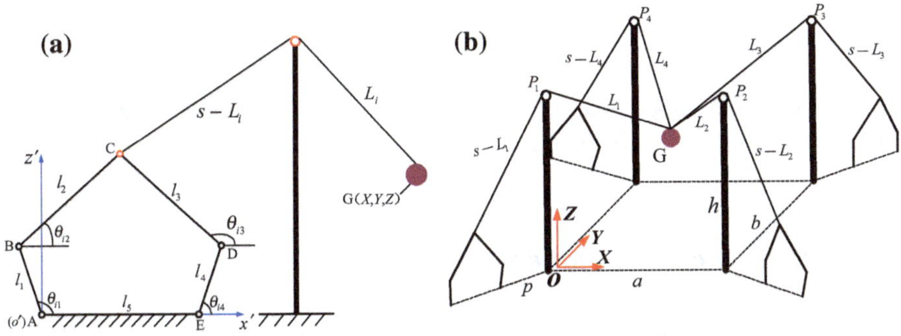

Fig. 6.3 Structure model **a** of the ith group of the HDPM and schematic sketch **b** of the HDCPR the HDPM and *i*th cable tower

In Eq. (6.4) θ_{i1} and θ_{i4} are known. Consequently θ_{i2} and θ_{i3} can be calculated by θ_{i1} and θ_{i4} specificity as follows

$$\theta_{i3} = \arctan\left(\frac{\pm\sqrt{A^2(\theta_{i1},\theta_{i4}) + B^2(\theta_{i1},\theta_{i4}) - C^2(\theta_{i1},\theta_{i4})}}{C(\theta_{i1},\theta_{i4})}\right) + \arctan\left(\frac{B(\theta_{i1},\theta_{i4})}{A(\theta_{i1},\theta_{i4})}\right) \tag{6.5}$$

$$\theta_{i2} = \arctan\left(\frac{\mu(\theta_{i1},\theta_{i4}) + l_3\sin\theta_{i3}}{\lambda(\theta_{i1},\theta_{i4}) + l_3\cos\theta_{i3}}\right) \tag{6.6}$$

where

$$A(\theta_{i1},\theta_{i4}) = 2l_3\lambda(\theta_{i1},\theta_{i4}), B(\theta_{i1},\theta_{i4}) = 2l_3\mu(\theta_{i1},\theta_{i4}),$$

$$\begin{aligned} C(\theta_{i1},\theta_{i4}) &= l_2^2 - l_3^2 - \lambda^2(\theta_{i1},\theta_{i4}) - \mu^2(\theta_{i1},\theta_{i4}), \lambda(\theta_{i1},\theta_{i4}) \\ &= l_4\cos\theta_{i4} - l_1\cos\theta_{i1} + l_5, \end{aligned}$$

$$\mu(\theta_{i1},\theta_{i4}) = l_4\sin\theta_{i4} - l_1\sin\theta_{i1}.$$

Let (x'_C, z'_C) be the coordinates of joint C in the local coordinate system, and it can be expressed as

$$\begin{cases} x'_C = l_1\cos\theta_{i1} + l_2\cos\theta_{i2} = l_5 + l_3\cos\theta_{i3} + l_4\cos\theta_{i4} \\ z'_C = l_1\sin\theta_1 + l_2\sin\theta_{i2} = l_3\sin\theta_{i3} + l_4\sin\theta_{i4} \end{cases} \tag{6.7}$$

According to the geometric relationship of the HDCPR, Li can be derived as

$$L_i = s - \sqrt{(p + l_5 - x'_C)^2 + (h - z'_C)^2} \tag{6.8}$$

where p is the distance $\overline{EO}$, s is the total length of the cable. Similarly, the end-effector coordinates (X, Y, Z) of the HDCPR can be formulated by Eq. (6.3).

6.2.2 *Dynamic Formulation*

In the following discussion, we will describe the CPR and HDCPR equations of motion by means of Lagrangian formulation.

6.2.2.1 Dynamic Model of the CPR

The generalized coordinates is the cable length $L_i(i = 1, 2, 3, 4)$ of the CPR as shown in Fig. 6.2. Differentiating Eq. (6.3) with respect to time

$$\begin{cases} \dot{X} = \sum\limits_{i=1}^{4} \frac{\partial X}{\partial L_i} \dot{L}_i \\ \dot{Y} = \sum\limits_{i=1}^{4} \frac{\partial Y}{\partial L_i} \dot{L}_i \; (i = 1, 2, 3, 4) \\ \dot{Z} = \sum\limits_{i=1}^{4} \frac{\partial Z}{\partial L_i} \dot{L}_i \end{cases} \tag{6.9}$$

Hence, the kinetic energy of the CPR can be written as

$$E = \frac{1}{2} m (\dot{X}^2 + \dot{Y}^2 + \dot{Z}^2) = \frac{1}{2} m \sum_{i=1}^{4} R_i \dot{L}_i^2 + m \sum_{i=1}^{4} \sum_{j=1, j \neq i}^{4} R_{ij} \dot{L}_i \dot{L}_j \tag{6.10}$$

where $R_i = \left(\frac{\partial X}{\partial L_i}\right)^2 + \left(\frac{\partial Y}{\partial L_i}\right)^2 + \left(\frac{\partial Z}{\partial L_i}\right)^2, R_{ij} = \frac{\partial X}{\partial L_i} \frac{\partial X}{\partial L_j} + \frac{\partial Y}{\partial L_i} \frac{\partial Y}{\partial L_j} + \frac{\partial Z}{\partial L_i} \frac{\partial Z}{\partial L_j}$.

The potential energy of the CPR is described by

$$U = mgZ \tag{6.11}$$

where g is the acceleration due to gravity.

Lagrange's equation can be written in the form of potential energy, kinetic energy, and generalized forces or torques in the following form

$$\frac{\partial}{\partial t}\left(\frac{\partial E}{\partial \dot{L}_i}\right) - \frac{\partial E}{\partial L_i} + \frac{\partial U}{\partial L_i} = Q_i \quad (i = 1, 2, 3, 4) \tag{6.12}$$

where

$$\frac{\partial}{\partial t}\left(\frac{\partial E}{\partial \dot{L}_i}\right) = \begin{bmatrix} mR_1 & mR_{12} & mR_{13} & mR_{14} \\ & mR_2 & mR_{23} & mR_{24} \\ & & mR_3 & mR_{34} \\ & & & mR_4 \end{bmatrix} \begin{bmatrix} \ddot{L}_1 \\ \ddot{L}_2 \\ \ddot{L}_3 \\ \ddot{L}_4 \end{bmatrix} = D(L_i)\ddot{L}_i \tag{6.13}$$

$$-\frac{\partial E}{\partial L_i} = -\left(\frac{1}{2}m\frac{\partial \sum_{i=1}^{4} R_i \dot{L}_i}{\partial L_i} + m\frac{\partial \sum_{i=1}^{4}\sum_{j=1, j\neq i}^{4} R_{ij}\dot{L}_j}{\partial L_i}\right)\dot{L}_i = C\left(L_i, \dot{L}_i\right)\dot{L}_i \tag{6.14}$$

$$\frac{\partial U}{\partial L_i} = \tilde{U}(L_i) \tag{6.15}$$

$$Q = \left[\tau_1/r \quad \tau_2/r \quad \tau_3/r \quad \tau_4/r\right]^T \tag{6.16}$$

where τ_1, τ_2, τ_3 and τ_4 represent the driving torque of the windlass, respectively; r is the radius of the windlass.

Based on Eqs. (6.12)–(6.16), through a series of transformations, substitutions and simplifying the resulting expression, here, dynamic equations of the CPR are written as the following general form

$$D(L_i)\ddot{L}_i + C\left(L_i, \dot{L}_i\right)\dot{L}_i + \tilde{U}(L_i) = Q_i \quad (i = 1, 2, 3, 4) \tag{6.17}$$

where $D(L_i)$ is the inertia matrix of the system, $C\left(L_i, \dot{L}_i\right)$ is the vector of Coriolis and centripetal terms, $\tilde{U}(L_i)$ is the vector of gravity terms.

6.2.2.2 Dynamic Model of the HDCPR

Angular displacement θ_{i1} and θ_{i4} (i = 1, 2, 3, 4) of the two driving links of four groups of HDPMs of the HDCPR are selected to be as generalized coordinates (see Fig. 6.3a), and the angular velocity of all links are given as

$$\dot{\theta}_{ij} = \frac{\partial \theta_{ij}}{\partial \theta_{i1}}\dot{\theta}_{i1} + \frac{\partial \theta_{ij}}{\partial \theta_{i4}}\dot{\theta}_{i4} \quad (\mathrm{j} = 1, 2, 3, 4) \tag{6.18}$$

The centroid velocity of each link in and directions can be formulated as follows

$$\begin{cases} \dot{x}'_{ij} = \dfrac{\partial x'_{ij}}{\partial \theta_{i1}} \dot{\theta}_{i1} + \dfrac{\partial x'_{ij}}{\partial \theta_{i4}} \dot{\theta}_{i4} \\ \dot{z}'_{ij} = \dfrac{\partial z'_{ij}}{\partial \theta_{i1}} \dot{\theta}_{i1} + \dfrac{\partial z'_{ij}}{\partial \theta_{i4}} \dot{\theta}_{i4} \end{cases} \quad (j = 1, 2, 3, 4) \tag{6.19}$$

From Eqs. (6.18) and (6.19), one can derive the kinetic energy of the ith group of the HDPM

$$\begin{aligned} E_{li} &= \frac{1}{2} \sum_{j=1}^{4} \left[m_j \left(\dot{x}'^2_{ij} + \dot{z}'^2_{ij} \right) + J_j \dot{\theta}^2_{ij} \right] \\ &= \frac{1}{2} d_{i1} \dot{\theta}^2_{i1} + d_{i2} \dot{\theta}_{i1} \dot{\theta}_{i4} + \frac{1}{2} d_{i3} \dot{\theta}^2_{i4} \quad (i = 1, 2, 3, 4) \end{aligned} \tag{6.20}$$

in which,

$$d_{i1} = \sum_{j=1}^{4} \left[m_j \left(\frac{\partial x'_{ij}}{\partial \theta_{i1}} \right)^2 + m_j \left(\frac{\partial z'_{ij}}{\partial \theta_{i1}} \right)^2 + J_j \left(\frac{\partial \theta_{ij}}{\partial \theta_{i1}} \right)^2 \right]$$

$$d_{i2} = \sum_{j=1}^{4} \left[m_j \left(\frac{\partial x'_{ij}}{\partial \theta_{i1}} \right) \left(\frac{\partial x'_{ij}}{\partial \theta_{i4}} \right) + m_j \left(\frac{\partial z'_{ij}}{\partial \theta_{i1}} \right) \left(\frac{\partial z'_{ij}}{\partial \theta_{i4}} \right) + J_j \left(\frac{\partial \theta_{ij}}{\partial \theta_{i1}} \right) \left(\frac{\partial \theta_{ij}}{\partial \theta_{i4}} \right) \right]$$

$$d_{i3} = \sum_{j=1}^{4} \left[m_j \left(\frac{\partial x'_{ij}}{\partial \theta_{i4}} \right)^2 + m_j \left(\frac{\partial z'_{ij}}{\partial \theta_{i4}} \right)^2 + J_j \left(\frac{\partial \theta_{ij}}{\partial \theta_{i4}} \right)^2 \right]$$

The sum of kinetic energy of four groups of the HDPM can be expressed by the following equation

$$E_l = \sum_{i=1}^{4} E_{li} \tag{6.21}$$

The velocity of the end-effector with respect to X, Y, and Z directions can be written as

$$\begin{cases} V_X = \sum\limits_{i=1}^{4} \left(\frac{\partial X}{\partial \theta_{i1}} \dot{\theta}_{i1} + \frac{\partial X}{\partial \theta_{i4}} \dot{\theta}_{i4} \right) \\ V_Y = \sum\limits_{i=1}^{4} \left(\frac{\partial Y}{\partial \theta_{i1}} \dot{\theta}_{i1} + \frac{\partial Y}{\partial \theta_{i4}} \dot{\theta}_{i4} \right) \\ V_Z = \sum\limits_{i=1}^{4} \left(\frac{\partial Z}{\partial \theta_{i1}} \dot{\theta}_{i1} + \frac{\partial Z}{\partial \theta_{i4}} \dot{\theta}_{i4} \right) \end{cases} \tag{6.22}$$

Hence, the kinetic energy of the end-effector can be obtained as follows

$$E_g = \frac{1}{2}m\left(V_X^2 + V_Y^2 + V_Z^2\right)$$
$$= \sum_{i=1}^{4}\left(\frac{1}{2}m\tilde{d}_{i1}\dot{\theta}_{i1}^2 + \frac{1}{2}m\tilde{d}_{i3}\dot{\theta}_{i4}^2\right) + m\tilde{d}\dot{\theta}_{i1}\dot{\theta}_{i4} \tag{6.23}$$
$$+ m\sum_{i=1}^{4}\sum_{j=1,j\neq i}^{4}\hat{d}_{ij}\dot{\theta}_{i1}\dot{\theta}_{j1} + m\sum_{i=1}^{4}\sum_{j=1,j\neq i}^{4}\bar{d}_{ij}\dot{\theta}_{i4}\dot{\theta}_{j4}$$

where

$$\tilde{d}_{i1} = \left(\frac{\partial X}{\partial\theta_{i1}}\right)^2 + \left(\frac{\partial Y}{\partial\theta_{i1}}\right)^2 + \left(\frac{\partial Z}{\partial\theta_{i1}}\right)^2$$

$$\tilde{d}_{i3} = \left(\frac{\partial X}{\partial\theta_{i4}}\right)^2 + \left(\frac{\partial Y}{\partial\theta_{i4}}\right)^2 + \left(\frac{\partial Z}{\partial\theta_{i4}}\right)^2$$

$$\tilde{d} = \sum_{i=1}^{4}\frac{\partial X}{\partial\theta_{i1}}\sum_{i=1}^{4}\frac{\partial X}{\partial\theta_{i4}} + \sum_{i=1}^{4}\frac{\partial Y}{\partial\theta_{i1}}\sum_{i=1}^{4}\frac{\partial Y}{\partial\theta_{i4}} + \sum_{i=1}^{4}\frac{\partial Z}{\partial\theta_{i1}}\sum_{i=1}^{4}\frac{\partial Z}{\partial\theta_{i4}}$$

$$\hat{d}_{ij} = \frac{\partial X}{\partial\theta_{i1}}\frac{\partial X}{\partial\theta_{j1}} + \frac{\partial Y}{\partial\theta_{i1}}\frac{\partial Y}{\partial\theta_{j1}} + \frac{\partial Z}{\partial\theta_{i1}}\frac{\partial Z}{\partial\theta_{j1}}$$

$$\bar{d}_{ij} = \frac{\partial X}{\partial\theta_{i4}}\frac{\partial X}{\partial\theta_{j4}} + \frac{\partial Y}{\partial\theta_{i4}}\frac{\partial Y}{\partial\theta_{j4}} + \frac{\partial Z}{\partial\theta_{i4}}\frac{\partial Z}{\partial\theta_{j4}}$$

Referring to Eqs. (6.21) and (6.23), the total kinetic energy of the system can be written as

$$\tilde{E} = E_l + E_g \tag{6.24}$$

The potential energy can be formulated as follows

$$\tilde{G} = m_1 g\sum_{i=1}^{4}z'_{i1} + m_2 g\sum_{i=1}^{4}z'_{i2} + m_3 g\sum_{i=1}^{4}z'_{i3} + m_4 g\sum_{i=1}^{4}z'_{i4} + mgZ \tag{6.25}$$

Dynamic equations of the HDCPR are given by using the following Lagrange's equation

$$\frac{\partial}{\partial t}\left(\frac{\partial\tilde{E}}{\partial\dot{\theta}}\right) - \frac{\partial\tilde{E}}{\partial\theta} + \frac{\partial\tilde{G}}{\partial\theta} = F \tag{6.26}$$

where $\theta = (\theta_{i1}\ \theta_{i4})^{\mathrm{T}}$(i = 1,2,3,4) denotes the vector of the generalized coordinates, F is the vector of the generalized forces corresponding to generalized coordinates.

Then, substituting the Eqs. (6.20)–(6.25) into Eq. (6.26), we obtain

$$F = \frac{\partial}{\partial t}\left(\frac{\partial E_l}{\partial \dot{\theta}}\right) - \frac{\partial E_l}{\partial \theta} + \frac{\partial}{\partial t}\left(\frac{\partial E_g}{\partial \dot{\theta}}\right) - \frac{\partial E_g}{\partial \theta} + \frac{\partial \tilde{G}}{\partial \theta} \tag{6.27}$$

$$\frac{\partial}{\partial t}\left(\frac{\partial E_l}{\partial \dot{\theta}}\right) - \frac{\partial E_l}{\partial \theta} = M_{li}\ddot{\theta} + C_{li}\dot{\theta} \quad (\mathrm{i} = 1, 2, 3, 4) \tag{6.28}$$

where

$$M_{li} = \begin{bmatrix} d_{i1} & d_{i2} \\ d_{i2} & d_{i3} \end{bmatrix}, C_{li} = \begin{bmatrix} \frac{1}{2}\frac{\partial d_{i1}}{\partial \theta_{i1}}\dot{\theta}_{i1} + \frac{\partial d_{i1}}{\partial \theta_{i4}}\dot{\theta}_{i4} & \left(\frac{\partial d_{i2}}{\partial \theta_{i4}} - \frac{1}{2}\frac{\partial d_{i3}}{\partial \theta_{i1}}\right)\dot{\theta}_{i4} \\ \left(\frac{\partial d_{i2}}{\partial \theta_{i1}} - \frac{1}{2}\frac{\partial d_{i1}}{\partial \theta_{i4}}\right)\dot{\theta}_{i1} & \frac{1}{2}\frac{\partial d_{i3}}{\partial \theta_{i4}}\dot{\theta}_{i4} + \frac{\partial d_{i3}}{\partial \theta_{i1}}\dot{\theta}_{i1} \end{bmatrix}.$$

$$\frac{\partial}{\partial t}\left(\frac{\partial E_g}{\partial \dot{\theta}}\right) - \frac{\partial E_g}{\partial \theta} = M_g\ddot{\theta} + C_g\dot{\theta} \tag{6.29}$$

where $M_g = \begin{bmatrix} m\tilde{d}_{i1} + m\hat{d}_{ij} & m\tilde{d} \\ m\tilde{d} & m\tilde{d}_{i3} + m\bar{d}_{ij} \end{bmatrix} \quad (\mathrm{i} = 1, 2, 3, 4;\ \mathrm{j} = 1, 2, 3, 4;\ \mathrm{i} \neq \mathrm{j})$

$$C_g = -\begin{bmatrix} \frac{m}{2}\frac{\partial \sum\limits_{i=1}^{4}\tilde{d}_{i1}}{\partial \theta_{i1}}\dot{\theta}_{i1} + m\frac{\partial \tilde{d}}{\partial \theta_{i1}}\dot{\theta}_{i4} + m\frac{\partial \sum\limits_{i=1}^{4}\sum\limits_{j=1, j\neq i}^{4}\hat{d}_{ij}\dot{\theta}_{j1}}{\partial \theta_{i1}} & \frac{m}{2}\frac{\partial \sum\limits_{i=1}^{4}\tilde{d}_{i3}}{\partial \theta_{i1}}\dot{\theta}_{i4} + m\frac{\partial \sum\limits_{i=1}^{4}\sum\limits_{j=1, j\neq i}^{4}\bar{d}_{ij}\dot{\theta}_{j4}}{\partial \theta_{i1}} \\ \frac{m}{2}\frac{\partial \sum\limits_{i=1}^{4}\tilde{d}_{i1}}{\partial \theta_{i4}}\dot{\theta}_{i1} + m\frac{\partial \sum\limits_{i=1}^{4}\sum\limits_{j=1, j\neq i}^{4}\hat{d}_{ij}\dot{\theta}_{j1}}{\partial \theta_{i4}} & \frac{m}{2}\frac{\partial \sum\limits_{i=1}^{4}\tilde{d}_{i3}}{\partial \theta_{i4}}\dot{\theta}_{i4} + m\frac{\partial \tilde{d}}{\partial \theta_{i4}}\dot{\theta}_{i1} + m\frac{\partial \sum\limits_{i=1}^{4}\sum\limits_{j=1, j\neq i}^{4}\bar{d}_{ij}\dot{\theta}_{j4}}{\partial \theta_{i4}} \end{bmatrix}$$

$$\frac{\partial \tilde{G}}{\partial \theta} = G \tag{6.30}$$

The generalized forces can be written as

$$F = (\tau_{i1}\ \tau_{i4})^{\mathrm{T}}(\mathrm{i} = 1, 2, 3, 4) \tag{6.31}$$

where τ_{i1}, τ_{i4} represents the input torque of all driving links of the HDCPR.

Through a series of transformations, substitutions and simplifying the resulting expression, here, dynamic equations of the HDCPR are written as

$$M(\theta)\ddot{\theta} + C\left(\theta, \dot{\theta}\right)\dot{\theta} + G(\theta) = F \tag{6.32}$$

Here, $M = M_{li} + M_g$, $M(\theta)$ is the symmetric positive definite inertia matrix; $C = C_{li} + C_g$, $C\left(\theta, \dot{\theta}\right)$ is the centrifugal force and Coriolis force matrix, and $\dot{M}(\theta) - 2C\left(\theta, \dot{\theta}\right)$ is skew symmetric; $G(\theta)$ is the gravity matrix.

6.2.3 Workspace Determination

6.2.3.1 Workspace of the CPR

Dynamic model of the CPR can also be formulated as

$$N_{3\times 4}T_{4\times 1} = \Theta_{3\times 1} \tag{6.33}$$

where

$$N_{3\times 4} = \begin{bmatrix} \frac{X_1 - X}{L_1} & \frac{X_2 - X}{L_2} & \frac{X_3 - X}{L_3} & \frac{X_4 - X}{L_4} \\ \frac{Y_1 - Y}{L_1} & \frac{Y_2 - Y}{L_2} & \frac{Y_3 - Y}{L_3} & \frac{Y_4 - Y}{L_4} \\ \frac{Z_1 - Z}{L_1} & \frac{Z_2 - Z}{L_2} & \frac{Z_3 - Z}{L_3} & \frac{Z_4 - Z}{L_4} \end{bmatrix}$$

$T_{4\times 1} = \begin{bmatrix} T_1 & T_2 & T_3 & T_4 \end{bmatrix}^{\mathrm{T}}$ and $\Theta_{3\times 1} = \begin{bmatrix} m\ddot{X} & m\ddot{Y} & mg + m\ddot{Z} \end{bmatrix}^{\mathrm{T}}$, and the driving force of cable ith exerted on the end-effector is defined as a tension cable force T_i (i = 1, 2, 3, 4).

Unlike rigid-link manipulators, the workspace of the CPR is always determined and characterized by positive tension status of driving cables due to the unilateral property of cables ($T_i > 0$). In addition, the points of the end-effector can be positioned in the cuboid space of the CPR. Therefore, the motion range can be expressed as:

$$0 < X < a,\; 0 < Y < b,\; 0 < Z < h \tag{6.34}$$

Hence, the flowchart of generating the workspace of the CPR is shown in Fig. 6.4.

6.2.3.2 Workspace of the HDCPR

The workspace of the HDCPR is characterized as the set of points where the end-effector can be positioned while all cables are in tension ($T_i > 0$). However, it is not sufficient to obtain the actual workspace of the HDCPR if it only depends on the tension condition; we should also consider the following constraints:

(1) The workspace of the HDPM. According to Eq. (6.3), the set of points (X, Y, Z) where the end-effector can be positioned are influenced by cable length L_i. Cable length L_i is calculated by the coordinates joint C $\left(x'_C, z'_C\right)$ from Eq. (6.8). Thus, the workspace of the joint C in $x' - z'$ plane is one of the constraints of the workspace of the HDCPR, and it can be expressed as

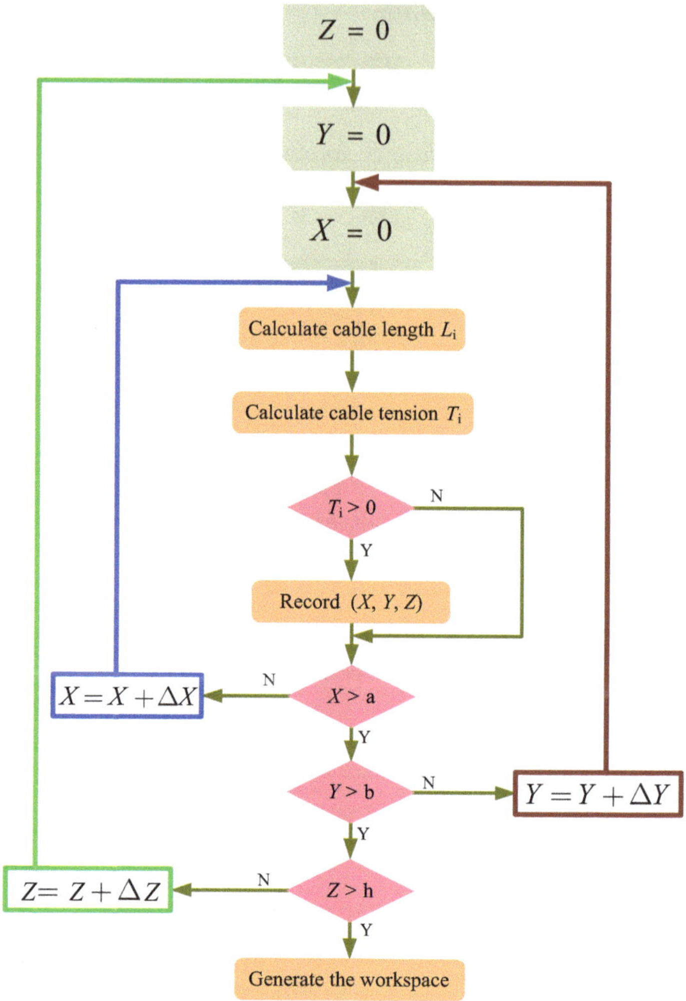

Fig. 6.4 Flowchart of generating the workspace of the CPR

$$\begin{cases} |l_1 - l_2| < L_{AC} < l_1 + l_2 \\ |l_3 - l_4| < L_{CE} < l_3 + l_4 \\ L_i = s - \sqrt{(p + l_5 - x'_C)^2 + (h - z'_C)^2} \end{cases} \tag{6.35}$$

where $L_{AC} = \sqrt{x'^2_C + z'^2_C}$, $L_{CE} = \sqrt{(x'_C - l_5)^2 + z'^2_C}$.

(2) The motion range of the end-effector can be expressed as

$$\begin{cases} 0<X=\frac{L_1^2-L_2^2+a^2}{2a}<a \\ 0<Y=\frac{L_1^2-L_4^2+b^2}{2b}<b \\ 0<Z=h-\sqrt{L_1^2-X^2-Y^2}<h \end{cases} \tag{6.36}$$

A general numerical workspace generation approach is employed here. The possible motion range of joint C in the HDPM is first discretized into a number of points. According to the Eq. (6.35), a series of the cable length Li can be calculated. Then, from Eq. (6.36), various combinations of Li can calculate the end-effector coordinates (X, Y, Z). Thus, a series of points in Cartesian coordinate system are determined. At the end, the points are checked and those satisfying the tension condition (T_i > 0) will generate the workspace. The flowchart of generating the workspace of the HDCPR is shown in Fig. 6.5.

6.2.4 *Stiffness Performance Analysis*

6.2.4.1 Stiffness Performance of the CPR

The mechanical component deflection of the parallel manipulator will be produced during the transmission process of the force and moment. Supposing the stiffness of the total drive system is K_p, the driving force of active joints is τ, and the deflection caused by active joints is Δq, the following equation can be obtained

$$\tau = K_p \Delta q \tag{6.37}$$

where $K_p = diag(k_{11}, k_{14}, k_{21}, k_{24}, k_{31}, k_{34}, k_{41}, k_{44})$, K_p represents the spring coefficient, and it is defined as unit matrix.

In order to study the stiffness of the CPR, the Jacobian matrix of the CPR should be worked out. According to the kinematics analysis above, the following equation is concluded

$$\dot{p} = J\dot{q} \tag{6.38}$$

where

$$\dot{p} = \begin{bmatrix} \dot{X} & \dot{Y} & \dot{Z} \end{bmatrix}^T, \dot{q} = \begin{bmatrix} \dot{L}_1 & \dot{L}_2 & \dot{L}_3 & \dot{L}_4 \end{bmatrix}^T, J = \begin{bmatrix} \frac{L_1}{a} & \frac{L_2}{a} & 0 & 0 \\ \frac{L_1}{b} & 0 & 0 & \frac{L_4}{b} \\ \frac{\partial Z}{\partial L_1} & \frac{\partial Z}{\partial L_2} & \frac{\partial Z}{\partial L_3} & \frac{\partial Z}{\partial L_4} \end{bmatrix}$$

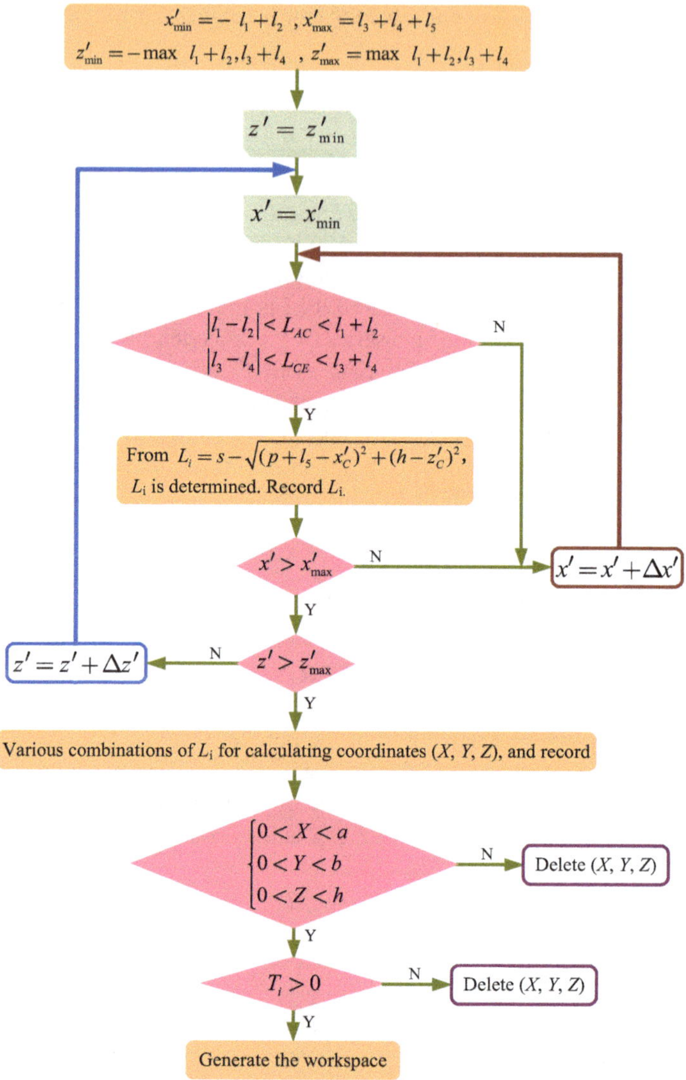

Fig. 6.5 Flowchart of generating the workspace of the HDCPR

Then, the stiffness matrix can be expressed as

$$S = (J^T)^{-1} K_p J^{-1} \tag{6.39}$$

where, J is the Jacobian matrix of the CPR.

In a similar way to derive the extremum of $\|F\|$, the Lagrange equation is constructed as follows

$$L_p = F^T S^{-T} S^{-1} F - \lambda_p (F^T F - 1) \tag{6.40}$$

where λ_P is the Lagrange multiplier. F is the norm of the external force that acts on the end-effector.

The necessary conditions that the extremum exists can be expressed as

$$\begin{cases} \frac{\partial L_p}{\partial \lambda_p} = F^T F - 1 = 0 \\ \frac{\partial L_p}{\partial F} = S^{-T} S^{-1} F - \lambda_p F = 0 \end{cases} \tag{6.41}$$

From Eq. (6.41), it is obviously that the Lagrange multiplier λ_P is the eigenvalue of the matrix $S^{-T}S^{-1}$. Combining Eqs. (6.40) and (6.41) leads to

$$\|\Delta p\|^2 = F^T S^{-T} S^{-1} F = \lambda_p \tag{6.42}$$

Equation (6.42) indicates that the extremum of $\|\Delta p\|^2$ is the eigenvalue of the matrix $S^{-T}S^{-1}$, thus the extremum of the norm of the deflection Δp can be represented as

$$\begin{cases} \|\Delta p_{max}\| = \sqrt{\max(|\lambda_p|)} \\ \|\Delta p_{min}\| = \sqrt{\min(|\lambda_p|)} \end{cases} \tag{6.43}$$

In principle, the less the norm of the deflection is, the better the stiffness performance is. It is expected that the minimum deflection should be smaller. Thus, the minimum of the norm of output force $\|\Delta p_{min}\|$ is referred to as the stiffness index (SI)

$$\mathrm{SI} = \sqrt{\min(|\lambda_p|)} \tag{6.44}$$

6.2.4.2 Stiffness Performance of the HDCPR

The stiffness performance analysis of HDCMP is similar to the CPR roughly, the only one difference is their Jacobi matrixes. So we merely study the Jacobi matrix of the HDCPR in this section.

According to Eq. (6.22), we can obtain

$$\begin{bmatrix} \dot{X}_i \\ \dot{Y}_i \\ \dot{Z}_i \end{bmatrix} = \begin{bmatrix} \frac{\partial X}{\partial \theta_{i1}} & \frac{\partial X}{\partial \theta_{i4}} \\ \frac{\partial Y}{\partial \theta_{i1}} & \frac{\partial Y}{\partial \theta_{i4}} \\ \frac{\partial Z}{\partial \theta_{i1}} & \frac{\partial Z}{\partial \theta_{i4}} \end{bmatrix} \begin{bmatrix} \dot{\theta}_{i1} \\ \dot{\theta}_{i4} \end{bmatrix} ,(i = 1, 2, 3, 4) \tag{6.45}$$

The equation can be expressed as

$$\dot{p}_i' = j_i \dot{q}_i', \quad (i = 1, 2, 3, 4) \tag{6.46}$$

Then, we can get the following equation

$$\dot{p}' = J' \dot{q}' \tag{6.47}$$

where J' is the Jacobi matrix of the HDCPR.

$$\dot{p}' = \begin{bmatrix} \dot{X} & \dot{Y} & \dot{Z} \end{bmatrix}^T \dot{q}' = [\theta_{11} \quad \theta_{14} \quad \theta_{21} \quad \theta_{24} \quad \theta_{31} \quad \theta_{34} \quad \theta_{41} \quad \theta_{44}]^T$$

$$J' = \begin{bmatrix} j_1 & j_2 & j_3 & j_4 \end{bmatrix}$$

6.2.5 Singularity Analysis

The singularity of the system is related to the stability and security of the institutions. When encounter the singular point, the agency will not be stable and controllable, and the institutional features is weak. Considering the constraint equations obtained above, analytical methodology based on the Jacobian matrix can be used to locate singularity regions within the workspace. In order to find the singular point of the CPR, we define the that:

$$\det(J) = \begin{vmatrix} \frac{L_1}{a} & \frac{L_2}{a} & 0 & 0 \\ \frac{L_1}{b} & 0 & 0 & \frac{L_4}{b} \\ \frac{\partial Z}{\partial L_1} & \frac{\partial Z}{\partial L_2} & \frac{\partial Z}{\partial L_3} & \frac{\partial Z}{\partial L_4} \end{vmatrix} = 0$$

In order to find the singular point of the HDCPR, we define the that:

$$\det(J') = \begin{vmatrix} \frac{\partial X}{\partial \theta_{11}} & \frac{\partial X}{\partial \theta_{14}} & \frac{\partial X}{\partial \theta_{21}} & \frac{\partial X}{\partial \theta_{24}} & \frac{\partial X}{\partial \theta_{31}} & \frac{\partial X}{\partial \theta_{34}} & \frac{\partial X}{\partial \theta_{41}} & \frac{\partial X}{\partial \theta_{44}} \\ \frac{\partial Y}{\partial \theta_{11}} & \frac{\partial Y}{\partial \theta_{14}} & \frac{\partial Y}{\partial \theta_{21}} & \frac{\partial Y}{\partial \theta_{24}} & \frac{\partial Y}{\partial \theta_{31}} & \frac{\partial Y}{\partial \theta_{34}} & \frac{\partial Y}{\partial \theta_{41}} & \frac{\partial Y}{\partial \theta_{44}} \\ \frac{\partial Z}{\partial \theta_{11}} & \frac{\partial Z}{\partial \theta_{14}} & \frac{\partial Z}{\partial \theta_{21}} & \frac{\partial Z}{\partial \theta_{24}} & \frac{\partial Z}{\partial \theta_{31}} & \frac{\partial Z}{\partial \theta_{34}} & \frac{\partial Z}{\partial \theta_{41}} & \frac{\partial Z}{\partial \theta_{44}} \end{vmatrix} = 0$$

The calculation of determinants will be through a program in matlab software, and the calculation process will not be shown in detail in this section. The results will be displayed below.

6.2.6 Results and Discussion

This section applies the previously mechanics performance analysis to carry out simulation studies. Four groups of the HDPMs of the HDCPR have the same structure parameters and the mechanisms have the symmetrical design in the three-dimensional space. The parameters of the CPR and HDCPR are listed in Table 6.1.

During simulation, the end-effector of the CPR and HDCPR is asked to move along the given spatial trajectory. The spatial trajectory is expressed as

$$\begin{cases} X = 0.2\cos 0.5\pi t + 0.5 \\ Y = 0.25\sin 0.5\pi t + 0.5 \quad 0 \le t \le 4(s) \\ Z = 0.5 \end{cases} \tag{6.48}$$

The end-effector of the CPR and HDCPR moves along the trajectory as shown in Fig. 6.6.

Figure 6.7 shows the spot demo interface of the CPR during following the trajectory of the end-effector in different time. Figure 6.8 shows the spot demo interface of the HDCPR during following the trajectory of the end-effector in different time. From the spot demo interface of Figs. 6.7 and 6.8, it can be noted that the end-effector runs smoothly, and the movement of all joints is fluency, so by which tracking the desired trajectory movement is accompanied well.

Figure 6.9 shows cable motion of the CPR and HDCPR in the process of simulation. The cable lengths Li are generated by Eq. (6.8) and vary symmetrically. Figure 6.10 shows the curves of cable tension exerted on the end-effector of the CPR and HDCPR related to the ellipse trajectory. Since the end-effector moves on a horizontal ellipse trajectory, the changes of cable tensions in Fig. 6.10 are reasonable and the transition is smooth.

Table 6.1 Parameters of the CPR and HDCPR

System parameters	Value
m	50 kg
g	9.81 m/s^2
h	1.2 m
a	1 m
b	1 m
s	2.5 m
r	0.2 m
l_1	0.2 m
l_2	0.5 m
l_3	0.5 m
l_4	0.28 m
l_5	0.51 m
Density of the links	7.85×10^{-3} kg/m^3
Cross sectional area of the links	2×10^{-4} m^2

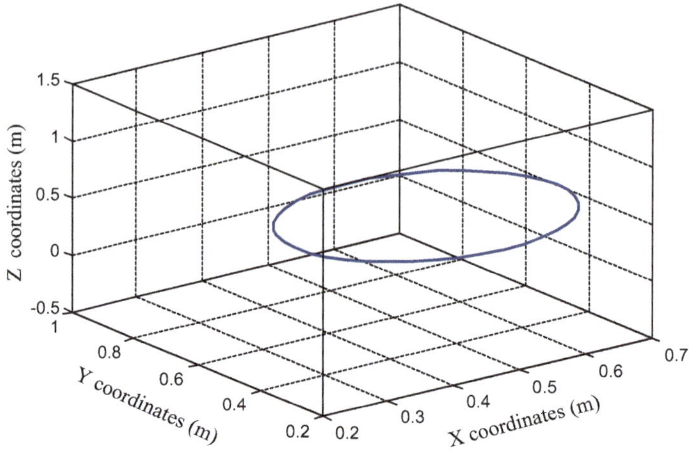

Fig. 6.6 Following trajectory of the ellipse motion workspace of the CPM the HDPM and *i*th cable tower

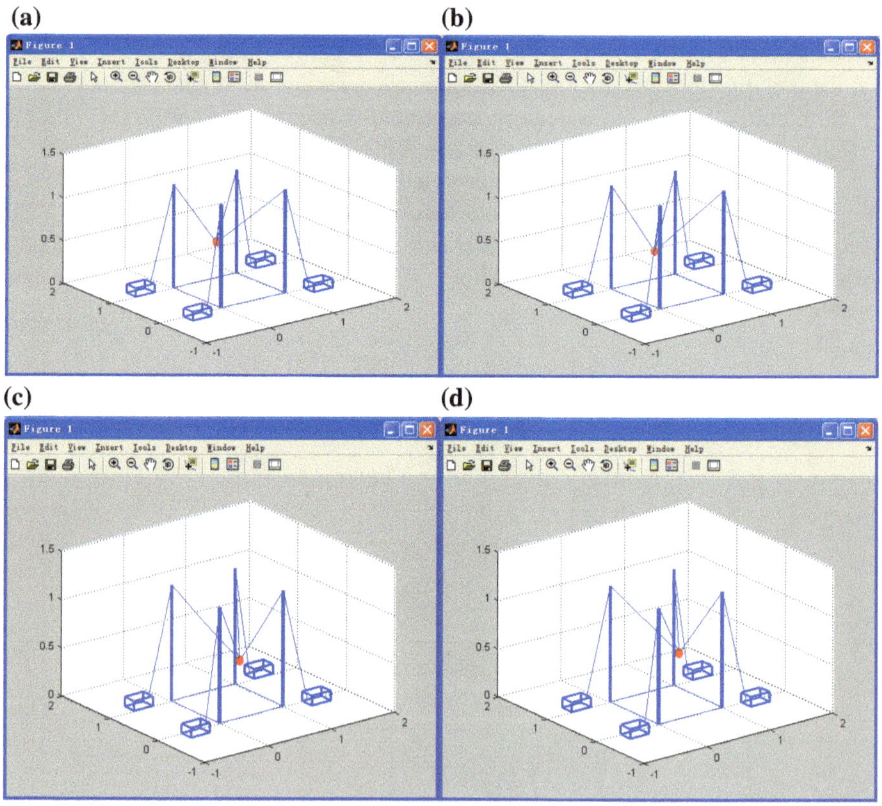

Fig. 6.7 Spot demo interface of the CPR: **a** t = 1 s, **b** t = 2 s, **c** t = 3 s and **d** t = 4 s

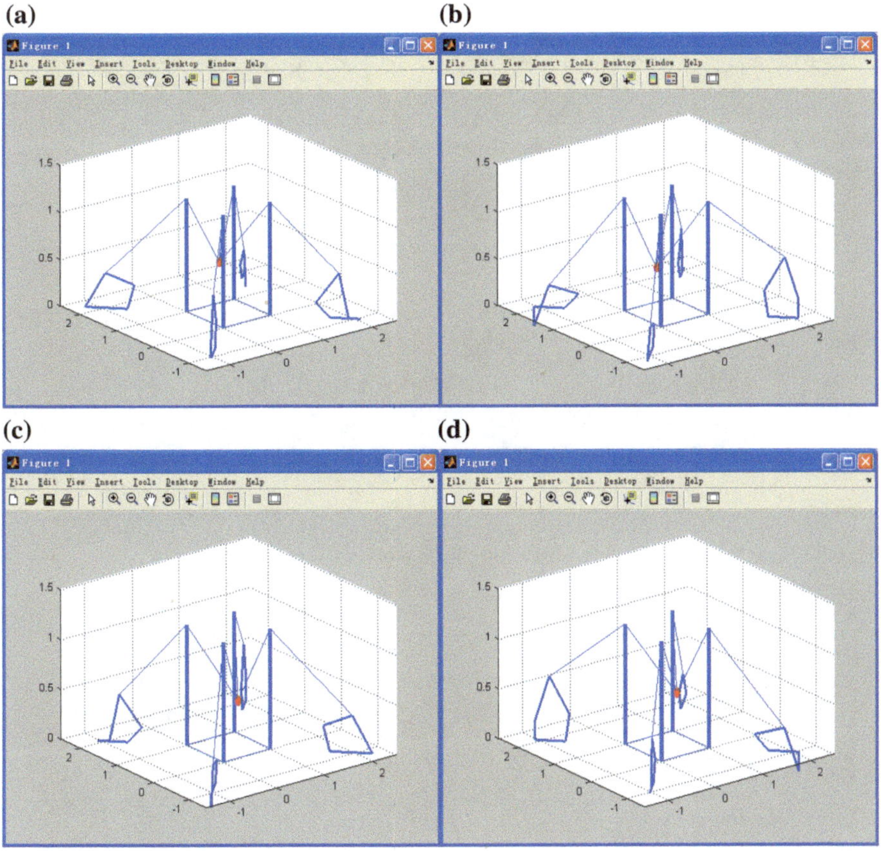

Fig. 6.8 Spot demo interface of the HDCPR: **a** t = 1 s, **b** t = 2 s, **c** t = 3 s and **d** t = 4 s

Figures 6.11 and 6.12 display the actuating torque curves of the CPR and HDCPR, respectively. From Figs. 6.11 and 6.12, it can be seen that the changes of the actuating torques are reasonable and the transition is smooth since the end-effector moves on a horizontal ellipse trajectory. Figure 6.13 illustrates comparison of torque curves between the four groups of the HDPMs of the HDCPR and the CPR. The maximum torque of the four groups of the HDPMs of the HDCPR is smaller than that of the CPR, and according to the HDCPR, the CV motors provide the majority of torque supply and the servomotors mainly act as a motion regulation device as depicted in Fig. 6.13.

The actuating power of the CPR can be obtained by

$$P_i = T_i \times \dot{L}_i \quad (1, 2, 3, 4) \tag{6.49}$$

where P_i is the ith servomotor power, T_i represents the ith cable tension, $\dot{L}_i$ denotes the ith cable velocity.

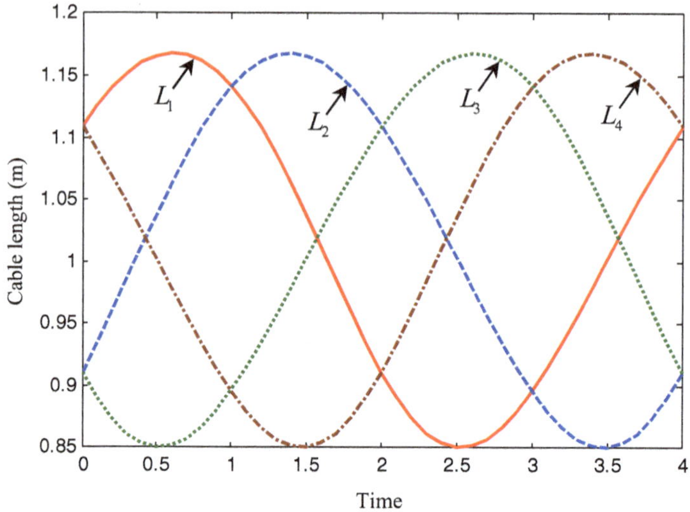

Fig. 6.9 Time behavior of cable length for ellipse motion

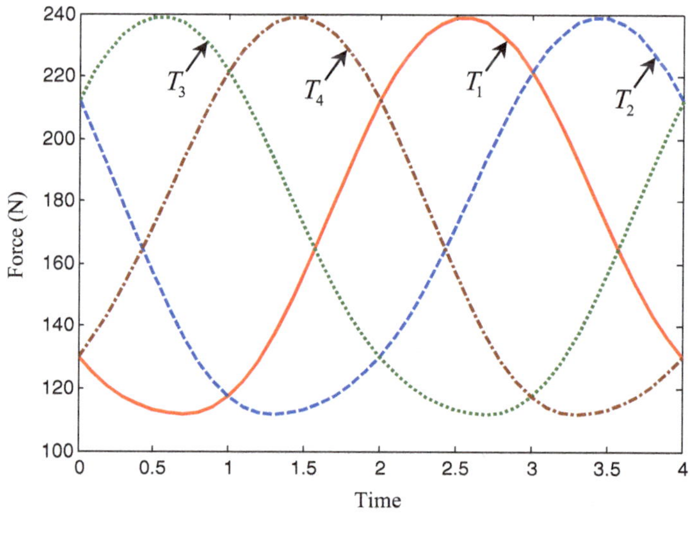

Fig. 6.10 Curves of cable tension

From Eq. (6.49), the absolute value actuating power curves of the CPR are shown in Fig. 6.14.

The CV motor actuating power and servomotor actuating power of the HDPMs of the HDCPR can be calculated by

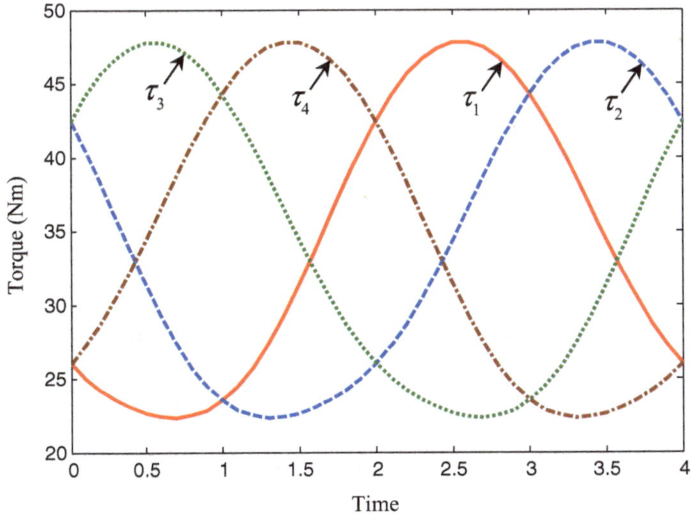

Fig. 6.11 Torque curves of the CPM

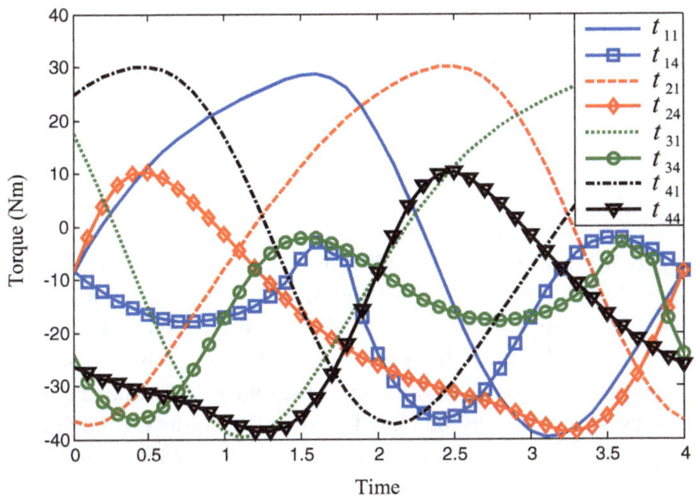

Fig. 6.12 Torque curves of the HDCPM

$$\begin{cases} P_{i1} = \tau_{i1} \times \dot{\theta}_{i1} \\ P_{i4} = \tau_{i4} \times \dot{\theta}_{i4} \end{cases} \tag{6.50}$$

where P_{i1} is the CV motor power of the ith group of the HDPM, P_{i4} is the servomotor power of the ith group of the HDPM.

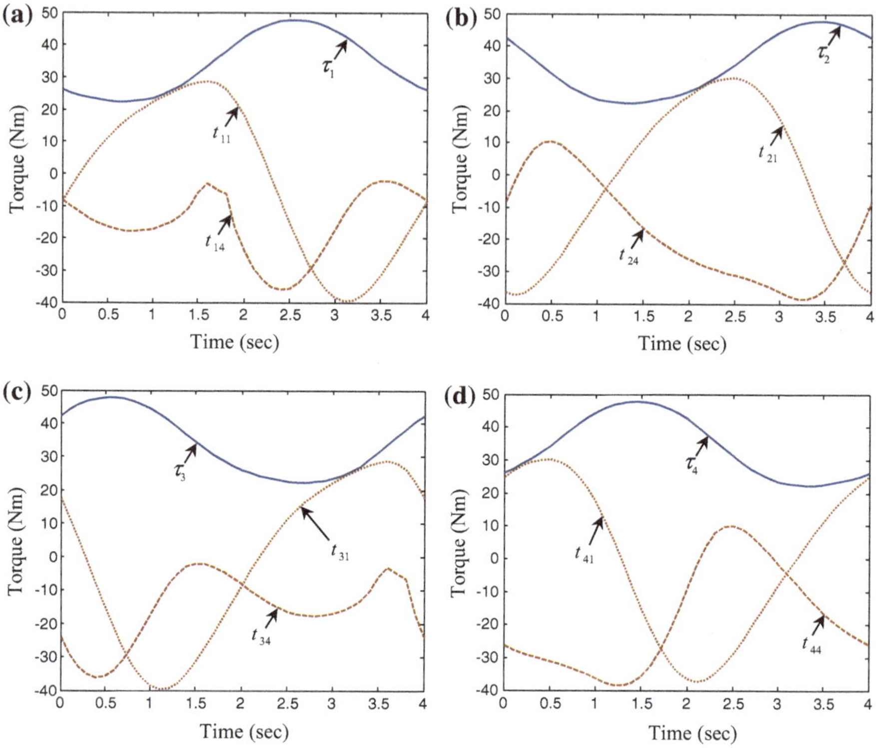

Fig. 6.13 Comparison of torque curves between four groups of the HDPMs of HDCPR and the CPR

From Eq. (6.50), Fig. 6.15 shows absolute value actuating power curves of four groups of the HDPMs of the HDCPR. Figure 6.16 illustrates comparison of power curves between the four groups of the HDPMs of the HDCPR and the CPR. From Figs. 6.14, 6.15 and 6.16, it can be seen that the changes of the actuating powers are reasonable and the transition is smooth since the end-effector moves on a horizontal ellipse trajectory. According to the HDCPR, the CV motors provide the majority of torque supply and the servomotors mainly act as a motion regulation device as depicted in Fig. 6.15.

From the flowchart of generating the workspace of the CPR as depicted in Fig. 6.4, the workspace can be given as shown in Fig. 6.17. According to Fig. 6.5, similarly, the simulation result of the reachable 3D workspace for the HDCPR is illustrated in Fig. 6.18 and the whole workspace is approximately an upside-down cone. It should be noted that most of the workspace volume is concentrated in the upper part of the tower cube, which is important for trajectory planning of the end-effector of the HDCPR. From Figs. 6.17 and 6.18, note that the workspace of the CPR is larger than that of the HDCPR in the case of the same simulation parameters of the CPR and HDCPR.

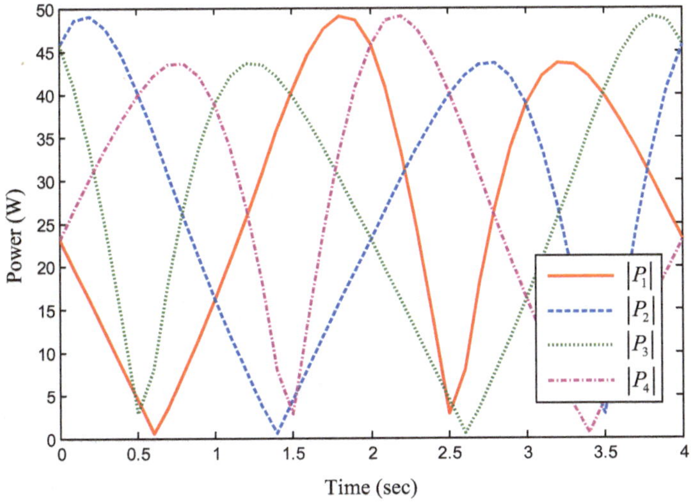

Fig. 6.14 Actuating power curves of the CPM

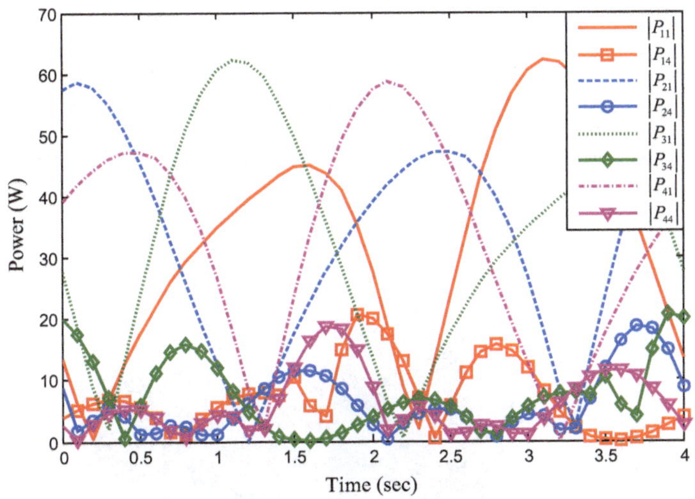

Fig. 6.15 Actuating power curves of the four groups of HDPMs of the HDCPM

Figures 6.19 and 6.20 illustrate the SI distributions of the CPR and HDCPR the at various x-y plane positions about different z-axis coordinate values, where z is 0.2, 0.4, 0.6, 0.8 m, respectively. Referring to Figs. 6.19 and 6.20, the SI distribution of the CPR is disciplinary, yet the SI distribution of the HDCPR is a little messy. In another, the SI of the HDCPR and the CPR has a gradually increasing tendency with the increase of z. That is to say, the higher the load is promoted, the worse stiffness performance is. The most important is that the SI of the HDCPR is much less than that of the CPR, and the stiffness performance of the HDCPR has a huge advantage than the CPR.

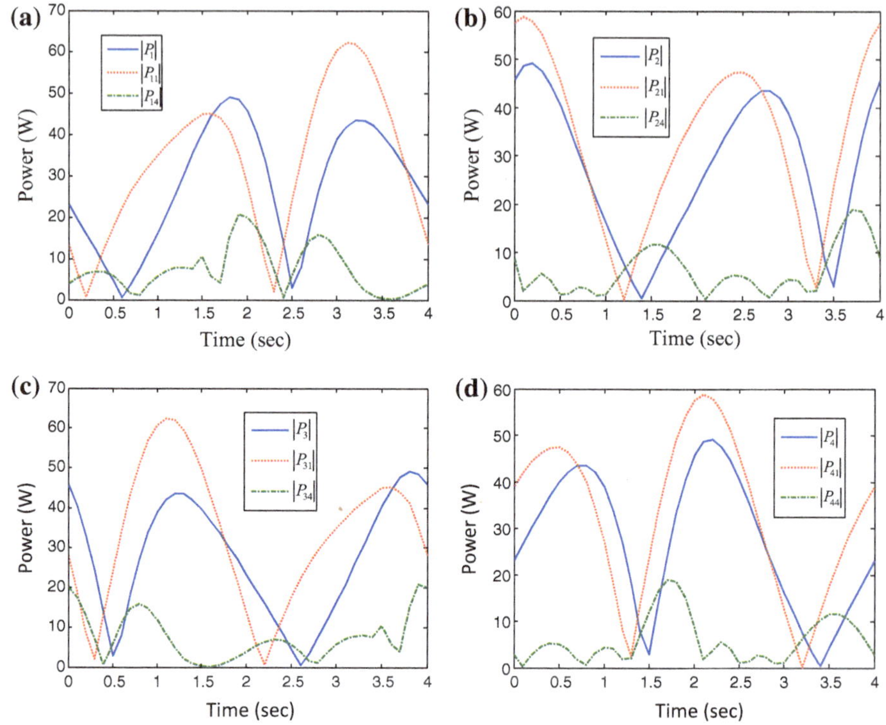

Fig. 6.16 Comparison of power curves between four groups of the HDPMs of HDCPM and the CPM

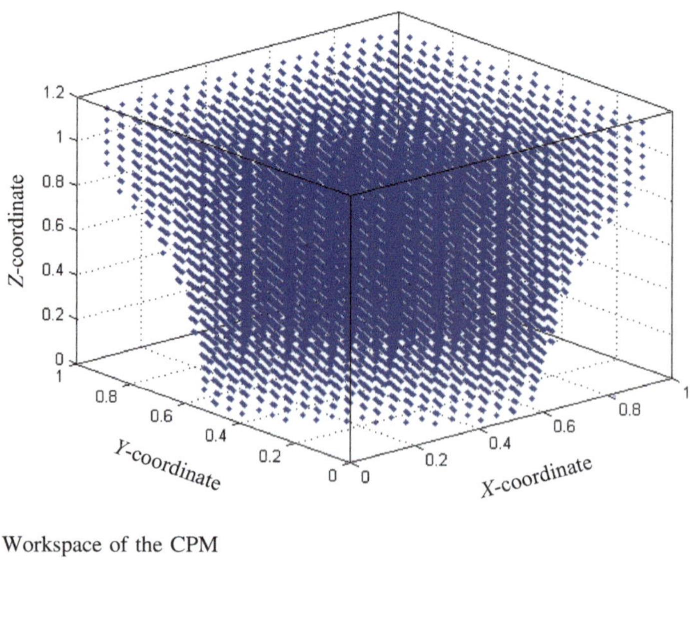

Fig. 6.17 Workspace of the CPM

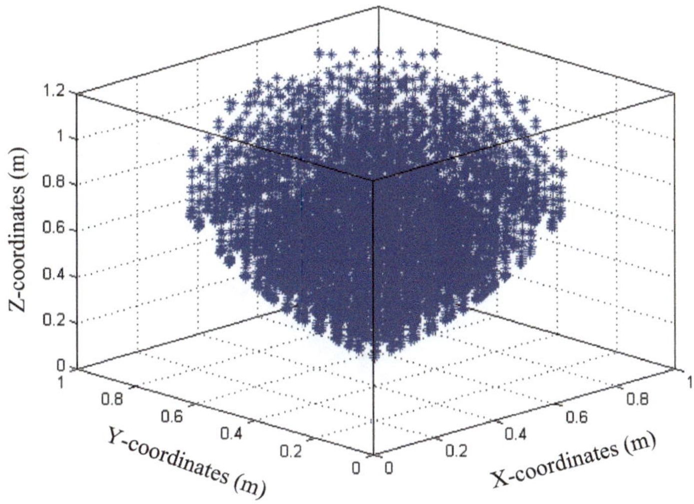

Fig. 6.18 Workspace of the HDCPM

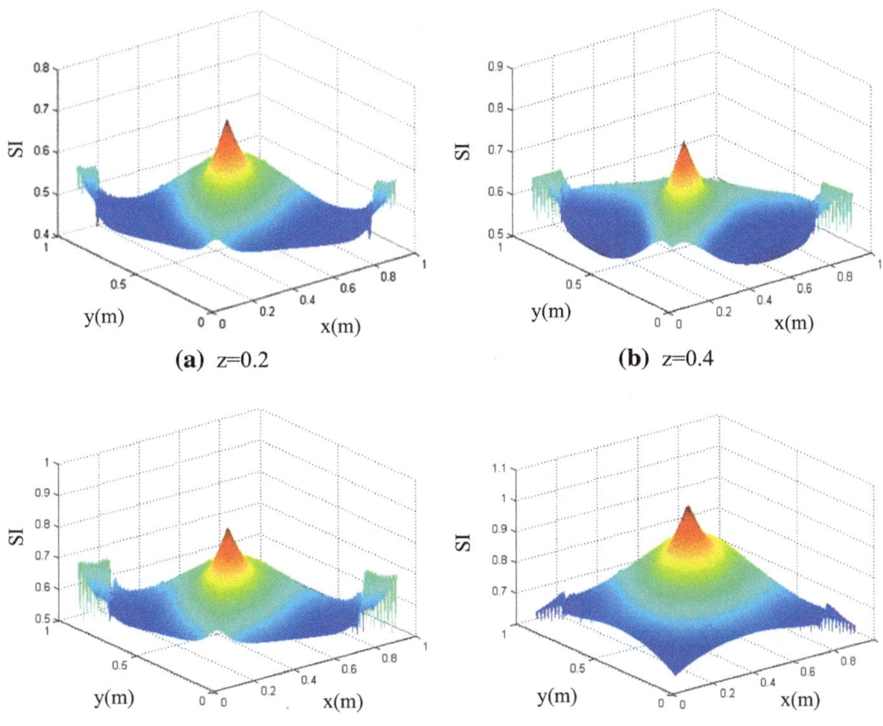

(a) z=0.2 **(b)** z=0.4 **(c)** z=0.6 **(d)** z=0.8

Fig. 6.19 SI distribution with different z of the CPR

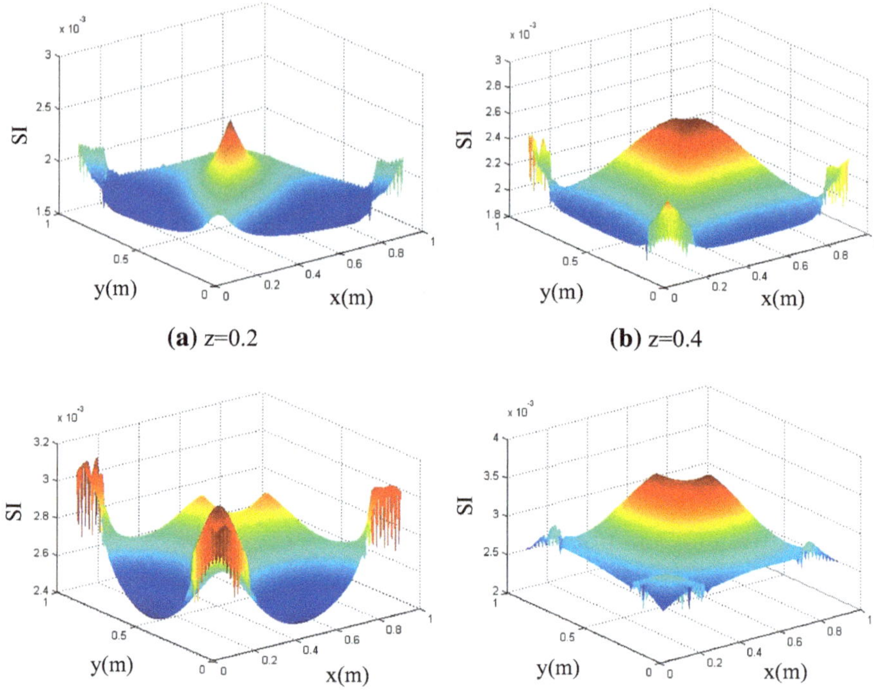

(a) z=0.2 (b) z=0.4 (c) z=0.6 (d) z=0.8

Fig. 6.20 SI distribution with different z of the HDCPR

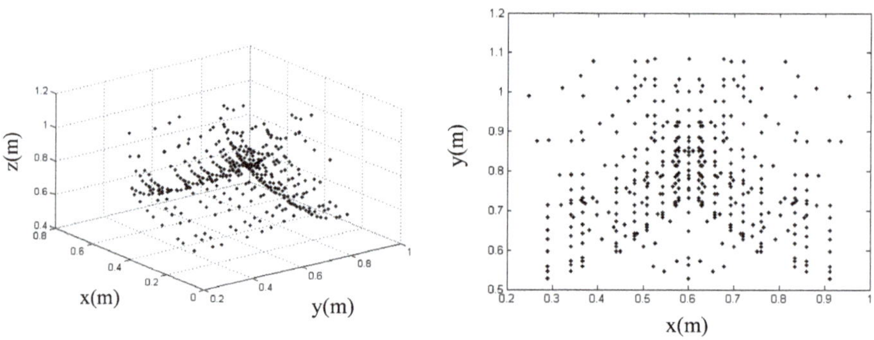

(a) 3D view (b) X-Y view

Fig. 6.21 The singularities within the workspace of the CPR

Figures 6.21 and 6.22 have shown singular points inside the workspace in 3-dimensional view and the projection in XOY plane. Figure 6.21 shows the singularity distribution of the CPR, and Fig. 6.22 shows the singularity distribution of

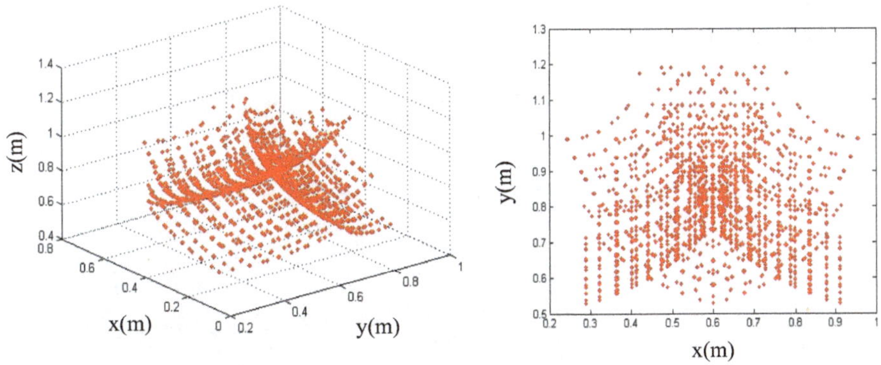

(a) 3D view (b) X-Y View

Fig. 6.22 The singularities within the workspace of the HDCPR

the HDCPR. It can be observed that the number of singular points in the CPR is much less than that in the HDCPR. Moreover, the singular configurations are nearly symmetrical for the identical structure of the four groups of the HDPMs. When cable parallel manipulators encounter the singular point position, the motor output of the system is uncertain, and effective measures should be taken to avoid this situation

6.3 Performance Evaluation and Comparison of a 3-Dof Hybrid-Driven Based Cable-Suspended Parallel Manipulator

In the design process of the manipulator, the performance index will be inevitably involved. The condition number of the Jacobian matrix has been commonly used in evaluating the performance of the manipulator within a workspace. In addition, in designing of parallel manipulators various criteria such as workspace, payload index and stiffness index should also be investigated.. The major contribution of this section is to propose the improved HDCPR, and then conditioning index, stiffness index, and payload index based on it are presented and compared.

6.3.1 Manipulator Model

The HDCPR suspends an end-effector by three cables and restrains all motion degrees of freedom for the object using the cables and gravitational force when the end-effector moves within the workspace, and the schematics of the HDCPR design

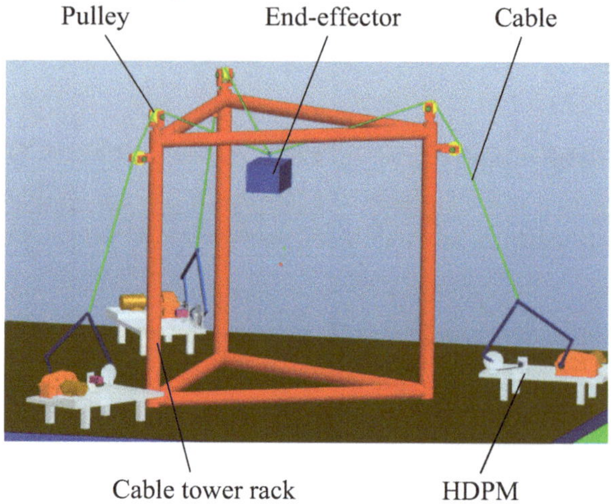

Fig. 6.23 Three-dimensional model of the HDCPR

is shown in Fig. 6.23. Referring to Fig. 6.23, the HDCPR consists of end-effector, cables, cable tower racks, pulleys, and HDPM and so on. One cable end is connected to the end-effector, and the other one rolls through a pulley fixed on the top of the relative cable tower rack and then is fed into the HDPM. The end-effector is supported in parallel by three cables controlled by three groups of HDPMs.

To find the position of the end-effector, assigned global coordinate frames (OXYZ) of the HDCPR are shown in Fig. 6.24. The assumptions were made for simplicity. The cables are treated as a massless body with no deformation, and the cable tower racks have the same height and are arrayed in a triangle on the ground, when deformation is ignored. We also assigned a local coordinate system $(o'x'z')$ at the bottom of the HDPM.

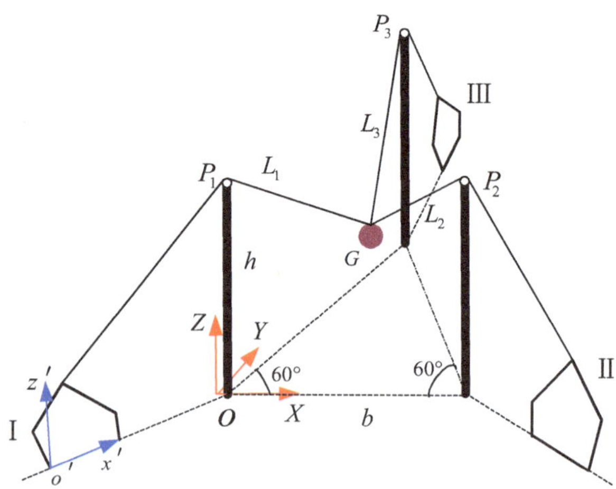

Fig. 6.24 Schematic sketch of the HDCPR

6.3.2 Kinematics and Jacobian Matrix

In the following we present kinematics and Jacobian matrix of the HDCPR. The kinematic schematic of the ith group of the HDPM and cable structure is described in Fig. 6.25. As can be observed in Fig. 6.25, the length of the links of the ith group of the HDPM are l_1, l_2, l_3, l_4, l_5, respectively. The angle between links and x'-axis of the HDPM are $\theta_{i1}, \theta_{i2}, \theta_{i3}, \theta_{i4}, 0°$, respectively. (x_{ic}, y_{ic}) is the position coordinates of joint C_i in the local coordinate system $(o'x'z')$ at the bottom of the HDPM.

As it can be clearly seen from Fig. 6.25, the closed loop polygon vector equation of the first group of the HDPM can be expressed as on the base of the method of vector chain

$$\overrightarrow{A_1B_1} + \overrightarrow{B_1C_1} = \overrightarrow{A_1E_1} + \overrightarrow{E_1D_1} + \overrightarrow{D_1C_1} \tag{6.51}$$

Equation (6.51) may also be expressed in the plural form

$$l_1 e^{i\theta_{11}} + l_2 e^{i\theta_{12}} = l_3 e^{i\theta_{13}} + l_4 e^{i\theta_{14}} + l_5 = x_{1c} + y_{1c} i \tag{6.52}$$

Developing Eq. (6.52) into real part and imaginary part, the following equations can be expressed as

$$\begin{cases} x_{1c} = l_1 c_{11} + l_2 c_{12} \\ y_{1c} = l_1 s_{11} + l_2 s_{12} \\ x_{1c} = l_3 c_{13} + l_4 c_{14} + l_5 \\ y_{1c} = l_3 s_{13} + l_4 s_{14} \end{cases} \tag{6.53}$$

For the sake of brevity, here, $s_{ij} = \sin\theta_{ij}$ and $c_{ij} = \cos\theta_{ij}$.

θ_{12} and θ_{13} may be eliminated from Eq. (6.53), then simplify and order it. Hence, the constraint problem of the HDPM can be formulated as follows

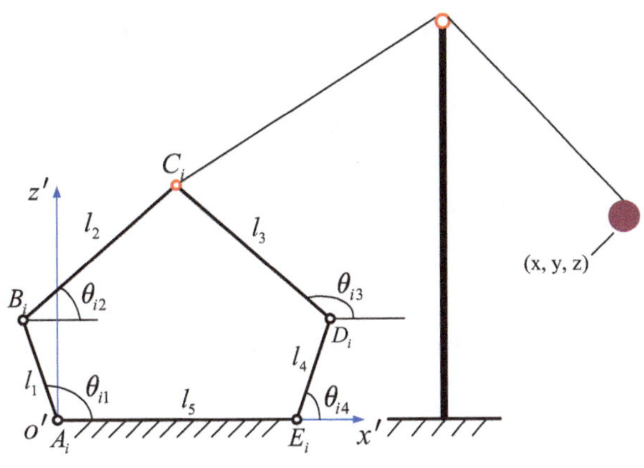

Fig. 6.25 Schematic of the ith group of the HDPM and cable

$$u_1 : 2l_1y_{1c}s_{11} + 2l_1x_{1c}c_{11} + l_2^2 - l_1^2 - x_{1c}^2 - y_{1c}^2 = 0 \tag{6.54}$$

$$u_2 : 2l_4y_{1c}s_{14} + 2l_4(x_{1c} - l_5)c_{14} + l_3^2 + 2l_5x_{1c} - l_4^2 - l_5^2 - x_{1c}^2 - y_{1c}^2 = 0 \tag{6.55}$$

Based on the aforementioned analysis, the constraint problem of the first group of the HDPMs have been formulated. Similarly, the constraint problem of the second and third groups of the HDPMs can also be formulated. According to Eqs. (6.54) and (6.55), the constraint equations of the HDPM may be expressed as

$$u_{ik}(\theta_{i1}, \theta_{i4}, x_{ic}, y_{ic}) = 0 \ i \in \{1, 2, 3\ \} \ k \in \{1, 2\} \tag{6.56}$$

Deriving Eq. (6.56) with respect to all the variables one gets the following relation

$$\frac{\partial u_{ik}}{\partial x_{ic}} dx_{ic} + \frac{\partial u_{ik}}{\partial y_{ic}} dy_{ic} = -\frac{\partial u_{ik}}{\partial \theta_{i1}} d\theta_{i1} - \frac{\partial u_{ik}}{\partial \theta_{i4}} d\theta_{i4} \tag{6.57}$$

With Eqs. (6.54)–(6.57) one receive

$$\begin{bmatrix} \dot{x}_{ic} \\ \dot{y}_{ic} \end{bmatrix} = \begin{bmatrix} D_{11} & D_{12} \\ D_{21} & D_{22} \end{bmatrix}^{-1} \begin{bmatrix} E_{11} & 0 \\ 0 & E_{22} \end{bmatrix} \begin{bmatrix} \dot{\theta}_{i1} \\ \dot{\theta}_{i4} \end{bmatrix} \tag{6.58}$$

where $D_{11} = l_1c_{i1} - x_{ic}$, $D_{12} = l_1s_{i1} - y_{ic}$, $D_{21} = l_4c_{i4} + l_5 - x_{ic}$, $D_{22} = l_4s_{i4} - y_{ic}$, $E_{11} = l_1x_{ic}s_{i1}$ - $l_1y_{ic}c_{i1}$, and $E_{22} = l_4(x_{ic} - l_5)s_{i4} - l_4y_{ic}c_{i4}$.

In a compact form, Eq. (6.58) may be rewritten as follows

$$\dot{p}_{ic} = J_{ic}\dot{q}_{ic} \tag{6.59}$$

where $\dot{p}_{ic} = [\dot{x}_{ic} \quad \dot{y}_{ic}]^T$, $\dot{q}_{ic} = [\dot{\theta}_{i1} \quad \dot{\theta}_{i4}]^T$. $\dot{p}_{ic}$ and $\dot{q}_{ic}$ represent the output velocity vector of joint C_i and the input velocity vector of the drive joint, respectively. Equation (6.59) describes the mapping relation from $\dot{q}_{ic}$ to $\dot{p}_{ic}$. $J_{ic}(i \in \{1, 2, 3\ \})$ is Jacobian matrix of the HDPMs, and from Eq. (6.58), one can write

$$J_{ic} = \begin{bmatrix} D_{11} & D_{12} \\ D_{21} & D_{22} \end{bmatrix}^{-1} \begin{bmatrix} E_{11} & 0 \\ 0 & E_{22} \end{bmatrix} \tag{6.60}$$

Referring to Fig. 6.2, the constraint problem of the CPR can be formulated as

$$f_i : L_i^2 = (x - x_i)^2 + (y - y_i)^2 + (z - z_i)^2, i \in \{1, 2, 3\ \} \tag{6.61}$$

Equation (6.61) may also be expressed as

$$f_i(x, y, z, L_i) = 0 \tag{6.62}$$

Deriving Eq. (6.62) with respect to all the variables one gets the following relation

$$\frac{\partial f_i}{\partial x}dx + \frac{\partial f_i}{\partial y}dy + \frac{\partial f_i}{\partial z}dz = -\frac{\partial f_i}{\partial L_1}dL_1 - \frac{\partial f_i}{\partial L_2}dL_2 - \frac{\partial f_i}{\partial L_3}dL_3 \tag{6.63}$$

Substituting Eq. (6.61) into Eq. (6.63) one has

$$\begin{bmatrix} \dot{x} \\ \dot{y} \\ \dot{z} \end{bmatrix} = \begin{bmatrix} x - x_1 & y - y_1 & z - z_1 \\ x - x_2 & y - y_2 & z - z_2 \\ x - x_3 & y - y_3 & z - z_3 \end{bmatrix}^{-1} \begin{bmatrix} L_1 & 0 & 0 \\ 0 & L_2 & 0 \\ 0 & 0 & L_3 \end{bmatrix} \begin{bmatrix} \dot{L}_1 \\ \dot{L}_2 \\ \dot{L}_3 \end{bmatrix} \tag{6.64}$$

Similarly, Eq. (6.64) may be rewritten as follows

$$\dot{p} = J_0 \dot{q}_0 \tag{6.65}$$

Here, $\dot{p} = [\dot{x} \quad \dot{y} \quad \dot{z}]^T$, $\dot{q}_0 = \left[\dot{L}_1 \quad \dot{L}_2 \quad \dot{L}_3\right]^T$. $\dot{p}$ is the velocity vector of the end-effector. $\dot{q}_0$ is the velocity vector of the drive cables. J_0 is the Jacobian matrix, and according to Eq. (6.64), one can write

$$J_0 = \begin{bmatrix} x - x_1 & y - y_1 & z - z_1 \\ x - x_2 & y - y_2 & z - z_2 \\ x - x_3 & y - y_3 & z - z_3 \end{bmatrix}^{-1} \begin{bmatrix} L_1 & 0 & 0 \\ 0 & L_2 & 0 \\ 0 & 0 & L_3 \end{bmatrix} \tag{6.66}$$

Referring to Fig. 6.3, the following equation describes the relation between the HDPMs and the CPR

$$p_i : (a - x_{ic})^2 + (h - y_{ic})^2 = (L_0 - L_i)^2, \ i \in \{1, 2, 3\} \tag{6.67}$$

where a is the distance between joint A_i and the bottom of the cable pillar. L_0 represents the length of the ith total cable. Equation (6.67) may be rewritten as follows

$$p_i(L_i, x_{ic}, y_{ic}) = 0 \tag{6.68}$$

The derivative of Eq. (6.68) with respect to all the variables can be computed by

$$\begin{aligned} \frac{\partial p_i}{\partial L_1}dL_1 + \frac{\partial p_i}{\partial L_2}dL_2 + \frac{\partial p_i}{\partial L_3}dL_3 = & -\frac{\partial p_i}{\partial x_{1c}}dx_{1c} - \frac{\partial p_i}{\partial y_{1c}}dy_{1c} - \frac{\partial p_i}{\partial x_{2c}}dx_{2c} - \frac{\partial p_{ii}}{\partial y_{2c}}dy_{2c} \\ & -\frac{\partial p_i}{\partial x_{3c}}dx_{3c} - \frac{\partial p_i}{\partial y_{3c}}dy_{3c} \end{aligned} \tag{6.69}$$

Substituting Eq. (6.67) into Eq. (6.69), we obtain

$$\begin{bmatrix} \dot{L}_1 \\ \dot{L}_2 \\ \dot{L}_3 \end{bmatrix} = \begin{bmatrix} L_1 - L_0 & 0 & 0 \\ 0 & L_2 - L_0 & 0 \\ 0 & 0 & L_3 - L_0 \end{bmatrix}^{-1} \begin{bmatrix} x_{1c} - a & y_{1c} - h & 0 & 0 & 0 & 0 \\ 0 & 0 & x_{2c} - a & y_{2c} - h & 0 & 0 \\ 0 & 0 & 0 & 0 & x_{3c} - a & y_{3c} - h \end{bmatrix} \left[\dot{x}_{1c}\ \dot{y}_{1c}\ \dot{x}_{2c}\ \dot{y}_{2c}\ \dot{x}_{3c}\ \dot{y}_{3c}\right]^T \tag{6.70}$$

Equation (6.70) can be rewritten in a compact form

$$\dot{p}_L = J_L \dot{q}_L \tag{6.71}$$

Here, $\dot{p}_L = \left[\dot{L}_1 \quad \dot{L}_2 \quad \dot{L}_3\right]^T$, $\dot{q}_L = \left[\dot{x}_{ic} \quad \dot{y}_{ic}\right]^T$. $\dot{p}_L$ is the velocity vector of the cables between the pulley center and the end-effector. $\dot{q}_L$ is the velocity vector of joint C_i. J_L is the Jacobian matrix, and one can obtain

$$J_L = \begin{bmatrix} L_1 - L_0 & 0 & 0 \\ 0 & L_2 - L_0 & 0 \\ 0 & 0 & L_3 - L_0 \end{bmatrix}^{-1} \begin{bmatrix} x_{1c} - a & y_{1c} - h & 0 & 0 & 0 & 0 \\ 0 & 0 & x_{2c} - a & y_{2c} - h & 0 & 0 \\ 0 & 0 & 0 & 0 & x_{3c} - a & y_{3c} - h \end{bmatrix} \tag{6.72}$$

Based on the foregoing kinematics of the three groups of the HDPMs and the CPR, the constraint equations of the HDCPR can be expressed as

$$\begin{cases} 2l_1 y_{ic} s_{i1} + 2l_1 x_{ic} c_{i1} + l_2^2 - l_1^2 - x_{ic}^2 - y_{ic}^2 = 0 \\ 2l_4 y_{ic} s_{i4} + 2l_4 (x_{ic} - l_5) c_{i4} + l_3^2 + 2l_5 x_{ic} \\ -l_4^2 - l_5^2 - x_{ic}^2 - y_{ic}^2 = 0 \\ L_i^2 = (x - x_i)^2 + (y - y_i)^2 + (z - z_i)^2 \\ (a - x_{ic})^2 + (h - y_{ic})^2 = (L_0 - L_i)^2 \end{cases} \tag{6.73}$$

According to Eqs. (6.59), (6.65) and (6.71), the end-effector velocity in global coordinates relates to the individual actuator inputs by the HDCPR Jacobian, given by

$$\dot{p} = J_0 \left[J_L \left[J_{1c}\dot{q}_{1c} \quad J_{2c}\dot{q}_{2c} \quad J_{3c}\dot{q}_{3c} \right]^T \right] \tag{6.74}$$

Equation (6.74) may be rewritten as follows

$$\dot{p} = J\dot{q} \tag{6.75}$$

Here $\dot{q} = \begin{bmatrix} \dot{\theta}_{11} & \dot{\theta}_{14} & \dot{\theta}_{21} & \dot{\theta}_{24} & \dot{\theta}_{31} & \dot{\theta}_{34} \end{bmatrix}^T$, the Jacobian matrix J of the HDCPR is the mapping from the input velocity vector of the drive joint to the output velocity vector of the end-effector.

6.3.3 Conditioning Performance Analysis

The condition number of the Jacobian matrix is regarded as the local performance index for evaluating the velocity and accuracy between the joint variables and the end-effector of the robots. The relationship between the condition number k and the Jacobian matrix J can be expressed by [2]

$$1 \leq k = \sigma_{max}/\sigma_{min} \leq \infty \tag{6.75}$$

where σ_{max} and σ_{min} are the maximum and minimum singular values of the Jacobian matrix J, respectively.

The conditioning index of the robots is discussed in detail in [5]. Here we give a short overview. The conditioning index (CI) may be defined by

$$0 \leq \mathrm{CI} = 1/k \leq 1 \tag{6.76}$$

When the CI is close to 1, the HDCPR acquires a highest degree of dexterity. On the contrary, when the CI is close to 0, the dexterity of the HDCPR is lowest.

Note that the CI varies with the configuration of the HDCPR, the global conditioning index (GCI) [5] is introduced to evaluate the global behavior of the manipulator on a workspace, and the GCI can be defined as

$$\mathrm{GCI} = \left(\int_{\mathrm{V}} \mathrm{CI}d\mathrm{V} \right) / \mathrm{V} \tag{6.77}$$

where V is the volume of the overall workspace.

In order to compare the performances of the HDCPR and the CPR, simulation study is carried out. It must be pointed out that the CPR here is driven by the common low power controllable motor. All the simulations are performed with MATLAB Version 7.10 software. Three groups of the HDPMs have the same structure parameters and the mechanisms have the symmetrical design in the three-dimensional space. The kinematics parameters of the HDCPR and the CPR are listed in Table 6.2.

Table 6.2 Kinematics parameters of the HDCPR and the CPR

Parameters	h	b	a	L_0	l_1	l_2	l_3	l_4	l_5
Value (m)	1.00	1.00	1.51	2.00	0.20	0.50	0.50	0.28	0.51

In view of the CI varies with the different position in the workspace, we choose different elevation of z-axis coordinate to analyze. Figures 6.26, 6.27, 6.28, 6.29 and 6.30 illustrate the CI distributions of the HDCPR and the CPR at various x-y plane positions about different z-axis coordinate values, where z is 0.3, 0.45, 0.6, 0.75, 0.9 m, respectively.

Figures 6.26, 6.27, 6.28, 6.29 and 6.30, demonstrate that when the end-effector moves at the middle levels in the workspace, the CI is generally larger than that is at the upper and lower levels. The end-effector moves more close to the central position of the middle levels, the CI is larger. In addition, the CI distributions of the HDCPR and the CPR have a similar trend. For the sake of the high dexterity, we should make the end-effector of the HDCPR move at the central position of the middle levels in the workspace.

When z is the average value of z-axis coordinate, the CI projection distribution of the HDCPR and the CPR at x-y plane is displayed in Fig. 6.31. It implies that the maximum value emerges in the central position of the workspace while minimum

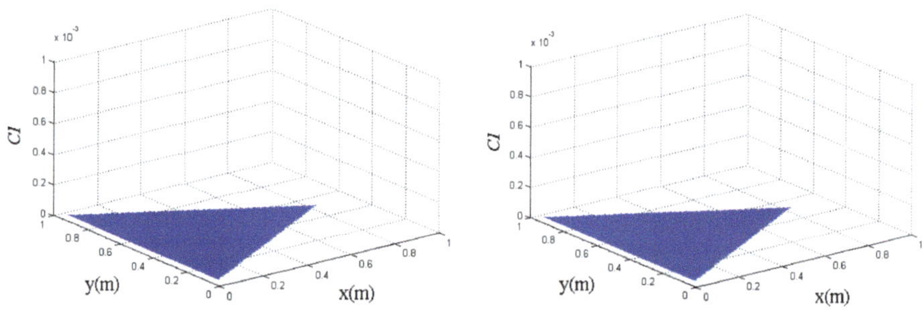

(**a**) HDCPR (**b**) CPR

Fig. 6.26 CI distribution with z = 0.3

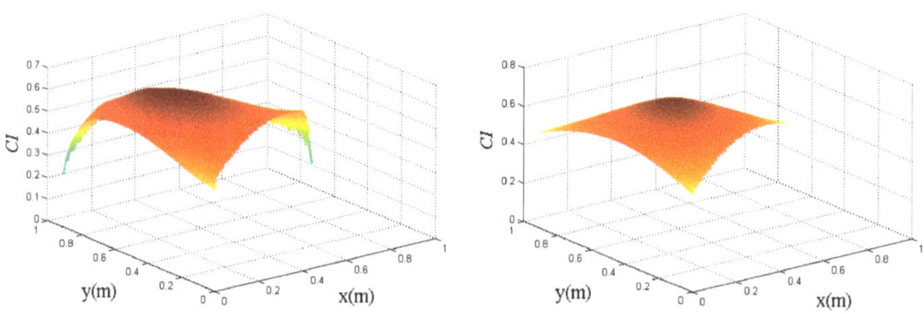

(**a**) HDCPR (**b**) CPR

Fig. 6.27 CI distribution with z = 0.45

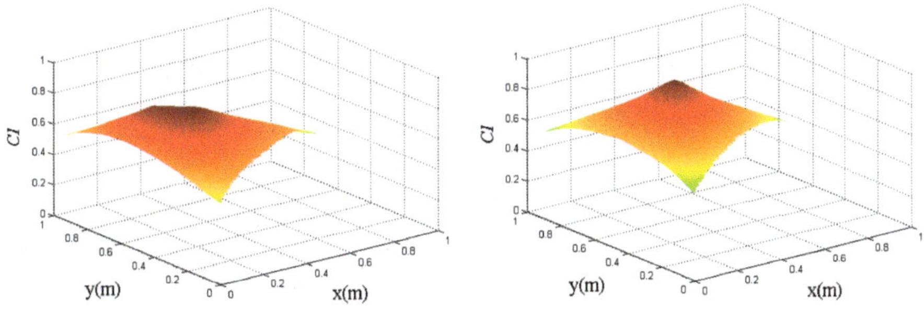

(a) HDCPR **(b)** CPR

Fig. 6.28 CI distribution with z = 0.6

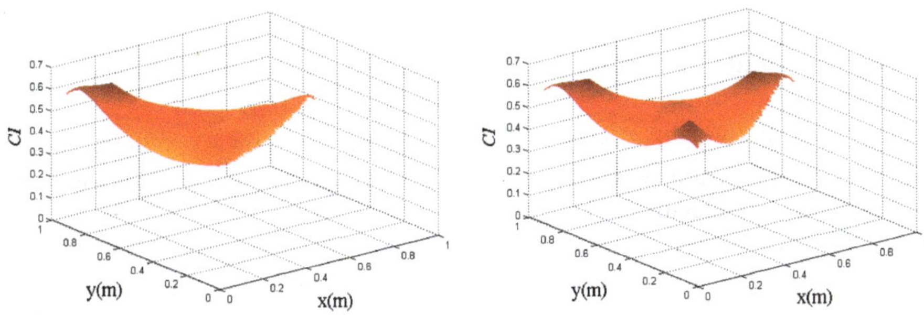

(a) HDCPR **(b)** CPR

Fig. 6.29 CI distribution with z = 0.75

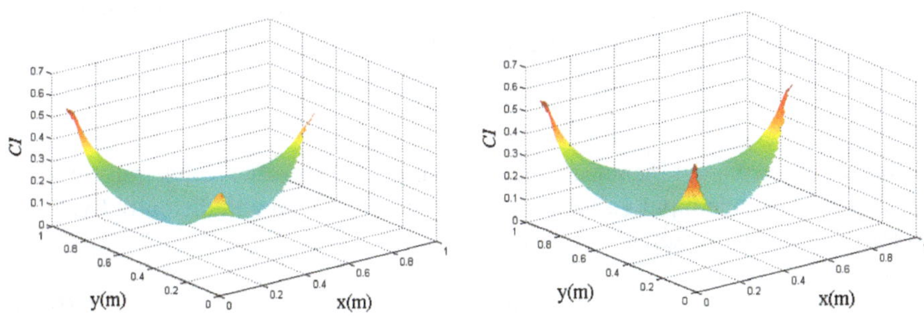

(a) HDCPR **(b)** CPR

Fig. 6.30 CI distribution with z = 0.9

value emerges at the edge position. The maximum values of the CI of the HDCPR and the CPR are 0.502 and 0.580. Thus, it can be concluded that the CPR has a bit better conditioning performance than the HDCPR.

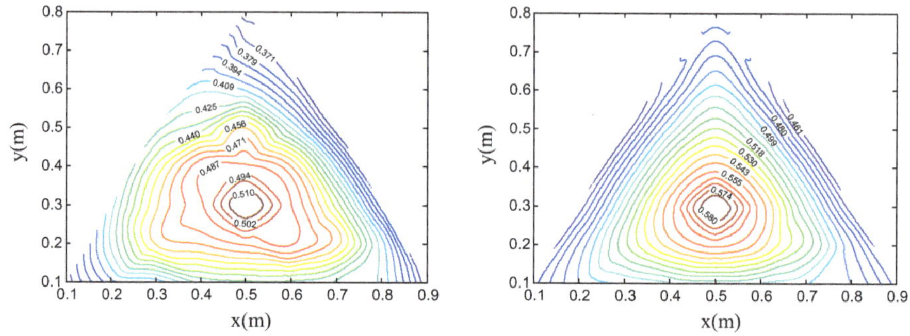

(**a**) HDCPR (**b**) CPR

Fig. 6.31 CI projection distribution

6.3.4 Payload Performance Analysis

Payload capability is one of the most important performance indices of the manipulator. It determines whether the manipulator can support the external force acting on the end-effector or not.

Let the driving force of active joints be τ, and the external force that acts on the end-effector be F, the following equation can be obtained

$$F = G\tau \tag{6.78}$$

where $\tau = \begin{bmatrix} \tau_{11} & \tau_{14} & \tau_{21} & \tau_{24} & \tau_{31} & \tau_{34} \end{bmatrix}^T$, G is the static Jacobian matrix of the HDCPR.

Let the deflection of active joints be Δq, and the deflection of the end-effector be Δq, based on the virtual work principle, one can derive

$$F^T \Delta p = \tau^T \Delta q \tag{6.79}$$

Since the forward kinematics Jacobian matrix J represents the transfer relationship between the input velocity vector of the drive joint and the output velocity vector of the end-effector, it will result in

$$\Delta p = J \Delta q \tag{6.80}$$

Substituting Eq. (6.80) into Eq. (6.79), we obtain

$$F = \left[J^T\right]^{-1} \tau \tag{6.81}$$

As the Jacobian matrix J is not a square matrix, the pseudo-inverse of can be expressed as $\left[J^T\right]^{-1}$.

Comparing Eqs. (6.78) and (6.81), the relationship between the static Jacobian matrix and the obverse kinematics Jacobian matrix can be written as

$$G = [J^T]^{-1} \tag{32}$$

The payload index of the HDCPR is defined as the extremum of the norm of output force vector F, while the input driving force vector of active joints τ is unity. In order to derive the extremum, the Lagrange equation is constructed as follows

$$L_F = \tau^T G^T G\tau - \lambda_F(\tau^T\tau - 1) \tag{33}$$

where λ_F is the Lagrange multiplier.

The necessary conditions that the extremum exists can be expressed as

$$\begin{cases} \frac{\partial L_F}{\partial \lambda_F} = \tau^T\tau - 1 = 0 \\ \frac{\partial L_F}{\partial \tau} = G^T G\tau - \lambda_F\tau = 0 \end{cases} \tag{6.84}$$

From Eq. (6.84), it is obviously that the Lagrange multiplier λ_F is the eigenvalue of the matrix $G^T G$.

According to Eqs. (6.81) and (6.84) on can obtain

$$\|F\|^2 = \tau^T G^T G\tau = \lambda_F \tag{6.85}$$

Equation (6.85) indicates that the extremum of $\|F\|^2$ is the eigenvalue of the matrix $G^T G$, thus the extremum of the norm of output force vector F can be represented as

$$\begin{cases} \|F_{\max}\| = \sqrt{\max(|\lambda_F|)} \\ \|F_{\min}\| = \sqrt{\min(|\lambda_F|)} \end{cases} \tag{6.86}$$

In principle, the larger the norm of output force is, the better the payload capability is. It is expected that the maximum force should be larger. Hence, the maximum of the norm of output force $\|F_{max}\|$ is referred to as the payload index (PI)

$$\text{PI} = \sqrt{\max(|\lambda_F|)} \tag{6.87}$$

Similarly, note that the PI varies with the configuration of the HDCPR, the global payload index (GPI) is introduced to evaluate the payload capability within the workspace can be defined as [7]

$$\text{GPI} = \left(\int_{\text{V}} \text{PI}d\text{V}\right)/\text{V} \tag{6.88}$$

where V is the volume of the overall workspace.

Figures 6.32, 6.33, 6.34, 6.35 and 6.36 illustrate the PI distributions of the HDCPR and the CPR at various x-y plane positions about different z-axis coordinate values, where z is 0.3, 0.45, 0.6, 0.75, 0.9 m, respectively.

Referring to Figs. 6.32, 6.33, 6.34, 6.35 and 6.36, one can see that the PI of the HDCPR is generally larger than that of the CPR. With the increase of z value, the PI of the HDCPR is larger gradually and approaches each other. Therefore, the payload capability of the HDCPR is better at the middle and upper levels than that is at the lower levels. In addition, the end-effector of the HDCPR moves more close to the central position of the middle levels, the PI is larger, and the payload capability is better.

When z is the average value of z-axis coordinate, the PI projection distribution of the HDCPR and the CPR at x-y plane is shown in Fig. 6.37. It implies that the PI of the HDCPR attains the maximum value around the central position while it attains

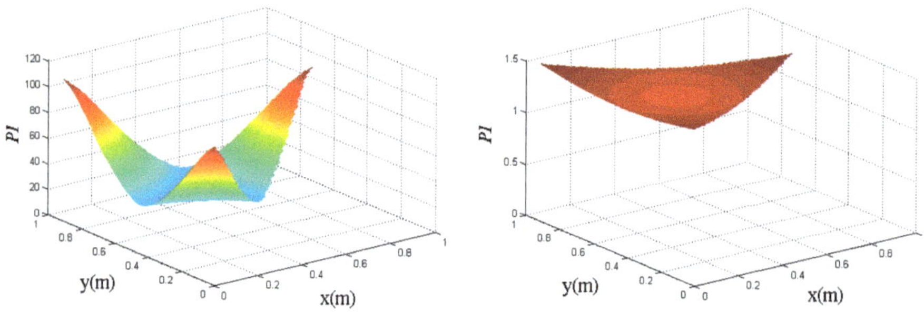

(a) HDCPR **(b)** CPR

Fig. 6.32 PI distribution with z = 0.3

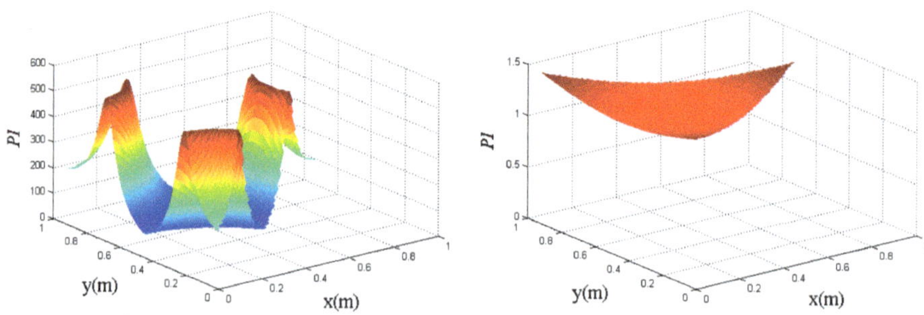

(a) HDCPR **(b)** CPR

Fig. 6.33 PI distribution with z = 0.45

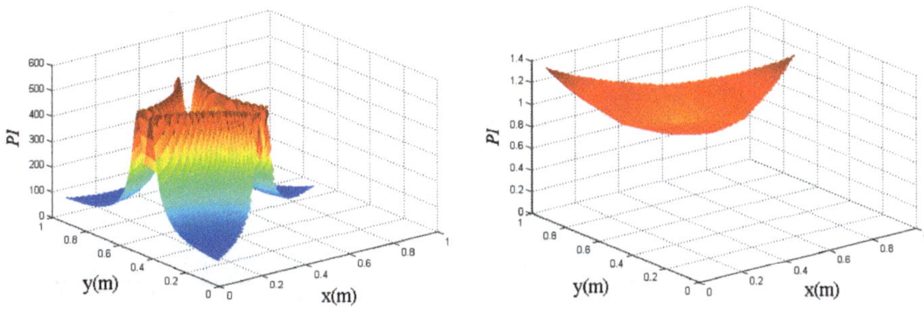

(a) HDCPR **(b)** CPR

Fig. 6.34 PI distribution with z = 0.6

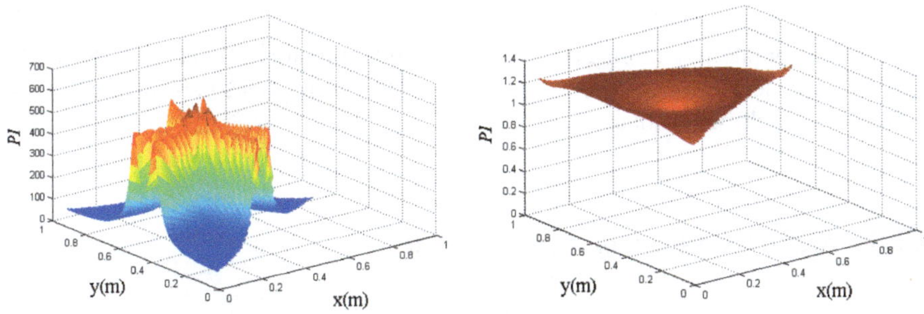

(a) HDCPR **(b)** CPR

Fig. 6.35 PI distribution with z = 0.75

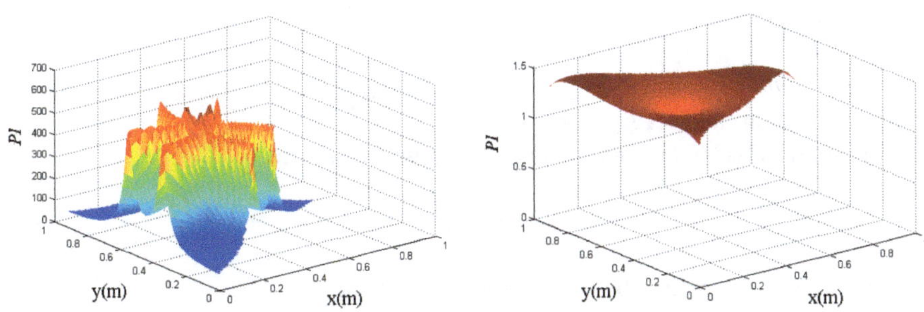

(a) HDCPR **(b)** CPR

Fig. 6.36 PI distribution with z = 0.9

minimum value at the edge position. However, the variations of the PI of the CPR are contrary. It may be influenced by the singularities that existed is the workspace. Note that the workspace of the CPR is larger than that of the HDCPR in the case of

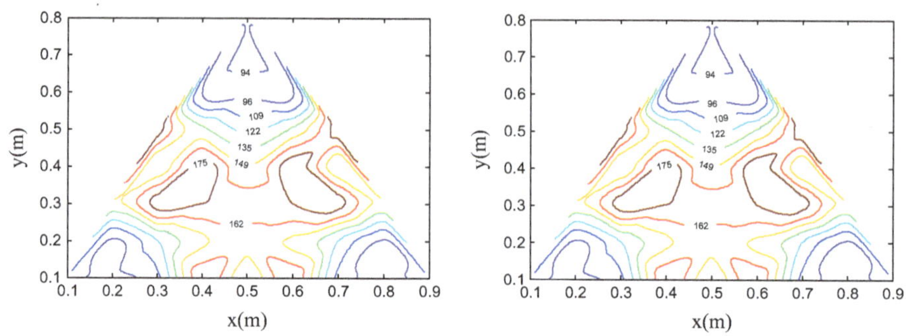

(a) HDCPR **(b)** CPR

Fig. 6.37 PI projection distribution

the same simulation parameters of the two manipulators. Since the PI of the HDCPR is generally larger than that of the CPR, the payload performance of the HDCPR is much better than that of the CPR.

6.3.5 Stiffness Performance Analysis

During the transmission process of the force and moment, the mechanical component of the parallel manipulator will produce deflection. Supposing the stiffness of the total drive system is K_p, the driving force of active joints is τ, and the deflection caused by active joints is Δq, the following equation can be obtained

$$\tau = K_p \Delta q \tag{6.89}$$

where $K_p = diag(k_{11}, k_{14}, k_{21}, k_{24}, k_{31}, k_{34})$, k_{i1}, k_{i4} ($i \in \{1, 2, 3\}$) represents the spring coefficient, and let $k_{i1} = k_{i4} = 1$.

Combining Eqs. (6.80), (6.81) and (6.89) will result in

$$F = (J^T)^{-1} K_p J^{-1} \Delta p \tag{6.90}$$

Let the stiffness matrix be expressed as

$$S = (J^T)^{-1} K_p J^{-1} \tag{6.91}$$

Hence, Eq. (6.90) can be simplified as

$$F = S \Delta p \tag{6.92}$$

The stiffness index of the HDCPR is defined as the extremum of the norm of the deflection Δp of the end-effector, while the norm of the external force F that acts on the end-effector is unity. In a similar way to derive the extremum of $\|F\|$, the Lagrange equation is constructed as follows

$$L_p = F^T S^{-T} S^{-1} F - \lambda_p (F^T F - 1) \tag{6.93}$$

where λ_P is the Lagrange multiplier.

The necessary conditions that the extremum exists can be expressed as

$$\begin{cases} \frac{\partial L_p}{\partial \lambda_p} = F^T F - 1 = 0 \\ \frac{\partial L_p}{\partial F} = S^{-T} S^{-1} F - \lambda_p F = 0 \end{cases} \tag{6.94}$$

From Eq. (6.94), it is obviously that the Lagrange multiplier λ_P is the eigenvalue of the matrix $S^{-T}S^{-1}$. Combining Eqs. (6.92) and (6.94) leads to

$$\|\Delta p\|^2 = F^T S^{-T} S^{-1} F = \lambda_p \tag{6.95}$$

Equation (6.95) indicates that the extremum of $\|\Delta p\|^2$ is the eigenvalue of the matrix $S^{-T}S^{-1}$, thus the extremum of the norm of the deflection Δp can be represented as

$$\begin{cases} \|\Delta p_{max}\| = \sqrt{\max(|\lambda_p|)} \\ \|\Delta p_{min}\| = \sqrt{\min(|\lambda_p|)} \end{cases} \tag{6.96}$$

In principle, the less the norm of the deflection is, the better the stiffness performance is. It is expected that the minimum deflection should be smaller. Thus, the minimum of the norm of output force $\|\Delta p_{min}\|$ is referred to as the stiffness index (SI)

$$\mathrm{SI} = \sqrt{\min(|\lambda_p|)} \tag{6.97}$$

Similarly, note that the SI varies with the configuration of the HDCPR, the global stiffness index (GSI) is introduced to evaluate the stiffness of the HDCPR within the workspace is defined by [4]

$$\mathrm{GSI} = \left(\int_{\mathrm{V}} \mathrm{SI} d\mathrm{V} \right) / \mathrm{V} \tag{6.98}$$

where V is the volume of the overall workspace.

Figures 6.38, 6.39, 6.40, 6.41 and 6.42 illustrate the SI distributions of the HDCPR and the CPR at various x-y plane positions about different z-axis

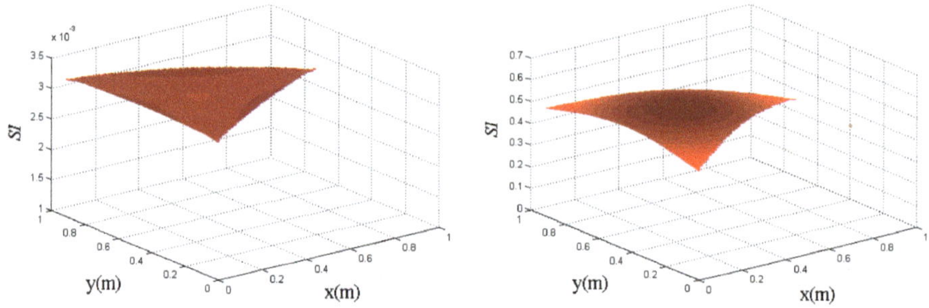

(a) HDCPR **(b)** CPR

Fig. 6.38 SI distribution with z = 0.3

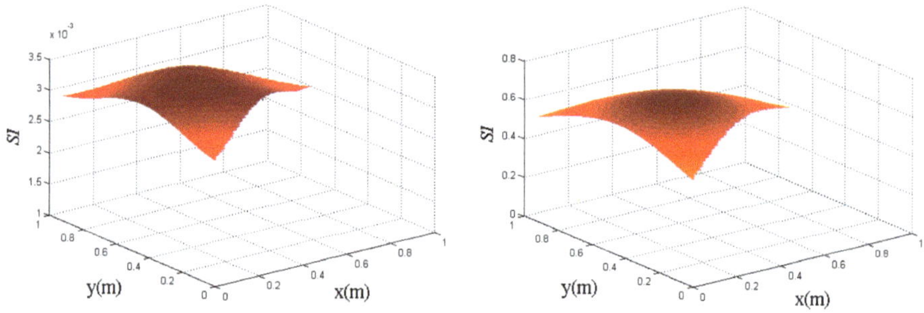

(a) HDCPR **(b)** CPR

Fig. 6.39 SI distribution with z = 0.45

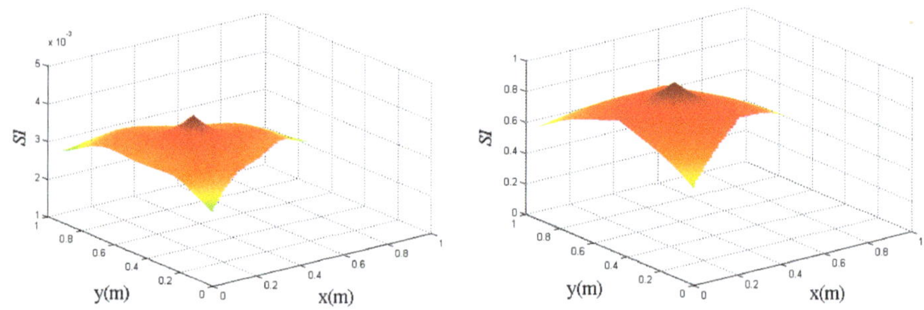

(a) HDCPR **(b)** CPR

Fig. 6.40 SI distribution with z = 0.6

coordinate values, where z is 0.3, 0.45, 0.6, 0.75, 0.9 m, respectively. As depicted in Figs. 6.38, 6.39, 6.40, 6.41 and 6.42, the SI of the HDCPR are much less than that of the CPR. With the increase of z value, the SI becomes less gradually.

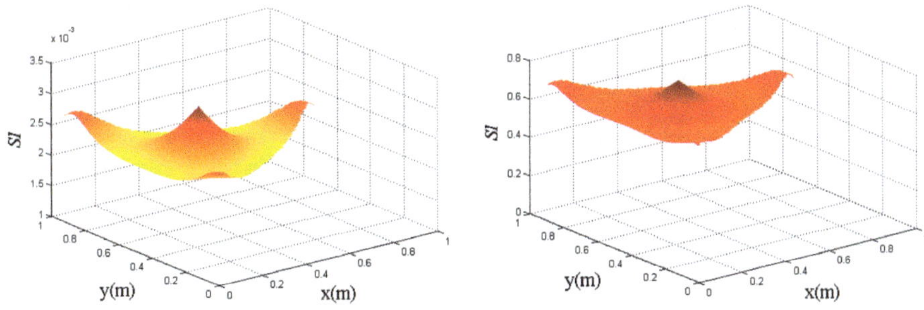

(**a**) HDCPR (**b**) CPR

Fig. 6.41 SI distribution with z = 0.75

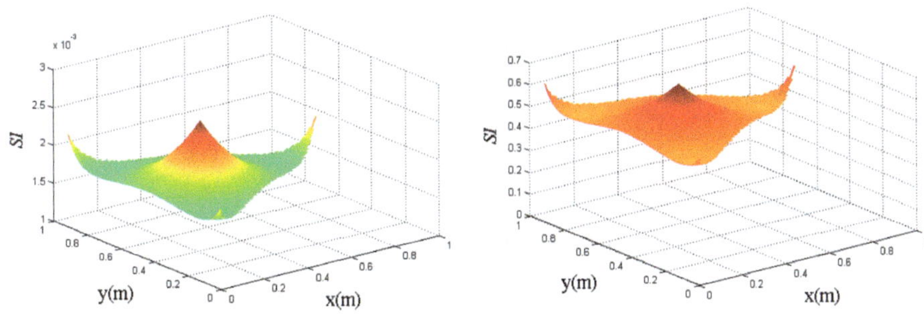

(**a**) HDCPR (**b**) CPR

Fig. 6.42 SI distribution with z = 0.9

The stiffness performance is better at the upper levels except at the central position than that is at the lower and middle levels.

When z is the average value of z-axis coordinate, the SI projection distribution of the HDCPR and the CPR at x-y plane is shown in Fig. 6.43. It implies that the SI attains the maximum value around the central position while it attains minimum value at the edge position. Besides, as a result of the symmetrical design of the mechanisms, the SI projection distribution is symmetrical. Comparing the minimum value of the SI shown in Fig. 6.43, it can be concluded that the stiffness performance of the HDCPR is much better than that of the CPR.

The global performance indices of the HDCPR and the CPR are shown in Table 6.3. Referring to Table 6.3, the values of the GCI of the HDCPR is less than that of the CPR, but they are close. The value of the GPI of the HDCPR is much larger than that of the CPR, while the value of the GSI of the HDCPR is much less than that of the CPR. Therefore, it can be concluded that the conditioning performance of the CPR is a bit better than that of the HDCPR. However, the HDCPR has much better payload performance and stiffness performance than the CPR.

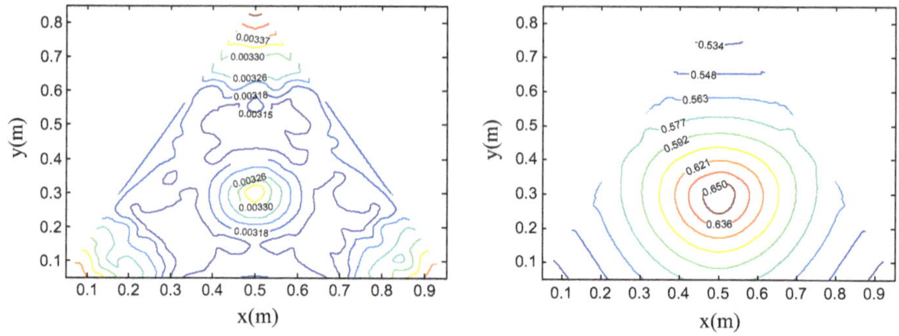

(**a**) HDCPR (**b**) CPR

Fig. 6.43 SI projection distribution

Table 6.3 The global performance indices of the HDCPR and the CPR

Manipulator	GCI	GPI	GSI
HDCPM	0.4308	136.9094	0.0032
CPM	0.5048	1.3310	0.5792

6.3.6 Improved HDCPR Description and Its Performance Analysis

In order to expand the workspace of HDCPR and improve its work performance, we put forward the improved HDCPR system. Figure 6.44 shows the design model of the improved HDCPR. Kinematical structure of the improved HDCPR is shown in Fig. 6.45. The improved HDCPR installs a twistable roulette at each of the HDPM which could put the rope around the roulette. Then the workplace of improved HDCPR will be greatly enlarged.

Fig. 6.44 Three-dimensional model of the improved HDCPR

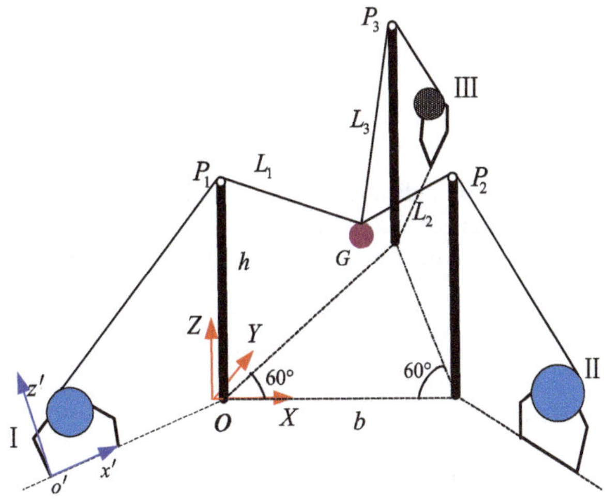

Fig. 6.45 Kinematic schematic of improved HDCPM

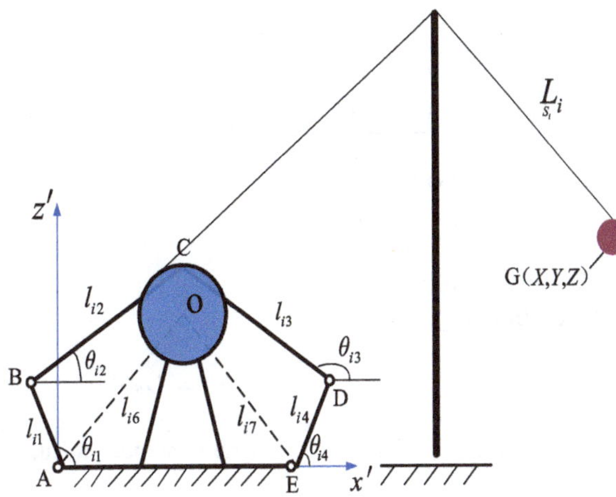

Fig. 6.46 The *i*th group of the improved HDPM

Figure 6.46 displays the schematic of the ith group of the improved HDPM and cable structure. Its basic parameters are similar with HDPM. The radius of roulette is r and the center is O. The length of links AO and EO are l_6 and l_7. The other parameters are the same.

From Fig. 6.45, it is well known that the closed loop polygon vector equation of the first group of the improved HDCPR can be described as

$$\left.\begin{array}{l}\vec{l}_{i1}+\vec{l}_{i2}=\vec{l}_{i6}+\vec{r}\\ \vec{l}_{i3}+\vec{l}_{i4}=\vec{r}+\vec{l}_{i7}\end{array}\right\} \tag{6.100}$$

Developing Eq. (6.100) into real part and imaginary part, the following equations can be expressed as

$$\left.\begin{aligned} l_{i1}\cos\theta_{i1}+l_{i2}\cos\theta_{i2} &= l_{i6x}+r\cos\phi \\ l_{i1}\sin\theta_{i1}+l_{i2}\sin\theta_{i2} &= l_{i6y}+r\sin\phi \\ l_{i3}\cos\theta_{i3}+l_{i4}\cos\theta_{i4} &= l_{i7x}+r\cos\phi \\ l_{i3}\sin\theta_{i3}+l_{i4}\sin\theta_{i4} &= l_{i7y}+r\sin\phi \end{aligned}\right\} \tag{6.101}$$

In which ϕ indicates the rotation angle of the roulette wheel, and r represents the radius of the roulette wheel.

θ_{i2} and θ_{i3} may be eliminated from Eq. (6.101), then simplify and order it. We obtain

$$A\sin\phi+B\cos\phi+C=0 \tag{6.102}$$

Here

$$A=r(l_{i1}\sin\theta_{i1}-l_{i6y}+l_{i4}\sin\theta_{i4}-l_{i7y}),\ B=r(l_{i1}\cos\theta_{i1}-l_{i6x}+l_{i4}\cos\theta_{i4}-l_{i7x}),$$
$$C=-l_{i4}^2-l_{i1}^2+l_{i1}(l_{i6y}\cos\theta_{i1}+l_{i6x}\cos\theta_{i1})+l_{i4}(l_{i7y}\cos\theta_{i4}+l_{i7x}\cos\theta_{i4})$$

The following equation is obtained by solving the Eq. (6.102)

$$\phi=2\arctan\frac{A-\sqrt{A^2+B^2-C^2}}{B-C},\ \theta_{i2}=\frac{r\sin\phi-l_{i1}\sin\theta_{i1}}{r\cos\phi+l_6-l_{i1}\cos\theta_{i1}},\ \theta_{i3}=\frac{r\sin\phi-l_{i4}\sin\theta_{i4}}{r\cos\phi+l_7-l_{i4}\cos\theta_{i4}} \tag{6.103}$$

According to the Fig. 6.44, it is well known that the equation of the cable length can be obtained as

$$L_i=L_{0i}-r\phi \tag{6.104}$$

The derivative of Eq. (6.104) with respect to all the variables can be computed by

$$\dot{L}_i=-\frac{l_{i1}\sin(\theta_{i1}-\theta_{i2})}{2\sin(\phi-\theta_{i2})}\dot{\theta}_{i1}-\frac{l_{i4}\sin(\theta_{i4}-\theta_{i3})}{2\sin(\phi-\theta_{i3})}\dot{\theta}_{i4} \tag{6.105}$$

Then, one can derive from Eqs. (6.105) and (6.65)

$$\dot{p}'=J_1J_0\dot{q}' \tag{6.106}$$

where, $p'=[x'\ \ y'\ \ z']$, $q'=\left[\dot{\theta}_{11}\ \ \dot{\theta}_{14}\ \ \dot{\theta}_{21}\ \ \dot{\theta}_{24}\ \ \dot{\theta}_{31}\ \ \dot{\theta}_{34}\right]$

$$J_1=\begin{vmatrix} -\frac{l_{11}\sin(\theta_{11}-\theta_{12})}{2\sin(\phi-\theta_{12})} & -\frac{l_{14}\sin(\theta_{14}-\theta_{13})}{2\sin(\phi-\theta_{13})} & 0 & 0 & 0 & 0 \\ 0 & 0 & -\frac{l_{21}\sin(\theta_{21}-\theta_{22})}{2\sin(\phi-\theta_{22})} & -\frac{l_{24}\sin(\theta_{24}-\theta_{23})}{2\sin(\phi-\theta_{23})} & 0 & 0 \\ 0 & 0 & 0 & 0 & -\frac{l_{31}\sin(\theta_{31}-\theta_{32})}{2\sin(\phi-\theta_{32})} & -\frac{l_{34}\sin(\theta_{34}-\theta_{33})}{2\sin(\phi-\theta_{33})} \end{vmatrix}$$

Table 6.4 Kinematics parameters of the improved HDCPR

Parameters	h	b	a	L_0	l_1	l_2	l_3	l_4	l_5	l_6	l_7	r
Value (m)	1.00	1.00	1.51	2.00	0.20	0.50	0.50	0.28	0.51	0.48	0.47	0.2

Equation (6.106) may be rewritten as follows

$$\dot{p}' = J'\dot{q}' \tag{6.107}$$

In order to compare the performances of the improved HDCPR and the HDCPR, we do simulation study again. The kinematics parameters of the improved HDCPR are listed in Table 6.4.

We also choose different elevation of z-axis coordinate to analyze. Figures 6.47, 6.48, 6.49, 6.50 and 6.51 illustrate the CI distributions of the improved HDCPR, where z is 0.3, 0.45, 0.6, 0.75, 0.9 m, respectively.

As shown in Figs. 6.47, 6.48, 6.49, 6.50 and 6.51, the CI distributions of the improved HDCPR and the CPR have a similar trend, and the difference between them is not obvious. When z is the average value of z-axis coordinate, the CI projection distribution of the improved HDCPR at x-y plane is displayed in Fig. 6.52. The maximum value is 0.552, which is a bit better conditioning performance than the HDCPR.

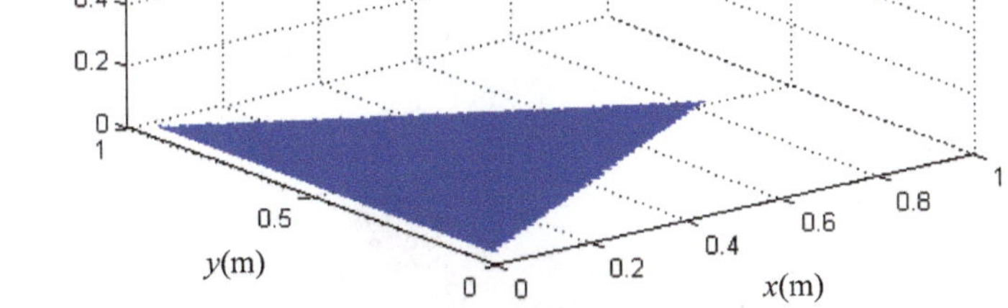

Fig. 6.47 *CI* distribution with z = 0.3

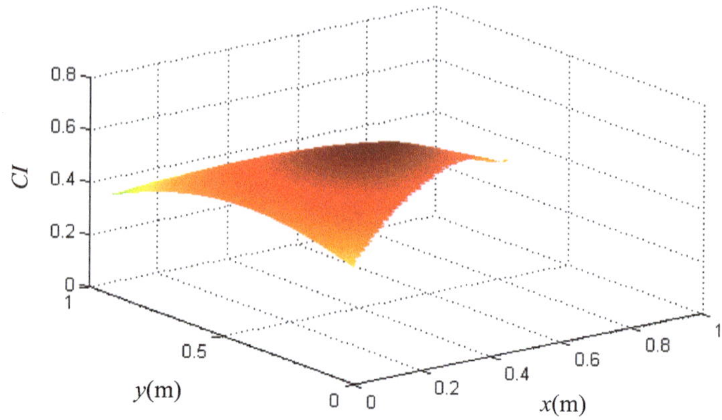

Fig. 6.48 *CI* distribution with z = 0.45

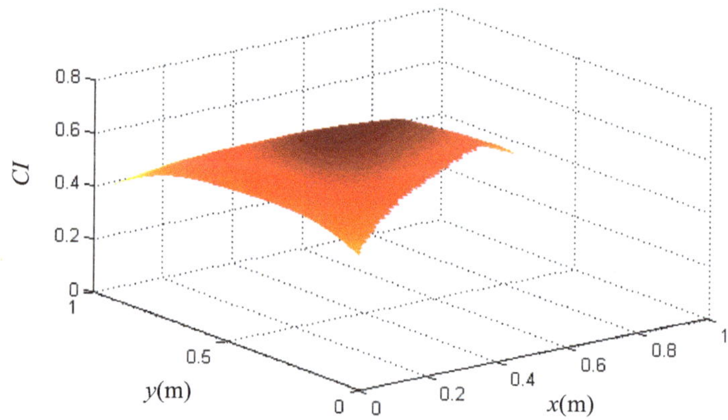

Fig. 6.49 *CI* distribution with z = 0.6

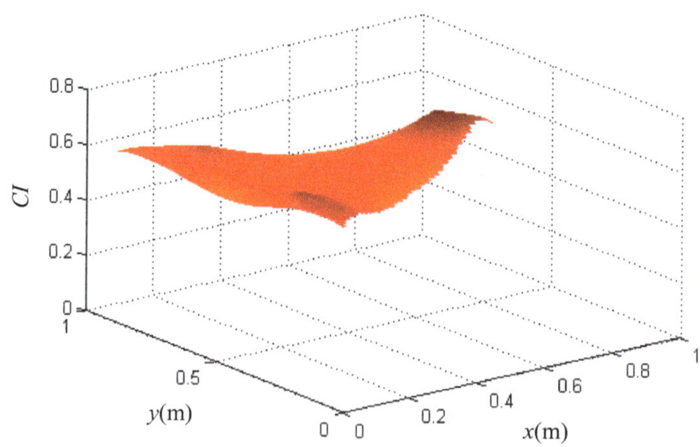

Fig. 6.50 *CI* distribution with z = 0.75

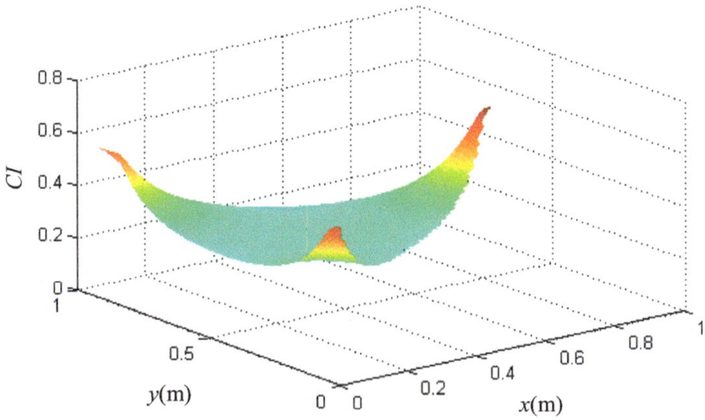

Fig. 6.51 *CI* distribution with z = 0.9

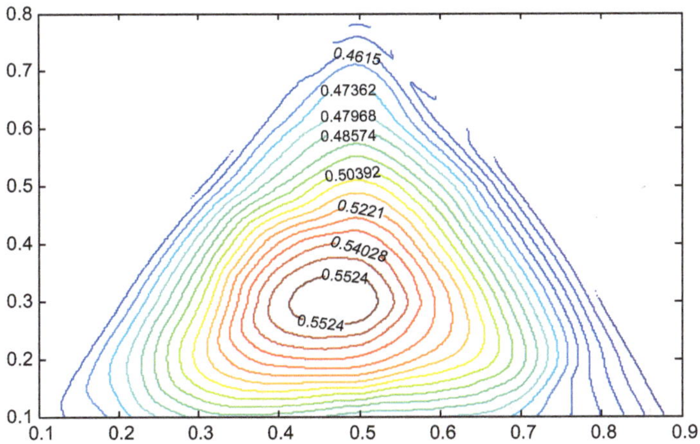

Fig. 6.52 *CI* projection distribution

In addition, the PI analysis of the improved HDCPR is also carried out. Figures 6.53, 6.54, 6.55, 6.56 and 6.57 illustrate the PI distributions of the improved HDCPR at various x-y plane positions about different z-axis coordinate values, where z is 0.3, 0.45, 0.6, 0.75, 0.9 m, respectively.

Referring to Figs. 6.53, 6.54, 6.55, 6.56 and 6.57, one can see that the PI of the improved HDCPR is a little larger than that of the HDCPR. When z is the average value of z-axis coordinate, the PI projection distribution of the improved HDCPR at x-y plane is illustrated in Fig. 6.58, and the PI distribution is more uniform. Therefore, the payload capability of the improved HDCPR is better.

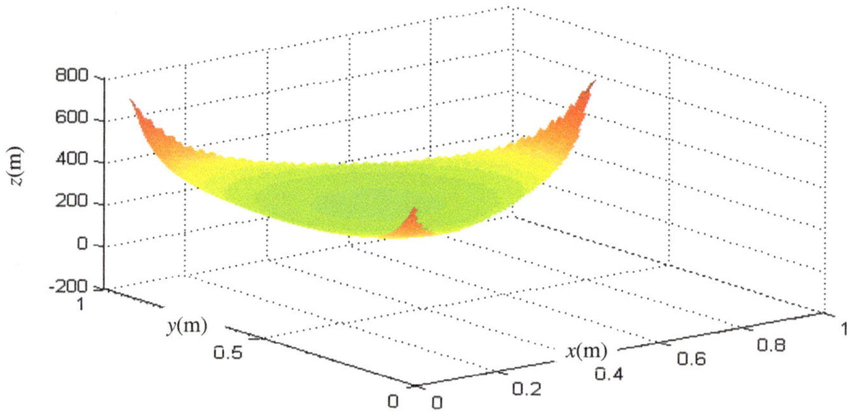

Fig. 6.53 PI distribution with z = 0.3

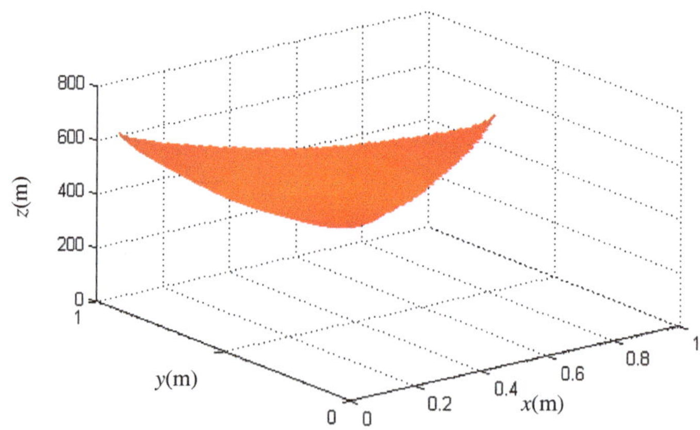

Fig. 6.54 PI distribution with z=0.45

Fig. 6.55 PI distribution with z=0.6

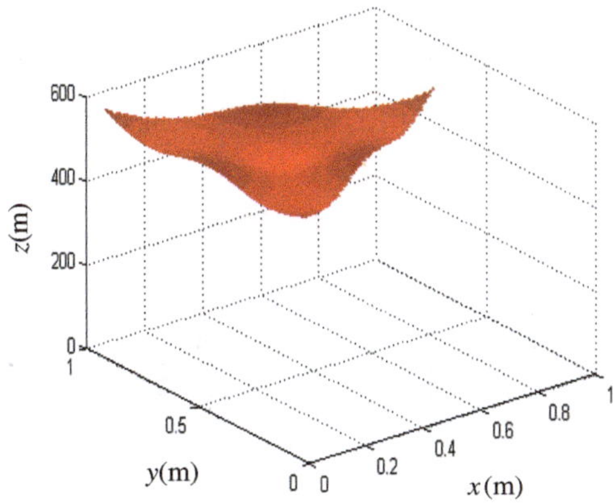

Fig. 6.56 PI distribution with z=0.75

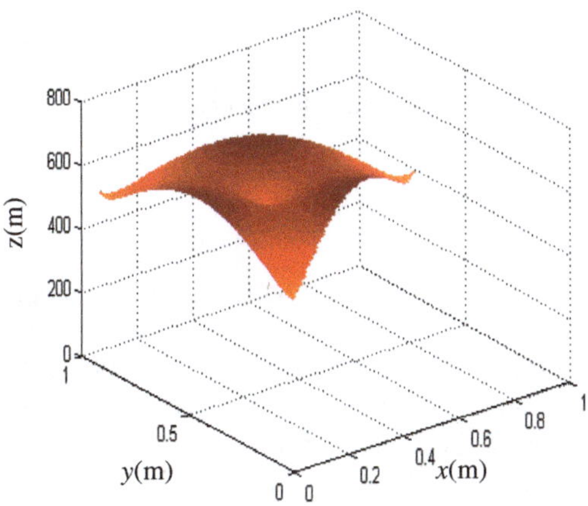

Fig. 6.57 PI distribution with z=0.9

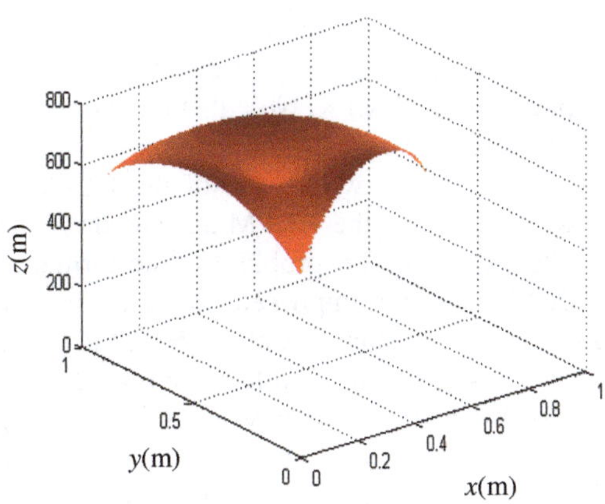

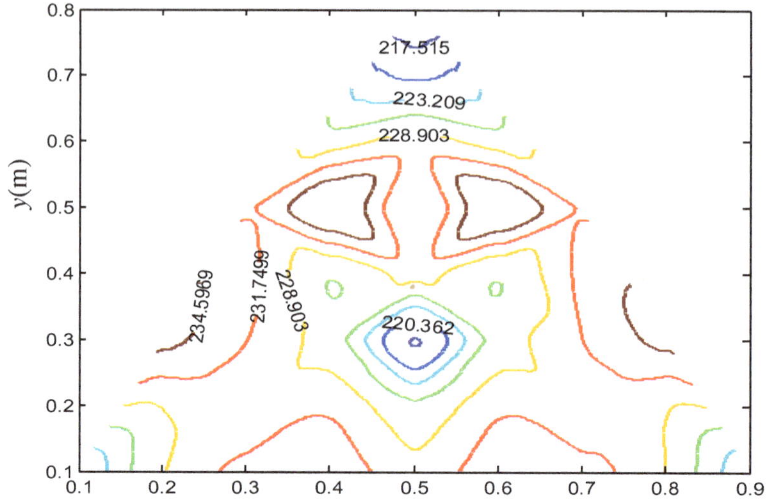

Fig. 6.58 PI projection distribution

6.4 Conclusions

This chapter presents the design of a completely restrained CPR with three translational motions and a new completely restrained HDCPR with three translational motions. Dynamic models of the CPR and HDCPR are established by using the Lagrangian formulation separately, and the inverse kinematic and dynamic problems of the CPR and HDCPR are resolved and compared on condition that an operation path of the end-effector has been planned. Workspace, drive torques, drive powers, stiffness performance and singularity are given to compare the main mechanics performance of the two manipulators. A case study is addressed for the validation of the proposed HDCPR. The results show the CPR and HDCPR have the similar mechanics performance, and the HDCPR has better load carrying capacity than the CPR. Note that the workspace of the CPR is larger than that of the HDCPR in the case of the same simulation parameters of the two manipulators. Nevertheless, the workspace of the HDCPR is enlarged by changing connection position between the HDPM and the drive cables of the CPR, which is a topic for further research.The problem of performance evaluation and comparison for the HDCPR and its improved mechanism has been investigated. Kinematic and Jacobian matrix analyses of the HDCPR are carried out by using closed loop vector conditions and direct differential method. Conditioning performance analysis, payload performance analysis, and stiffness performance analysis of the HDCPR are provided to evaluate performance variation of the HDCPR by means of Jacobian matrix. A case study is addressed for the validation of the proposed HDCPR. The results demonstrate the payload performance and stiffness performance of the HDCPR are much better than that of the CPR, while the conditioning performance of the CPR is a bit better than that of the HDCPR. In addition, the improved HDCPR is designed in order to enlarge the workspace of the HDCPR. From comparisons between the improved HDCPR and CPR, the simulations show that the conditioning performance of the improved HDCPR is as well as that of the CPR, and compared with the HDCPR, the improved HDCPR has better payload performance.

References

1. Gao F, Liu X, Gruver WA (1998) Performance evaluation of two-degree-of-freedom planar parallel robots. Mech Mach Theory 33(6):661–668
2. Zi B, Cao J, Zhu H et al (2014) Comparative study of cable parallel manipulators with and without hybrid-driven planar five-bar mechanism. Appl Math Model 38(24):5994–6017
3. Gao Z, Zhang D (2010) Design, analysis and fabrication of a multidimensional acceleration sensor based on fully decoupled compliant parallel mechanism. Sens Actuators A Phys 163 (1):418–427
4. Liu XJ, Wang J, Pritschow G (2006) Performance atlases and optimum design of planar 5R symmetrical parallel mechanisms. Mech Mach Theory 41(2):119–144
5. Gosselin C (2014) Cable-driven parallel mechanisms: state of the art and perspectives. Mech Eng Rev 1(1):DSM0004-DSM0004
6. Zi B, Zhu Z, Du J (2011) Analysis and control of the cable-supporting system including actuator dynamics. Control Eng Pract 19(5):491–501
7. Otis MJD, Perreault S, Nguyen-Dang TL et al (2009) Determination and management of cable interferences between two 6-DOF foot platforms in a cable-driven locomotion interface. IEEE Trans Syst Man Cybern-Part A Syst Hum 39(3):528–544
8. Lim WB, Yang G, Yeo SH et al (2011) A generic force-closure analysis algorithm for cable-driven parallel manipulators. Mech Mach Theory 46(9):1265–1275
9. Zi B, Wu X, Lin J, et al (2012) Inverse kinematics and singularity analysis for a 3-DOF hybrid-driven cable-suspended parallel robot. Int J Adv Robot Syst 9
10. Gosselin C, Angeles J (1991) A global performance index for the kinematic optimization of robotic manipulators. J Mech Des 113(3):220–226
11. Pond G, Carretero JA (2006) Formulating Jacobian matrices for the dexterity analysis of parallel manipulators. Mech Mach Theory 41(12):1505–1519
12. Shneor Y, Portman VT (2010) Stiffness of 5-axis machines with serial, parallel, and hybrid kinematics: Evaluation and comparison. CIRP Ann-Manuf Technol 59(1):409–412
13. Mackay AB, Smith DG, Magleby SP et al (2012) Metrics for evaluation and design of large-displacement linear-motion compliant mechanisms. J Mech Des 134(1):011008
14. Rezaei A, Akbarzadeh A, Akbarzadeh-T MR (2012) An investigation on stiffness of a 3-PSP spatial parallel mechanism with flexible moving platform using invariant form. Mech Mach Theory 51:195–216

Chapter 7
Design and Dynamics of the Cooperative Cable Parallel Manipulators for Multiple Mobile Cranes

7.1 Preamble

The design, dynamic modelling, and workspace are presented in this chapter concerning cooperative cable parallel manipulators for multiple mobile cranes (CPRMC). The CPRMC can handle complex tasks that are more difficult or even impossible for a single mobile crane. Kinematics and dynamics of the CPRMC are studied on the basis of geometric methodology and d'Alembert's principle, and a mathematical model of the CPRMC is developed and presented with dynamic simulation. The constant orientation workspace analysis of the CPRMC is carried out additionally. As an example, a cooperative cable parallel manipulator for triple mobile cranes with 6 Degrees of Freedom is investigated on the basis of the above design objectives.

Mobile cranes are widely used for loading, mounting, carrying large heavy-duty loads and for work performed in the presence of obstacles of various kinds such as civil engineering for the construction of buildings, power lines and similar technological installations [1]. With the development of the manufacturing technology and demand of the modern engineering, cooperative operation of multiple mobile cranes has become more common in modern construction projects and manipulation of heavy-duty objects in recent years [2], such as the one shown in Fig. 7.1.

Multiple mobile cranes can be treated as cooperative multiple manipulators [3]. Multiple manipulators have become a mature research field, in which the problem of the implementation of cooperative manipulation on conventional set-ups has attracted an increasing interest of the research community [4]. Multiple robot manipulators have been studied as having many possible applications [5, 6]. This is due to the extended capabilities that the multiple robot manipulators have to offer compared to the use of a single manipulator for the same task [7]. In the literature there is a variety of studies about multiple robot manipulators.

Similar to cable parallel manipulators (CPRs), the cooperative multiple mobile cranes can be viewed as the CPRs. The CPRs have potential advantages in terms of

B. Zi and S. Qian, *Design, Analysis and Control of Cable-Suspended Parallel Robots and Its Applications*, DOI 10.1007/978-981-10-1753-7_7

Fig. 7.1 Multiple mobile cranes

simple and light-weight structure, large reachable workspace, high acceleration capability, and easy reconfigurability [8–10]. For the preceding characteristics, the CPRs play an important role in many engineering fields, such as manipulation of heavy payloads for manufacturing and cargo handling, coordinate measurement, aircraft testing, super antenna, haptic devices, rehabilitation and welfare engineering [11–16]. For the past several decades, the CPRs have been extensively studied.

Recently, cooperative cable parallel manipulators for multiple mobile cranes (CPRMC) are finding increased use in a wide variety of modern engineering applications. In addition, the CPRMC can handle complex tasks that are more difficult or even impossible for a single crane. In this investigation, the design, dynamic modelling, and workspace of the CPRMC are presented. The remainder of this chapter is organized as follows: Section 7.2 describes the design model of the CPRMC. Then, kinematics and dynamics of the CPRMC are performed in Sect. 7.3. Workspace analysis of the CPRMC is provided in Sect. 7.4. In Sect. 7.5, illustrative simulation studies highlight its performances. Finally, some concluding remarks are summarized in Sect. 7.6.

7.2 System Description

In this chapter, cooperative cable parallel manipulators for triple mobile cranes are taken as an example, and the three-dimensional model of the CPRMC are established for the purpose of numerical analysis (see Fig. 7.2). The CPRMC consist of three mobile cranes and suspend the payload by three cables within the workspace through coordinate the action of each mobile crane. Each single mobile crane is mainly composed of telescopic subsystem, slewing subsystem, luffing subsystem and hoisting subsystem. For each cable, one end is connected to the hoisting point of payload, the other one rolls through a pulley fixed on the top of telescopic boom and then is fed into the drum of hoisting subsystem.

Fig. 7.2 Three-dimensional model of the CPRMC

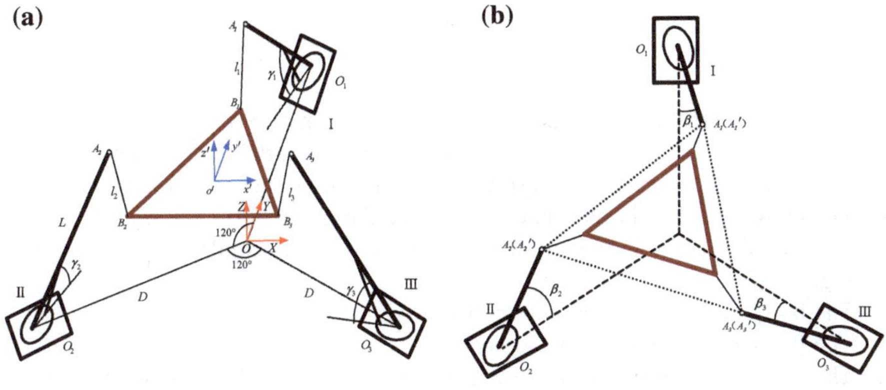

Fig. 7.3 **a** Schematic sketch of the CPRMC **b** a top view of the CPRMC

A simple schematic sketch of the CPRMC structure model is shown in Fig. 7.3, which provides both a three dimentional view and a top view of CPRMC with the associated coordinate systems. The three mobile cranes are arrayed at the vertex of triangle $O_1O_2O_3$ on the ground.

At the center of the triangle $O_1O_2O_3$, a global coordinate system O (XYZ) is established. The distance between the endpoint of boom A_i and the corresponding hoisting point of payload B_i is L_i(i = 1, 2, 3). At the center of the gravity of the payload, a local coordinate system O' $(x'y'z')$ is established. The mechanical configuration of a QY100k mobile crane is illustrated in Fig. 7.4.

For simplicity, we make the following assumptions.

1. Three mobile cranes of the same type QY100k are chosen in the CPRMC, made in Xuzhou Heavy Machinery Co., Ltd of XCMG.
2. In order to suspend the payload smoothly and flexibly, the mobile cranes are expected to be able to luff, slew and hoist simultaneously as manipulators,

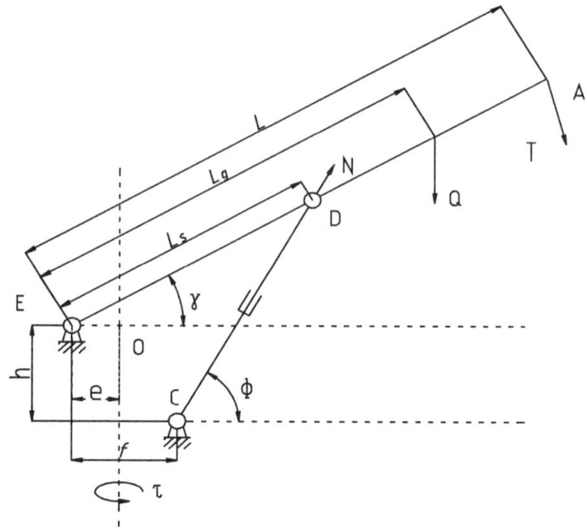

Fig. 7.4 Schematic sketch A a mobile crane

which can be realized by cooperative control system of mobile cranes. However, the mobile cranes are expected to move with a certain regulation to avoid collision with each other during the cooperation and simplify calculation. Therefore, the hypothesis is put forward that the triangle $A'_1A'_2A'_3$ whose vertexes are the projection points of the endpoints A_i is a equilateral triangle (see Fig. 7.3).

3. The body of mobile crane is regarded as rigid.
4. The payload is homogeneous object.
5. The cable is treated as a massless body and the flexibility is neglected.
6. Inertia force due to the body of the mobile crane is neglected because of the slow luffing and slewing speed.
7. The maximum acceleration in the vertical direction is smaller than the acceleration due to gravity.

7.3 Kinematics and Dynamics

According to the geometric relationship of the CPRMC in Fig. 7.3, the coordinates of three mobile cranes can be derived as

$$\begin{cases} O_1 = [0, D, 0]^T \\ O_2 = [-D\cos(\pi/6), \quad -D\sin(\pi/6), 0]^T \\ O_3 = [D\cos(\pi/6), \quad -D\sin(\pi/6), 0]^T \end{cases} \tag{7.1}$$

where D is the distance between each mobile crane and the reference point O.

The coordinates $P \ni R^6$ describe the spatial position and orientation of the payload. xp, yp, zp represent three Cartesian coordinates and α, o, n represent three Bryant angles.

$$P = [r\,\theta]^T, r = [xp, yp, zp]^T, \theta = [\alpha, o, n]^T \tag{7.2}$$

Hence, the coordinates of hoisting point of payload $B_i(x_{Bi}, y_{Bi}, z_{Bi})$ can be obtained by

$$B_i = Rb_i' + r \tag{7.3}$$

where R is the rotation matrix going from the local frame O' to the global frame O (as depicted in Fig. 7.3). Additionally, $b_i'$is the coordinates of hoisting points relative to O'.

$$R = \begin{bmatrix} C\alpha Co & C\alpha SoSn - S\alpha Cn & C\alpha SoCn + S\alpha Sn \\ S\alpha Co & S\alpha SoSn - C\alpha Cn & S\alpha SoCn - C\alpha Sn \\ -So & CoSn & CoCn \end{bmatrix} \tag{7.4}$$

$$\begin{cases} b_1' = [0, s, 0]^T \\ b_2' = [-s\ \cos(\pi/6), -s\ \sin(\pi/6), 0]^T \\ b_3' = [s\ \cos(\pi/6), -s\ \sin(\pi/6), 0]^T \end{cases} \tag{7.5}$$

where s represents the distance between the hoisting point B_i and the reference point O'. Additionally, we denote cos by C, with similar substitution S for sin.

According to Fig. 7.3, we can obtain the coordinates of $A_i(x_{Ai}, y_{Ai}, z_{Ai})$ as follows

$$\begin{cases} (x_{A1}, y_{A1}, z_{A1}) = [X - d\ \sin\delta, d\ \cos\delta + Y, \sqrt{L^2 - (X - d\ \sin\delta)^2 - (d\ \cos\delta + Y - D)^2}] \\ (x_{A2}, y_{A2}, z_{A2}) = [X - d\ \cos(\delta + \pi/6),\ Y - d\ \sin(\delta + \pi/6), \\ \qquad \sqrt{L^2 - (X - d\ \cos(\delta + \pi/6) + D\ \cos(\pi/6))^2 - (Y - d\ \sin(\delta + \pi/6) + D\ \sin(\pi/6))^2}] \\ (x_{A3}, y_{A3}, z_{A3}) = [d\ \sin(\delta + \pi/3) + X,\ Y - d\ \cos(\delta + \pi/3), \\ \qquad \sqrt{L^2 - (X + d\ \sin(\delta + \pi/3) - D\ \cos(\pi/6))^2 - (Y - d\ \cos(\delta + \pi/3) + D\ \sin(\pi/6))^2}] \end{cases} \tag{7.6}$$

where (X, Y, δ) represents the coordinates of the center and the rotation angle of equilateral triangle $A_1'A_2'A_3'$. L represents the boom length of mobile crane. d represents the distance between the center of equilateral triangle $A_1'A_2'A_3'$ and the vertex A_i'.

The relationship between the cable length l_i and the endpoint of boom A_i as well as the corresponding hoisting point of payload B_i, can be easily obtained as follows

$$l_i = \sqrt{(x_{Ai} - x_{Bi})^2 + (y_{Ai} - y_{Bi})^2 + (z_{Ai} - z_{Bi})^2} \tag{7.7}$$

Hence the cables exert zero-pitch wrenches on the payload which take the following form [16]

$$w_i = \frac{1}{l_i}\begin{bmatrix} \vec{B_i}\, A_i \\ \vec{O'}\, B_i \times \vec{B_i}\, A_i \end{bmatrix} \tag{7.8}$$

The gravity wrench takes the form as follows

$$\vec{G} = -mg[e_3 \quad r \times e_3]^T, e_3 = [0 \quad 0 \quad 1]^T \tag{7.9}$$

where g is the acceleration due to gravity.

According to Eq. (7.2), the translational velocity v of the payload relative to O is first order time derivative of its Cartesian coordinates r.

$$v = \dot{r} = [\dot{x} \quad \dot{y} \quad \dot{z}]^T \tag{7.10}$$

The angular velocity ω can be generated as

$$\omega = K^{-1}\dot{\theta} \tag{7.11}$$

where $\dot{\theta}$ is first order time derivative of Bryant angles θ of the payload. Additionally, K^{-1} is the transformation matrix going from Bryant angles θ to the angular velocity of the payload.

$$\dot{\theta} = [\dot{\alpha} \quad \dot{o} \quad \dot{n}]^T \tag{7.12}$$

$$K^{-1} = \begin{bmatrix} 1 & S\alpha So/Co & C\alpha So/Co \\ 0 & C\alpha & S\alpha \\ 0 & -S\alpha/Co & C\alpha/Co \end{bmatrix} \tag{7.13}$$

On the basis of geometric methodology and d'Alembert's principle, the dynamic equations of the payload can be formulated as

$$\begin{bmatrix} M & \\ & I \end{bmatrix}\begin{bmatrix} \dot{v} \\ \dot{\omega} \end{bmatrix} + [w_1 \quad w_2 \quad w_3]\begin{bmatrix} T_1 \\ T_2 \\ T_3 \end{bmatrix} + \vec{G} = 0 \tag{7.14}$$

where M is the mass matrix of the payload relative to its center of gravity, while I is inertia matrix of the payload relative to its center of gravity relative to O'. $T_i (i = 1, 2, 3)$ is the tension of the ith cable, that is the force acting on the hoisting subsystem.

Decomposing the cable tension T_i in the vertical direction, radial direction and tangential direction, three component forces can be formulated as

$$\begin{cases} f_{vi} = \frac{T_i(Z_{Ai} - Z_{Bi})}{l_i} \\ f_{ri} = \frac{T_i}{l_i\left|\overrightarrow{O_iA_i'}\right|}\overrightarrow{B_iA_i} \cdot \overrightarrow{O_iA_i'} \\ f_{ti} = \sqrt{T_i^2 - f_{xi}^2 - f_{zi}^2} \end{cases} \tag{7.15}$$

where f_{vi} is the vertical component force, f_{ri} is the radial component force and f_{ti} is the tangential component force, respectively.

$$\gamma_i = \text{a} \sin(z_{A_i}/L) \tag{7.16}$$

According to Eq. (7.15) and Fig. 7.4, the force N_i acting on the luffing subsystem and the torque τ_i acting on the slewing subsystem can be easily obtained by

$$N_i = (f_{ri}(Z_{Ai} - h) + (f_{vi} + L_g Q/L)(\left|\overrightarrow{O_iA_i'}\right| + e))/(L_S \sin(\phi_i - \gamma_i)) \tag{7.17}$$

$$\tau_i = f_{ti}\left|\overrightarrow{O_iA_i'}\right| \tag{7.18}$$

where γ is the luffing angle of mobile crane and ϕ is the angle between the cylinder and ground. h represents the vertical distance between the lower hinge joints of boom and cylinder. e is the horizontal distance between the lower hinge joint of boom and the slewing axle of mobile crane. L_g is the distance between the lower hinge joints of boom and the center of gravity of boom. L_S represents the distance between the lower hinge joint of boom and the upper hinge joint of cylinder.

$$\phi_i = \text{a} \tan\frac{L_s Z_{Ai}/L + h}{L_s\left(\left|\overrightarrow{O_iA_i'}\right| + e\right)/L - f} \tag{7.19}$$

where f is the horizontal distance between the lower hinge joints of boom and cylinder.

$$\gamma_i = \text{a} \tan\frac{Z_{Ai}}{\left(\left|\overrightarrow{O_iA_i'}\right| + e\right)} \tag{7.20}$$

The dynamic equations of the payload and the force acting on the subsystems of the CPRMC are described by the following formulation

$$\begin{cases} \begin{bmatrix} M & \\ & I \end{bmatrix} \begin{bmatrix} \dot{v} \\ \dot{\omega} \end{bmatrix} + [w_1 \quad w_2 \quad w_3] \begin{bmatrix} T_1 \\ T_2 \\ T_3 \end{bmatrix} + \vec{G} = 0 \\ \tau_i = f_{ti} \left| \overrightarrow{O_i A_i'} \right| \\ N_i = (f_{ri}(Z_{Ai} - h) + (f_{vi} + L_g Q/L)(\left| \overrightarrow{O_i A_i'} \right| + e))/(L_S \sin(\phi_i - \gamma_i)) \end{cases} \tag{7.21}$$

7.4 Workspace Determination

The workspace of the CPRMC is characterized as the set of points where the payload can be positioned while all cables are in tension ($T_i > 0$). Because of the mechanical structure limitation, the payload can not be suspended a circle around the reference point. In other words, there is no dexterous workspace for CPRMC. Therefore in this work, the constant orientation workspace of CPRMC is generated, which is the set of points where the payload can be positioned while its Bryant angles $\theta = [\alpha, o, n]^T$ is fixed.

The constant orientation workspace of CPRMC not only depends on the tension condition ($T_i > 0$), we should also consider the following constraints:

1. The main searching scope. Taking the engineering practice into consideration, the triangular prism is chosen as the main searching scope, which undersurface is the triangle $O_1O_2O_3$ and height is the length of the boom L. The main searching scope can be expressed as follows

$$\begin{cases} L < z < 0 \\ -D\ \sin(\pi/3) < x < D\ \sin(\pi/3) \\ -D\ \cos(\pi/3) < y < \sqrt{3}(D\ \sin(\pi/3) - |X|) - D\ \cos(\pi/3) \end{cases} \tag{7.22}$$

2. The working radius R. According to the Lifting capability of QY100k mobile crane (see Table 7.1), the working radius R of each mobile crane is limited after the fix of boom length and payload.

$$R_{\min} < R_i < R_{\max} \quad (i = 1, 2, 3) \tag{7.23}$$

$$R_i = \sqrt{(X_{Ai} - X_{Oi})^2 + (Y_{Ai} - Y_{Oi})^2} \tag{7.24}$$

3. The working area. For a mobile crane, there are three working areas in terms of the slewing angle: left area, right area and back area. Generally speaking, the

Table 7.1 Lifting capability of QY100k

Boom length (m) working radius (m)	12.8	17.4	22.1	26.7	30.4	36	40.7	45.3	50
3	100	80							
4	88	70	61						
5	70	61	54.3	42	40				
6	57	54	48.7	42	36.5	31.1			
7	47	47.5	44.1	38.8	33.6	29	24.5		
8	40.5	40	40.2	35.3	31.1	27.2	23.3	19.3	
9	34.5	35	35	32.3	28.7	25.4	21.8	18.4	14
10	30	30	30	29.9	26.6	23.6	20.6	17.5	13.4
12		23	22.5	24.1	23.3	20.9	18.5	15.7	11.8
14		16.6	16.4	17.7	18.6	18.5	16.4	14.2	11.3
16			12.2	13.4	14.2	14.8	14.3	14	9.2
18			9.2	10.4	11.2	11.8	12.3	11.6	10.1
20				8.1	8.9	9.5	10	10.2	8.3
22				6.3	7.1	7.7	8.2	8.4	7.7
24					5.7	6.2	6.8	7	7.1
26					4.5	5.1	5.7	5.8	6.1
28						4.6	4.7	4.8	5.1
30						3.3	3.9	4	4.2
32							3.2	3.3	3.5
34							2.5	2.7	2.9
36								2.1	2.4
38								1.6	1.9

stability when the crane works in the back area is more excellent than works in two side area. Hence the constraints of slewing angle of each mobile crane can be expressed as follows

$$\begin{cases} -\pi/4 < \beta_1 = ATAN2(x_{A_1}/(D - y_{A_1})) < \pi/4 \\ -\pi/4 < \beta_2 = ATAN2((y_{A_2} - y_{O_2})/(x_{A_2} - x_{O_2})) - \pi/6 < \pi/4 \\ -\pi/4 < \beta_3 = ATAN2((y_{A_3} - y_{O_3})/(x_{O_3} - x_{A_3})) - \pi/6 < \pi/4 \end{cases} \tag{7.25}$$

4. The cable tension. According to Table 7.1, the cable tension should be smaller than the lifting capability when the working radius and boom length are fixed.

$$0 < T_i < [T_i] \tag{7.26}$$

A general numerical workspace generation approach is employed here. The main searching scope is fixed at first. Thus, a series of points in Cartesian coordinate system are determined. According to the Eq. (7.24), a series of the working radius R_i can be calculated. From Eq. (7.25), the slewing angle β_1 of each crane can be calculated. Then, the cable tension T_i can be obtained from Eq. (7.14). At the end, the points are checked and those satisfying all the constraints will generate the constant orientation workspace of CPRMC. The flowchart of generating the constant orientation workspace of the CPRMC is shown in Fig. 7.5.

7.5 Dynamic Loading Capacity Analysis

The last decades have seen the rapid development of world economy. Due to the ever-increasing quantity of cargo, hazardous terrain and weather, some complex tasks are impossible to handle with a single traditional crane, inevitably. The cooperative multiple cranes attracts more attention recently for the redesigning cost is saved. Thus, as more reconfigurable and cost-effective cable parallel crane-type robots, the cooperative multiple cranes have broad application prospects.

Dynamic loading capacity is one of the most important character of the CPRMC, since the CPRMC is designed for carrying large heavy load [10]. A general numerical generation approach for dynamic loading capacity analysis is employed here. The predefined trajectory of the payload is determined at first. Thus the desire trajectory is divided into several elements and a prediction is assumpted for dynamic loading capacity analysis of the CPRMC. Then simulation of the system along the desire trajectory under full-load and none-load working condition, on the basis of the dynamic modeling and trajectory tracking controller. Thus the dynamic loading capacity can be derived based on the torque coefficient and accuracy coefficient. If the error between the simulation and predicted results are acceptable, the simulation value can be defined as the dynamic loading capacity of the CPRMC, otherwise the average value of the simulation and predicted values is calculated for next iteration until the simulation and predicted value are the same with an accepted accuracy.

The linear relation between the motors torque and the load is a required factor for using the algorithm explained for dynamic loading capacity, it can be seen that the same situation is also exists for the CPRMC. Thus the dynamic equation of the CPRMC an be explained [10]

$$\begin{aligned} &\dot{X} = DX; \ddot{X} = D^2X \\ &\Rightarrow rmD^2X + S(X)J(\frac{\partial\beta}{\partial X})D^2X + S(X)(J\frac{d}{dt}(\frac{\partial\beta}{\partial X}) + C(\frac{\partial\beta}{\partial X}))DX = S(X)\tau \end{aligned} \tag{7.27}$$

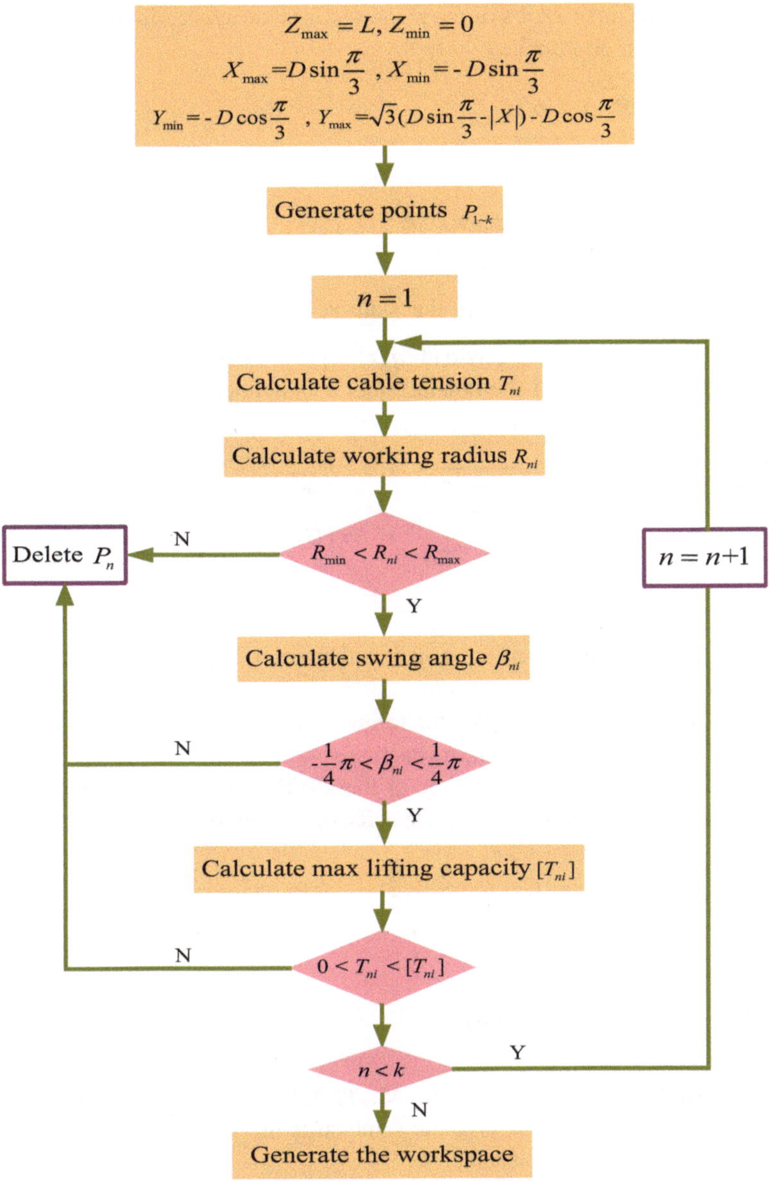

Fig. 7.5 Flowchart of generating the workspace of the CPRMC

where T is the tension of the cables, ψ is the torque exerted by the motors, r is the radius of the pulleys, ψ is the angle of the pulleys, c is the damping coefficient of the motors, J is the rotary inertia of the pulleys, S is the Jacobean matrix and ψ is the angle of the cables.

Considering the three DOFs of the payload, the state space of the dynamic model of the CPRMC can be obtained

$$\begin{cases} if : i = odd : \\ \dot{z_i} = z_{i+1} \\ if : i = even : \\ \dot{z_i} = \left\{ \frac{1}{D + (S^T \frac{J}{r})(\frac{\partial \beta}{\partial X})} (-g - S^T (\frac{1}{r} (\tau - J(\frac{d}{dt} (\frac{\partial \beta}{\partial X}) \dot{X})))) \right. \end{cases} \tag{7.28}$$

where $S = \left[\frac{\partial q_i}{\partial x_j} \right]_{i \times j}$, $z = \{X, \dot{X}\}$ is the state space variables of the system, $X = \{x, y, z\}$ is the position of the payload. $q(x)$ represents the cable length.

The torque coefficient of the hoisting motor is defined as follows

$$\begin{aligned} if(\tau_e)_i \geq 0 \rightarrow (C_a)_i &= \frac{\tau_i^+}{(\tau_e)_i - (\tau_n)_i}; \tau_i^+ = (U^+)_i - (\tau_e)_i, U^+ = \tau_s - \frac{\tau_s}{\omega_n} \dot{q} \\ if(\tau_e)_i \leq 0 \rightarrow (C_a)_i &= \frac{\tau_i^-}{(\tau_e)_i - (\tau_n)_i}; \tau_i^- = (U^-)_i - (\tau_e)_i, U^- = -\tau_s - \frac{\tau_s}{\omega_n} \dot{q} \end{aligned} \tag{7.29}$$

where τ_s is the stall torque of the hoisting motor, ω_n represents the free running angular velocity of the hoisting motor. τ_i^+, τ_i^- are the upper and lower bounds of the motor torque. $(\tau_e)_i$ and $(\tau_n)_i$ are the torque of the hoisting motor under full-load and none-load working condition, respectively.

Thus the dynamic loading capacity of the CPRMC can be defined as the minimum value of the dynamic loading capacity of the motors

$$C = \min\left\{ (C_a)_j \right\}; j = 1, 2, \cdots, 6 \Rightarrow m_{load} = C * m_e \tag{7.30}$$

7.6 Results and Discussion

Simulation study is carried out with MATLAB Version 7.10 software. Three mobile cranes of the CPRMC have the same structure parameters. The parameters of the CPRMC are listed in Table 7.2.

According to the flowchart of generating the workspace of the CPRMC as depicted in Fig. 7.5, the constant orientation workspace can be shown in Figs. 7.6 and 7.7. The whole workspace is approximately an axially symmetric column. The cable tensions become larger and working radiuses of mobile cranes become shorter with the increase of vertical coordinate of the payload. Consequently, the lifting capability of each single mobile crane decreases, according to the indications of Table 7.1, which may cause the result that the upper part of the column is cone conformation as shown in Fig. 7.6.

Table 7.2 Main parameters of the CPRMC

System parameters	Value
Mass of the payload m	1 x 10^4 kg
Acceleration due to gravity g	9.81 m/s^2
Gravity of boom Q	1.8 x 10^5 N
Distance between each mobile crane and the reference point *O* *D*	20 m
Distance between the center of equilateral triangle $A_1{}' A_2{}' A_3{}'$ and the vertex $A_i{}'$ d	8 m
Distance between the hoisting point B and the reference point *O'* s	7 m
Length of boom L	30.4 m
Vertical distance between the lower hinge joints of boom and cylinder *h*	1.5 m
Horizontal distance between the lower hinge joints of boom and cylinder *f*	2.45 m
Horizontal distance between the lower hinge joint of boom and the slewing axle e	0.25 m
Distance between the lower hinge joints and the center of gravity of boom *Lg*	10 m
Distance between the lower hinge joint of boom and the upper hinge joint of cylinder *Ls*	7 m

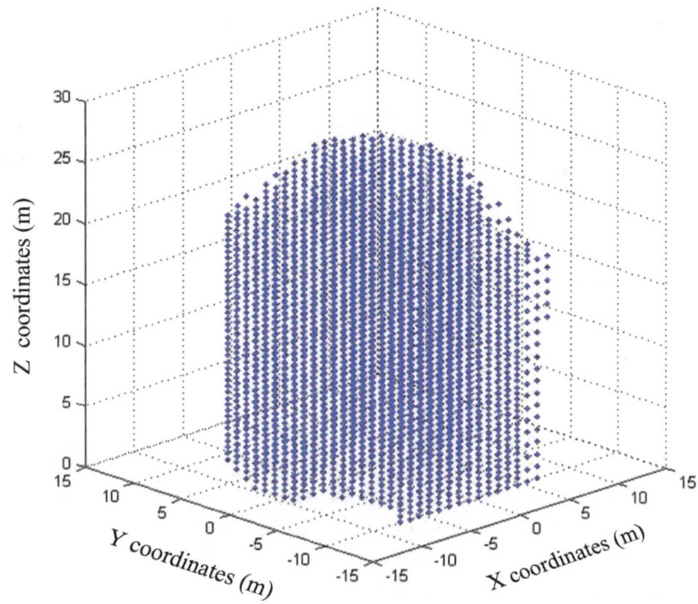

Fig. 7.6 Workspace of the CPRMC

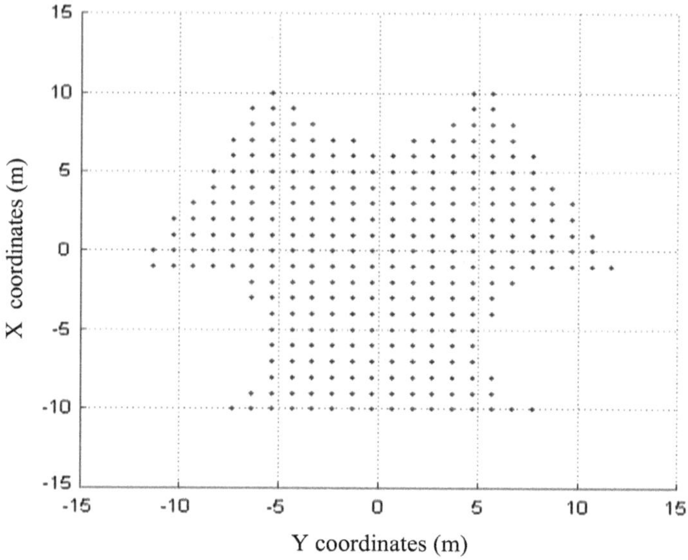

Fig. 7.7 Top view of Workspace of the CPRMC

During the following simulation, the payload of the CPRMC is expected to move along the given spatial trajectory. The spatial trajectory can be expressed as

$$\begin{cases} xp = 4 \times \sin(t) \\ yp = 4 \times \cos(t) \\ zp = 0.5 \times t \\ \alpha = \pi \times t/120 \\ o = \pi \times t/120 \\ n = \pi \times t/120 \end{cases} \quad 0 \le t \le 12(s) \tag{7.27}$$

The payload of the CPRMC moves along the trajectory as shown in Fig. 7.8.

It can be seen from Fig. 7.9 that the cable lengths of mobile cranes l_i showed downtrend with operation time during following the trajectory of the payload, which is reasonable since the payload rises during the simulation. Figs. 7.10 and 7.11 display the curves of slewing and luffing angles of mobile cranes, respectively. From Figs. 7.9, 7.10 and 7.11, it should be noted that the cable lengths as well as the slewing and luffing angles transit smoothly and periodically, obviously this is because that the payload moves along the spiral trajectory.

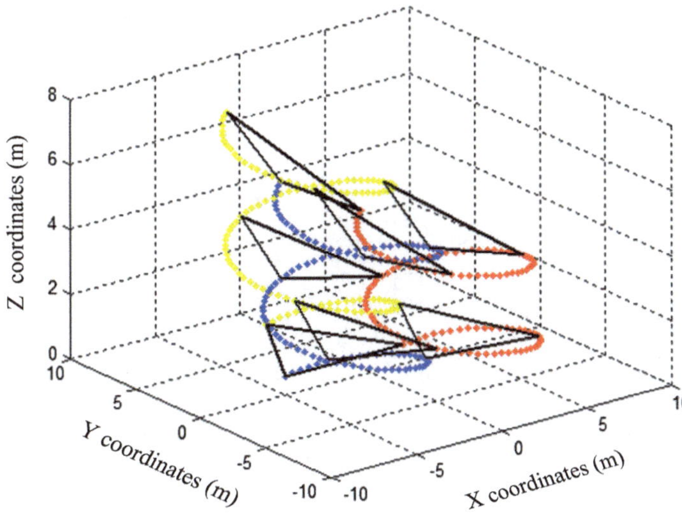

Fig. 7.8 Trajectory of the payload

Fig. 7.12 shows the curves of cable tension exerted on the payload related to the trajectory. One can observe that the tension of each cable waves smoothly nearby 3.5×10^4 N, which is slightly larger than one third of the gravity of the payload. Compared with a single mobile crane, the CPRMC have much more excellent lift capability. In addition, the fluctuation of cable tension increases during the motion.

Figs. 7.13, 7.14 and 7.15 show the curves of vertical component force f_{vi}, the radial component force f_{ri} and the tangential component force f_{ti}, respectively. Referring to Figs. 7.13, 7.14 and 7.15, it can be seen that the vertical component force f_{vi} is the maximum component of cable tension. In addition, the vertical

Fig. 7.9 Cable length of the CPRMC

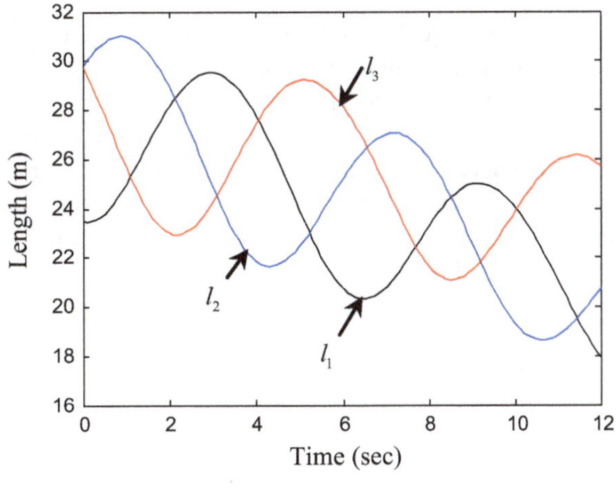

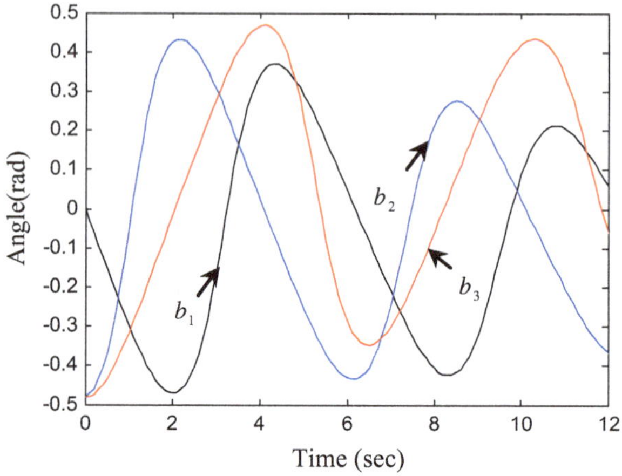

Fig. 7.10 Slewing angle of mobile cranes

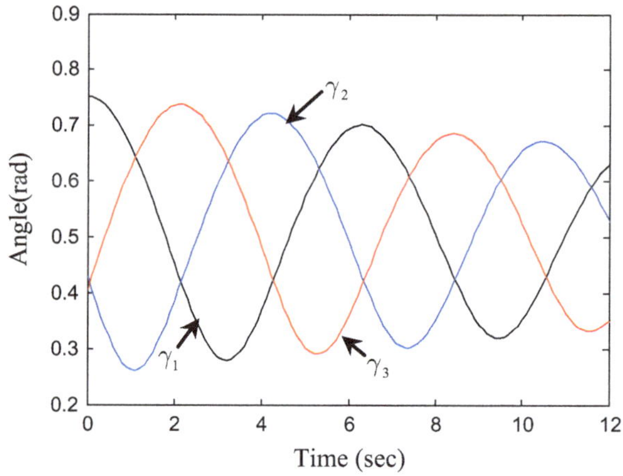

Fig. 7.11 Luffing angle of mobile cranes

component force has a larger fluctuation than any other component force with operation time, which is probably the primary factor for the fluctuation of cable tension T_i as illustrated in Fig. 7.12.

Fig. 7.16 illustrates the cylinder pressure of the mobile cranes. It should be noted that the curves of the cylinder pressure transit periodically in the main. From the simulations, we can see that the cylinder pressure has a close relationship with *xp* and *yp* coordinates of the payload, which are periodic functions of time as expressed in Eq. (7.27).

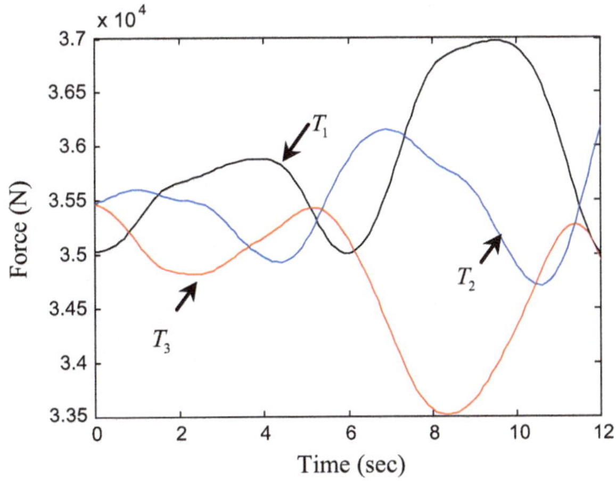

Fig. 7.12 Cable tension

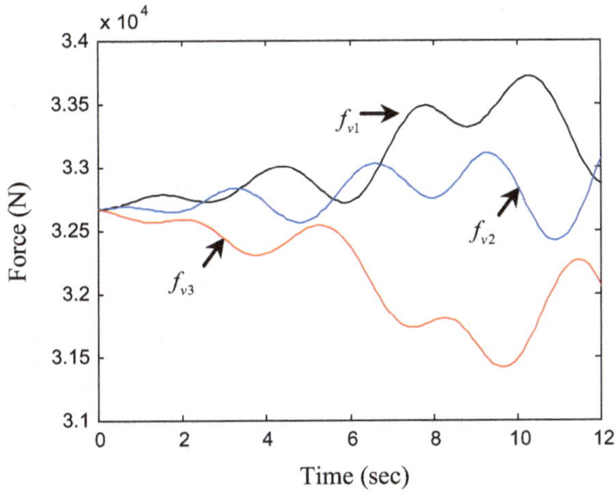

Fig. 7.13 Vertical component of cable tension

Fig. 7.17 displays the curves of the torque acting on slewing subsystem of the CPRMC. As it is seen, the curves of the torque is basically in accordance with the curves of the tangential component force f_{ti} of the cable tension.

The saturation torque of the motors are shown in Figs. 7.18, 7.19 and 7.20. According to the simulational results, the maximum dynamic loading capacity of the CPRMC is 20.4t, which causes the first saturation torque in the second hoisting motor as shown in Fig. 7.19.

Fig. 7.14 Radial component of cable tension

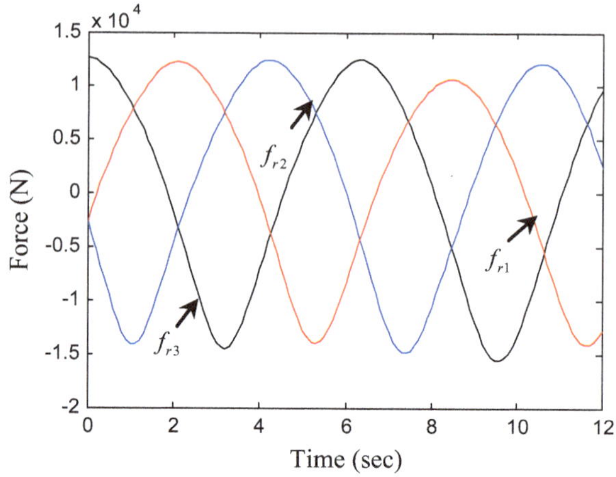

Fig. 7.15 Tangential component of cable tension

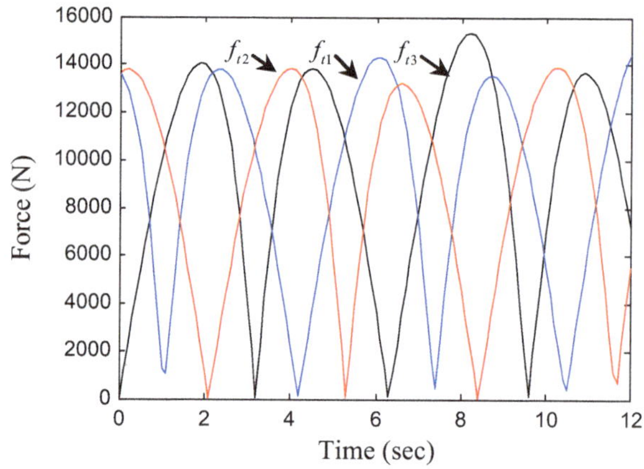

Fig. 7.16 Cylinder pressure of mobile cranes

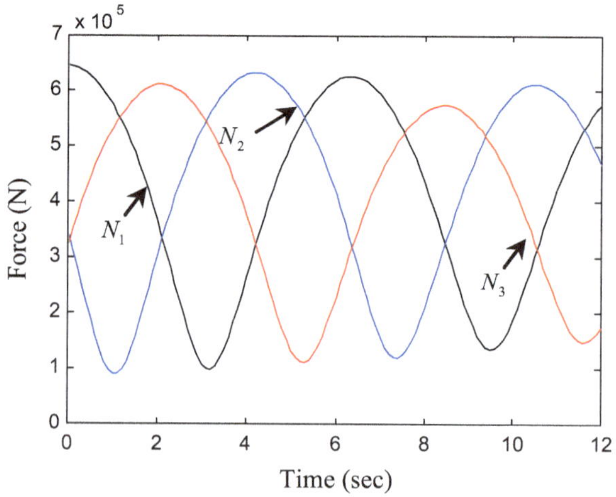

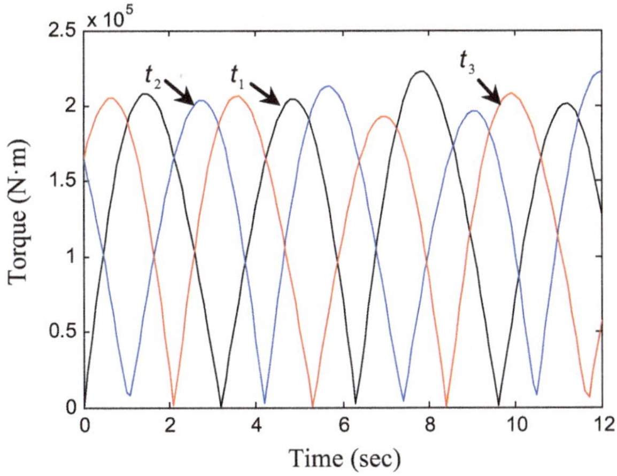

Fig. 7.17 Torque acting on slewing subsystem

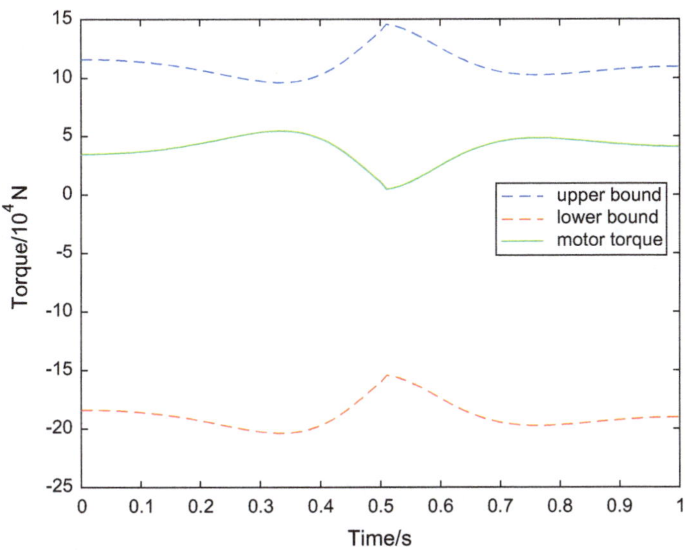

Fig. 7.18 Saturation torque of the first hoisting motor

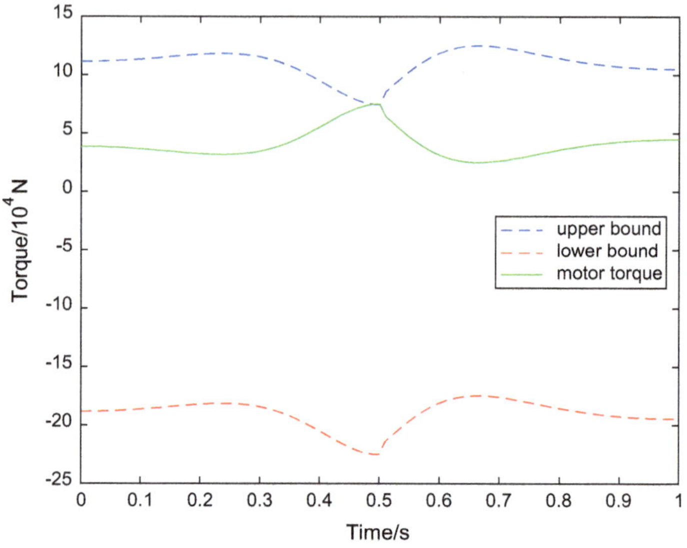

Fig. 7.19 Saturation torque of the second hoisting motor

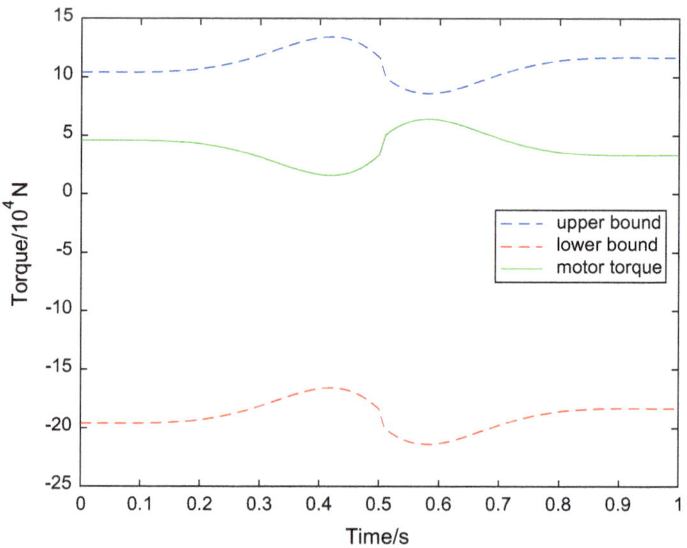

Fig. 7.20 Saturation torque of the third hoisting motor

7.7 Conclusions

This chapter presents the design and analysis of the CPRMC with 6 Degrees of Freedom. Detailed design and implementation of a virtual prototype model of the CPRMC is performed as well. Dynamic modelling of the CPRMC is addressed and presented together with dynamic simulation. The inverse kinematic and dynamic problems of the CPRMC system are resolved on condition that an operation path of the end-effector has been planned. Then, workspace analysis of the CPRMC is provided based on the structure constraints of the CPRMC. Finally, illustrative simulation studies highlight its performances, which lay a foundation for the further research on optimization and real-time control. In addition, the methodology presented in this work could be used as a guide for the design of cooperative multiple manipulators.

References

1. Kłosiński J (2005) Swing-free stop control of the slewing motion of a mobile crane. Control Eng Pract 13(4):451–460
2. Zi B, Qian S, Ding H et al (2012) Design and analysis of cooperative cable parallel manipulators for multiple mobile cranes. Int J Adv Rob Syst 9
3. Caccavale F, Chiacchio P, Marino A et al (2008) Six-DOF impedance control of dual-arm cooperative manipulators. IEEE/ASME Trans Mechatron 13(5):576–586
4. Basile F, Caccavale F, Chiacchio P et al (2012) Task-oriented motion planning for multi-arm robotic systems. Robot Comput-Integr Manuf 28(5):569–582
5. Mehrjerdi H, Saad M, Ghommam J (2011) Hierarchical fuzzy cooperative control and path following for a team of mobile robots. IEEE/ASME Trans Mechatron 16(5):907–917
6. Zi B, Sun H, Zhu Z et al (2012) The dynamics and sliding mode control of multiple cooperative welding robot manipulators. Int J Adv Robot Syst 9
7. Renzaglia A, Doitsidis L, Martinelli A et al (2012) Multi-robot three dimensional coverage of unknown areas. Int J Robot Res 0278364912439332
8. Otis MJD, Perreault S, Nguyen-Dang TL et al (2009) Determination and management of cable interferences between two 6-DOF foot platforms in a cable-driven locomotion interface. IEEE Trans Syst Man Cybern-Part A Syst Hum 39(3):528–544
9. Zi B, Zhu Z, Du J (2011) Analysis and control of the cable-supporting system including actuator dynamics. Control Eng Pract 19(5):491–501
10. Korayem MH, Tourajizadeh H, Bamdad M (2010) Dynamic load carrying capacity of flexible cable suspended robot: robust feedback linearization control approach. J Intel Robot Syst 60 (3–4):341–363
11. Duan BY, Qiu YY, Zhang FS et al (2009) On design and experiment of the feed cable-suspended structure for super antenna. Mechatronics 19(4):503–509
12. Pham CB, Yeo SH, Yang G et al (2009) Workspace analysis of fully restrained cable-driven manipulators. Robot Auton Syst 57(9):901–912
13. Lau D, Oetomo D, Halgamuge SK (2011) Wrench-closure workspace generation for cable driven parallel manipulators using a hybrid analytical-numerical approach. J Mech Des 133 (7):071004
14. Taniguchi S, Kino H, Ozawa R et al (2012) Inverse dynamics of human passive motion based on iterative learning control. IEEE Trans Syst Man Cybern-Part A Syst Hum 42(2):307–315

15. Baser O, Konukseven EI (2010) Theoretical and experimental determination of capstan drive slip error. Mech Mach Theory 45(6):815–827
16. Fahham HR, Farid M, Khooran M (2011) Time optimal trajectory tracking of redundant planar cable-suspended robots considering both tension and velocity constraints. J Dyn Syst Meas Control 133(1):011004

Chapter 8
Obstacle Avoidance Planning of the Cooperative Cable Parallel Manipulators for Multiple Mobile Cranes

8.1 Preamble

This chapter addresses the cooperative problems in terms of localization, obstacle avoidance planning and automatic leveling control for cable parallel robots for multiple mobile cranes (CPRMC). The design model of the CPRMC is elaborated on. The three-dimensional grid map method is utilized to plot the environment map based on the operation environment model. Combined the relative localization method with the absolute localization method, a cooperative localization scheme of the CPRMC is developed, and an improved localization algorithm is designed on the basis of multilateration method. Then, according to the artificial potential field method with the grid method, a global path planning of the CPRMC is performed, and consider the possible collision of the single mobile crane, the sensor technology is applied to the cooperative obstacle avoidance. In addition, a four-point collaborative leveling method is adopted for automatic leveling control of the platform of the CPRMC. Finally, the effectiveness of the CPRMC system is verified through simulations.

With the fast development of the manufacturing technology, mobile cranes are playing a more and more important role in different modern construction projects due to their excellent mobility performance and load carrying capacity, such as loading, mounting, carrying large heavy-duty loads for the construction of the factory, railway, port, mine, power equipment, etc. The increasing requirements for construction quality, safety and efficiency create new challenges for the comprehensive performance and lifting strategy of single mobile cranes. In recent years, cooperative multiple robotics are an active research domain, in which the problem of the implementation of cooperative manipulation on conventional set-ups has attracted an increasing interest of the researchers [1].

As a matter of fact, cooperative operation of multiple mobile crane cranes has become more common in modern construction projects, such as the one shown in Fig. 8.1, which can be treated as the CPRMC. Therefore, the CPRMC can take the

B. Zi and S. Qian, *Design, Analysis and Control of Cable-Suspended Parallel Robots and Its Applications*, DOI 10.1007/978-981-10-1753-7_8

Fig. 8.1 Multiple mobile cranes

advantages of the characteristics of both mobile cranes and cable parallel robots (CPRs).

For the purpose of designing a cost-effective, safe, and compliant solution for the operation of the CPRMC, there are several key problems to be solved. In particular, cooperative localization technology is fundamental to cooperative robotics within the unknown environment [2], which has been widely used in different applications including but not limited to tracking automated guided vehicle, robot navigating and large scale metrology [3–5]. Considering the positioning control problem in robot applications, localization is meaningful for data aggregation, velocity estimation, and geographic-aware routing, so robots cannot effectively carry out a task without the location information [6, 7]. Then, based on cooperative localization, the CPRMC are expected to traverse the path to desired goals within an obstacle-ridden environment [8]. Taking the safety factors for the CPRMC into consideration, there are two crucial requirements in terms of obstacle avoidance planning and automatic leveling control. Thus, the robots can avoid collision with obstacles that exist in its environment [9–11]. In addition, automatic leveling control can help the CPRMC to prevent the overturning of cargo platform caused by the asynchronous of each single mobile crane, and keep the stability and improve the accuracy of the whole system [12–14].

The major contribution of this chapter includes three aspects: (1) Design model and cooperative localization scheme of the CPRMC are presented, and the improved localization algorithm based on multilateration method is designed; (2) The global path planning of the CPRMC is performed based on the artificial potential field method with the grid method, and the sensor technology is applied to the cooperative obstacle avoidance, and combined with Matlab and LabVIEW, a co-simulation platform is built for cooperative obstacle avoidance analysis; (3) A four-point collaborative leveling method is adopted for automatic leveling control of the platform for the CPRMC.

The rest of this chapter is structured as follows. Section 8.2 briefly presents the design model of the CPRMC. Then, localization analysis of the CPRMC are reported in Sect. 8.3. Sections 8.4 and 8.5 describe path planning and obstacle avoidance, and automatic leveling control of the CPRMC, respectively. Finally, the last section summarizes this chapter.

8.2 Design Model

The CPRMC mainly consist of mobile cranes, cable driving devices, platform, cables and booms. The mobile cranes have an excellent mobility performance and load carrying capacity. The booms can implement luffing, slewing and telescoping action, which expand the operating radius of the mobile cranes. The platform moves along the workspace path by controlling the booms' action and the four-cable lengths, and four cables are linked to the platform, which has four degrees of freedom (see Fig. 8.2).

The structure of the platform is shown in Fig. 8.3. The platform is composed of the uprights, lugs, counterweights, top plate and bottom plate. The lugs are connected to the cables and top plate, and fixed at four vertices of the top plate. Four counterweights are placed at four vertices of the bottom plate, which are utilized to adjust the centroid of the platform.

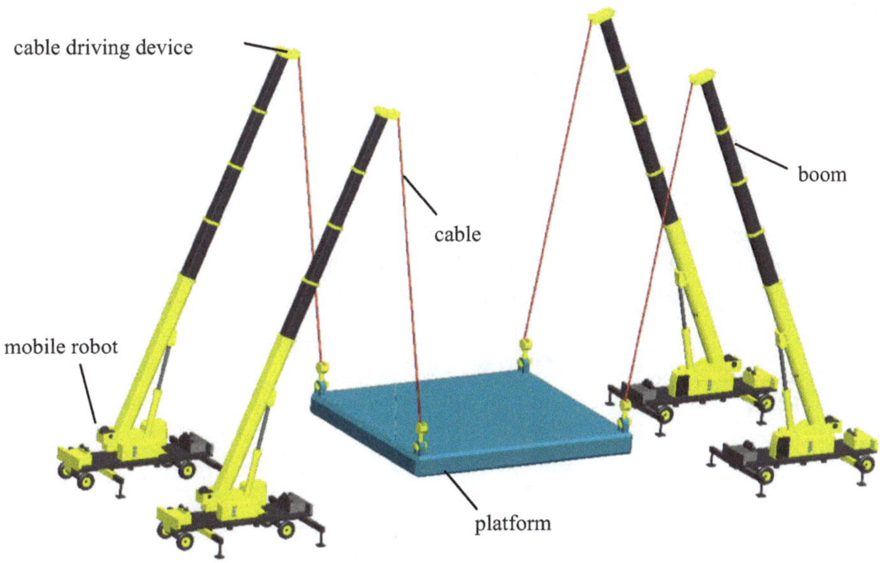

Fig. 8.2 Three-dimensional model of hoisting multi-mobile robots system

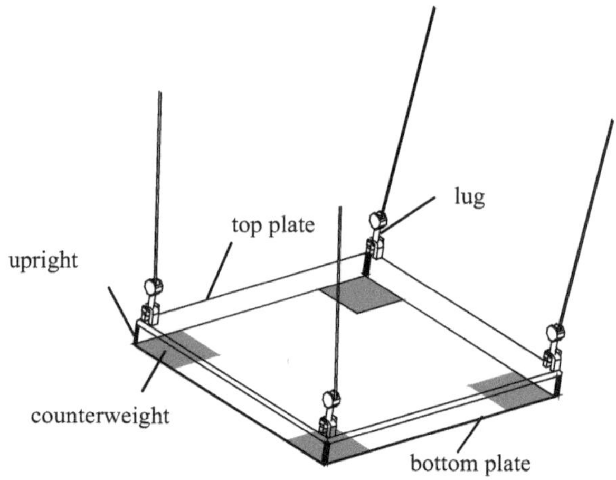

Fig. 8.3 Structure of the platform

8.3 Localization Analysis

In the following discussion, we will describe kinematics model of the CPRMC, map characterization of the operation environment, and the improved localization algorithm.

8.3.1 Localization

The pose information of the CPRMC can be represented as follows

$$p = (x, y, z, \theta)$$

where OXYZ is a Cartesian coordinate frame fixed to the base, (x, y, z) represents the coordinate of the geometric center of the mobile robots, and θ denotes the posture direction of the robots. The kinematics model of the CPRMC in the coordinate system OXYZ is shown in Fig. 8.4.

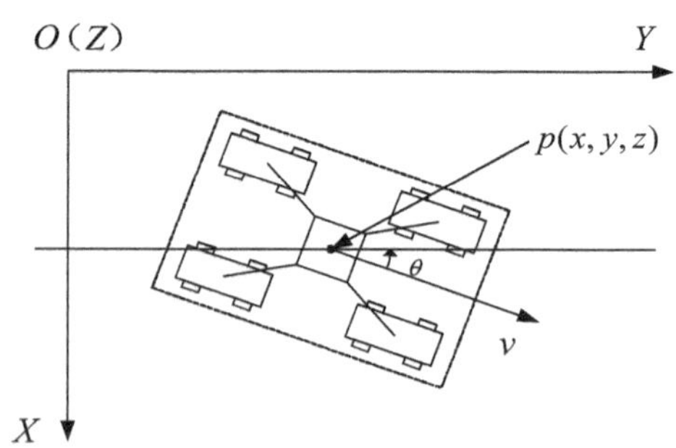

Fig. 8.4 Kinematics model of the CPRMC

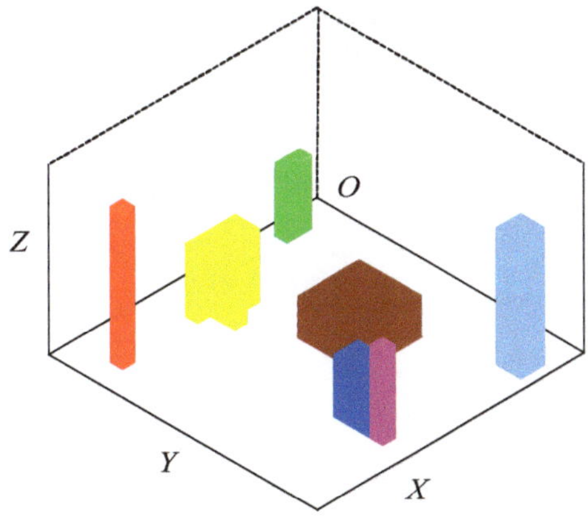

Fig. 8.5 Three-dimensional characteristic pattern of the working environment

In order to describe the various characteristics of the CPRMC in the environment, the characteristic pattern of the working environment should be considered. The environmental information is abstracted as a feature model so as to further represent the map characterization. The three-dimensional grid method is adopted in this section.

Figures 8.5 and 8.6 show three-dimensional characteristic pattern and XOY plane characteristic pattern of the working environment, respectively. Three-dimensional grid map and XOY plane grid map of the working environment are presented in Figs. 8.7 and 8.8. From Figs. 8.7 and 8.8, it can be seen that, the colorless squares represent the free space, while the colored squares show the obstacle space. The value in Z-axis direction within the unit square, represents the height information of the environment space. The environmental characteristics are discretized into a large number of grids based on modeling the operating environment as a grid map. Each grid covers the corresponding area of the environment. If the ratio of the individual grid that occupies by the obstacle is equal or greater

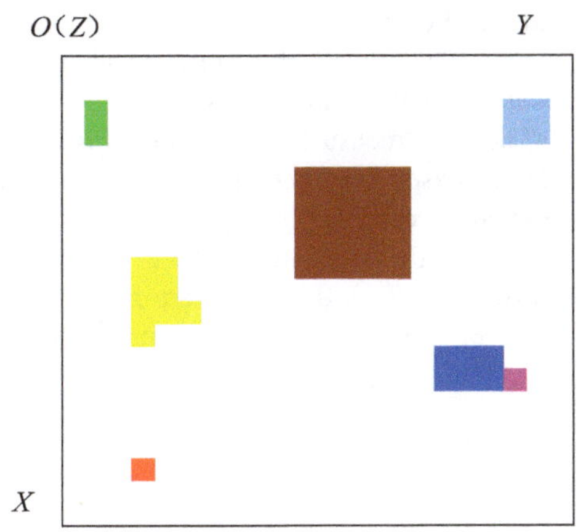

Fig. 8.6 *XOY* plane characteristic pattern of the working environment

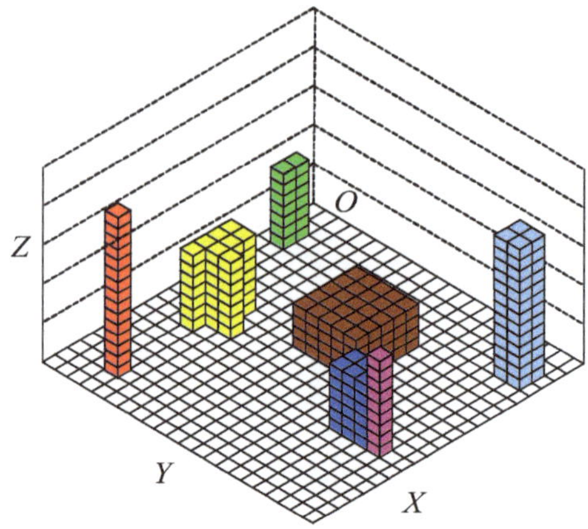

Fig. 8.7 Three-dimensional grid map of the working environment

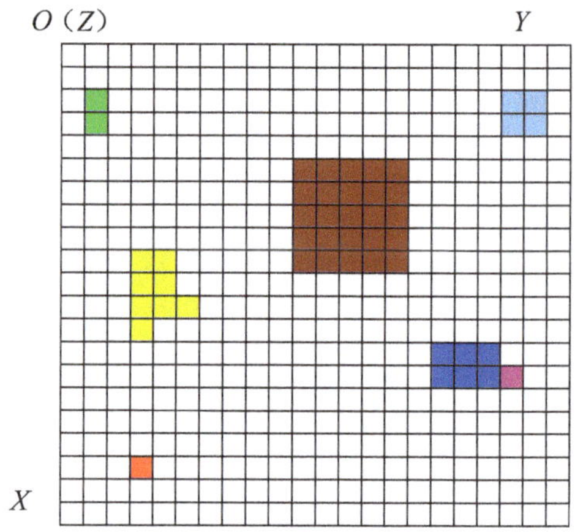

Fig. 8.8 *XOY* plane grid map of the working environment

than 50%, the grid area is regarded as obstacle space, otherwise free space. Any grid can be expressed as $Grid_i = (x_i, y_i, z_i)$ with respect to the global coordinate system. The higher the unit coordinates resolution of the grid is, the more the accuracy of the grid map is.

The pose measurement of the CPRMC can be realized with relative localization method based on fiber optic gyroscope. The fiber optic gyroscope is fixed at the geometric center of the system. In the global coordinate system, the CPRMC is located at the starting position, when the moving speed v = 0 and the posture information $\theta_0 = 0$. The rotational angle change of the fiber optic gyroscope is $\Delta\theta_i = \theta_i - \theta_{(i-1)}$ in the *i*th sampling period, and the posture information of the CPRMC is given as

$$\theta_n = \sum_{i=0}^{n} \Delta\theta_i \tag{8.1}$$

The multi-ultrasonic sensor information fusion network is made up of several group of ultrasonic transmitters and receivers. As shown in Fig. 8.9, the ultrasonic transmitters are fixed regularly in the global coordinate system, and the corresponding ultrasonic receivers are fixed around a rotary table, which is actuated by a constant speed motor installed on the bottom of the platform of the CPRMC.

The distance between the *i*th ultrasonic receiver and the ultrasonic transmitter D_i can be shown by $D_i = 1/2C \cdot \Delta t_i$. Here, C is the propagation velocity of the ultrasound, which can also be expressed by $C = 331.4\sqrt{1 + T/273}$ (m/s). T is the ambient temperature and Δt_i represents the time interval of the *i*th ultrasonic receiver and transmitter.

From the aforementioned methodology, the orientation information of the mobile crane is θ_n. The distances between the transmitter and the receiver measured by the multi-ultrasonic sensor information fusion network, are $D_1, D_2, D_3, \ldots, D_n$, respectively. The coordinates of the receivers in the global coordinate frame are (x_1, y_1, z_1), (x_2, y_2, z_2), (x_3, y_3, z_3), …, (x_n, y_n, z_n), respectively.

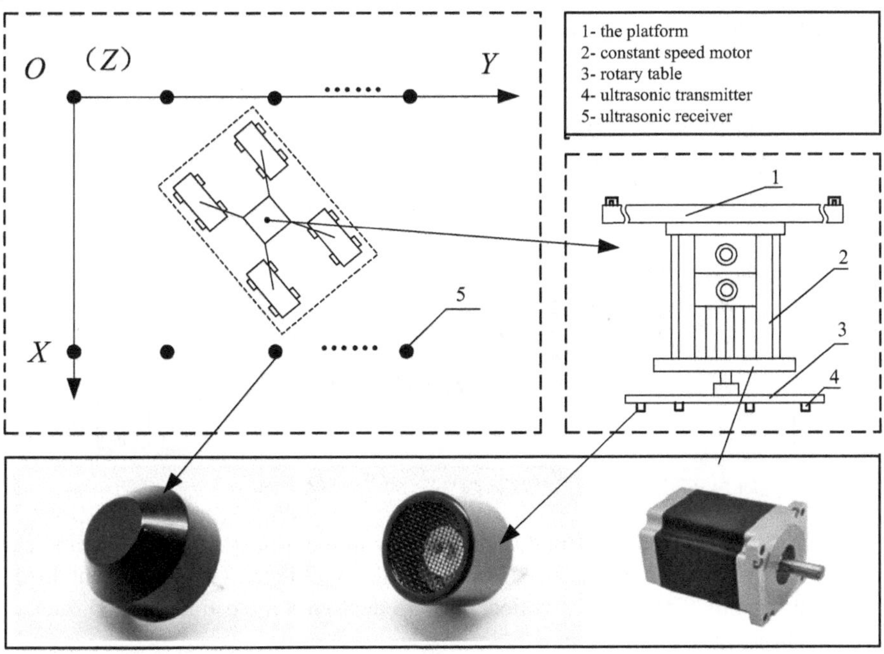

Fig. 8.9 Ultrasonic transmitter and receiver

On the basis of the multilateral measurement localization algorithm [13], the distances between the estimated position of mobile crane and the points (x_1, y_1, z_1), (x_2, y_2, z_2), (x_3, y_3, z_3), …, (x_n, y_n, z_n), are given as

$$\begin{cases} D_1 = \sqrt{(x_1 - x_m)^2 + (y_1 - y_m)^2 + (z_1 - z_m)^2} \\ D_2 = \sqrt{(x_2 - x_m)^2 + (y_2 - y_m)^2 + (z_2 - z_m)^2} \\ D_3 = \sqrt{(x_3 - x_m)^2 + (y_3 - y_m)^2 + (z_3 - z_m)^2} \\ \vdots \\ D_n = \sqrt{(x_n - x_m)^2 + (y_n - y_m)^2 + (z_n - z_m)^2} \end{cases} \tag{8.2}$$

where (x_m, y_m, z_m) is the estimated position coordinates.

Using Eq. (8.2), one can write

$$\begin{cases} x_1^2 - x_n^2 - 2(x_1 - x_n)x_m + y_1^2 - y_n^2 - 2(y_1 - y_n)y_m + z_1^2 - z_n^2 - 2(z_1 - z_n)z_m = D_1^2 - D_n^2 \\ x_2^2 - x_n^2 - 2(x_2 - x_n)x_m + y_2^2 - y_n^2 - 2(y_2 - y_n)y_m + z_2^2 - z_n^2 - 2(z_2 - z_n)z_m = D_2^2 - D_n^2 \\ \vdots \\ x_{n-1}^2 - x_n^2 - 2(x_{n-1} - x_n)x_m + y_{n-1}^2 - y_n^2 - 2(y_{n-1} - y_n)y_m + z_{n-1}^2 - z_n^2 - 2(z_{n-1} - z_n)z_m = D_{n-1}^2 - D_n^2 \end{cases} \tag{8.3}$$

Substituting Eq. (8.2) into Eq. (8.3), the linear equation sets can be expressed as in matrix form, $AX = b$, in which

$$\begin{aligned} \boldsymbol{A} &= \begin{bmatrix} 2(x_1 - x_n) & 2(y_1 - y_n) & 2(z_1 - z_n) \\ 2(x_2 - x_n) & 2(y_2 - y_n) & 2(z_2 - z_n) \\ 2(x_3 - x_n) & 2(y_3 - y_n) & 2(z_3 - z_n) \\ \vdots & \vdots & \vdots \\ 2(x_{n-1} - x_n) & 2(y_{n-1} - y_n) & 2(z_{n-1} - z_n) \end{bmatrix} \\ \boldsymbol{b} &= \begin{bmatrix} x_1^2 - x_n^2 + y_1^2 - y_n^2 + z_1^2 - z_n^2 + D_n^2 - D_1^2 \\ x_2^2 - x_n^2 + y_2^2 - y_n^2 + z_2^2 - z_n^2 + D_n^2 - D_2^2 \\ x_3^2 - x_n^2 + y_3^2 - y_n^2 + z_3^2 - z_n^2 + D_n^2 - D_3^2 \\ \vdots \\ x_{n-1}^2 - x_n^2 + y_{n-1}^2 - y_n^2 + z_{n-1}^2 - z_n^2 + D_n^2 - D_{n-1}^2 \end{bmatrix}, \quad \boldsymbol{X} = \begin{bmatrix} x_m \\ y_m \\ z_m \end{bmatrix} \end{aligned} \tag{8.4}$$

Hence, the estimated coordinates (x_m, y_m, z_m) in the global coordinate frame can be obtained. However, there are errors with the multilateral measurement localization algorithm. Therefore, it is necessary to deal with the estimated coordinates. The function can be defined as

$$\begin{cases} \Delta l_1 = D_1' - \sqrt{(x_1 - x_m)^2 + (y_1 - y_m)^2 + (z_1 - z_m)^2} \\ \Delta l_2 = D_2' - \sqrt{(x_2 - x_m)^2 + (y_2 - y_m)^2 + (z_2 - z_m)^2} \\ \Delta l_3 = D_3' - \sqrt{(x_3 - x_m)^2 + (y_3 - y_m)^2 + (z_3 - z_m)^2} \\ \vdots \\ \Delta l_n = D_n' - \sqrt{(x_n - x_m)^2 + (y_n - y_m)^2 + (z_n - z_m)^2} \end{cases} \tag{8.5}$$

where Δl_i is the difference between the measured distance and calculated distance. For decreasing the errors, Δl_i should be taken as small as possible. Thus, we can get

$$F(x_m, y_m, z_m) = \mathrm{Min}(\sum_{i=1}^{n} \Delta l_i) \tag{8.6}$$

$$\begin{cases} D_1' = \sqrt{(x_1 - x_r)^2 + (y_1 - y_r)^2 + (z_1 - z_r)^2} \\ D_2' = \sqrt{(x_2 - x_r)^2 + (y_2 - y_r)^2 + (z_2 - z_r)^2} \\ D_3' = \sqrt{(x_3 - x_r)^2 + (y_3 - y_r)^2 + (z_3 - z_r)^2} \\ \vdots \\ D_n' = \sqrt{(x_n - x_r)^2 + (y_n - y_r)^2 + (z_n - z_r)^2} \end{cases} \tag{8.7}$$

Equation (8.7) is the nonlinear equations, where (x_r, y_r, z_r) is the exact coordinates of the ultrasonic receivers. D_i' is the distance between the exact coordinates and the known coordinates of the ultrasonic receivers.

The error between the estimate coordinates (x_m, y_m, z_m) and the exact coordinates is $(\Delta x_{r-m}, \Delta y_{r-m}, \Delta z_{r-m})$. Therefore, the exact coordinates of the system in the global coordinate frame can be obtained as

$$\begin{cases} x_r = x_m + \Delta x_{r-m} \\ y_r = y_m + \Delta y_{r-m} \\ z_r = z_m + \Delta z_{r-m} \end{cases} \tag{8.8}$$

The offset between the exact position and the estimated position can be described as

$$F_1(x_r, y_r, z_r) = F_1(x_m + \Delta x_{r-m}, y_m + \Delta y_{r-m}, z_m + \Delta z_{r-m}) \tag{8.9}$$

The following equation can be derived by using Taylor Series Expansion

$$
\begin{aligned}
& F_1(x_m + \Delta x_{r-m}, y_m + \Delta y_{r-m}, z_m + \Delta z_{r-m}) \\
& \approx F_1(x_m, y_m, z_m) + \frac{\partial F_1(x_m, y_m, z_m)}{\partial x_m} \cdot \Delta x_{r-m} + \frac{\partial F_1(x_m, y_m, z_m)}{\partial y_m} \cdot \Delta y_{r-m} \\
& + \frac{\partial F_1(x_m, y_m, z_m)}{\partial z_m} \cdot \Delta z_{r-m}
\end{aligned} \tag{8.10}
$$

Linearizing Eq. (8.10) with respect to x_m, y_m and z_m, one can obtain

$$
\begin{cases}
\frac{\partial F_1(x_m, y_m, z_m)}{\partial x_m} = -\frac{x_i - x_m}{\sqrt{(x_i - x_m)^2 + (y_i - y_m)^2 + (z_i - z_m)^2}} \\
\frac{\partial F_1(x_m, y_m, z_m)}{\partial y_m} = -\frac{y_i - y_m}{\sqrt{(x_i - x_m)^2 + (y_i - y_m)^2 + (z_i - z_m)^2}} \\
\frac{\partial F_1(x_m, y_m, z_m)}{\partial z_m} = -\frac{z_i - z_m}{\sqrt{(x_i - x_m)^2 + (y_i - y_m)^2 + (z_i - z_m)^2}}
\end{cases} \tag{8.11}
$$

Substituting Eqs. (8.9)–(8.11) into Eq. (8.8) leads to

$$
D_i = D_i' - \frac{x_i - x_m}{D_i'} \cdot \Delta x_{r-m} - \frac{y_i - y_m}{D_i'} \cdot \Delta y_{r-m} - \frac{z_i - z_m}{D_i'} \cdot \Delta z_{r-m} \tag{8.12}
$$

which can be rewritten as

$$
-\Delta l_i = D_i' - D_i = \frac{x_i - x_m}{D_i'} \cdot \Delta x_{r-m} + \frac{y_i - y_m}{D_i'} \cdot \Delta y_{r-m} + \frac{z_i - z_m}{D_i'} \cdot \Delta z_{r-m} \tag{8.13}
$$

For simplicity, the following assumptions were made

$$
\begin{cases}
\Delta D_i = -\Delta l_i \\
a = \frac{x_i - x_m}{D_i'} \\
b = \frac{y_i - y_m}{D_i'} \\
c = \frac{z_i - z_m}{D_i'}
\end{cases} \tag{8.14}
$$

Equation (8.13) can also be rewritten as

$$
\Delta D_i = a \cdot \Delta x_{r-m} + b \cdot \Delta y_{r-m} + c \cdot \Delta z_{r-m} \tag{8.15}
$$

Thus

$$
\begin{cases}
\Delta D_1 = a_1 \cdot \Delta x_{r-m} + b_1 \cdot \Delta y_{r-m} + c_1 \cdot \Delta z_{r-m} \\
\Delta D_2 = a_2 \cdot \Delta x_{r-m} + b_2 \cdot \Delta y_{r-m} + c_2 \cdot \Delta z_{r-m} \\
\Delta D_3 = a_3 \cdot \Delta x_{r-m} + b_3 \cdot \Delta y_{r-m} + c_3 \cdot \Delta z_{r-m} \\
\vdots \\
\Delta D_n = a_n \cdot \Delta x_{r-m} + b_n \cdot \Delta y_{r-m} + c_n \cdot \Delta z_{r-m}
\end{cases} \tag{8.16}
$$

To solve Eq. (8.16), the following function can be defined as

$$F_2(\Delta x_{r-m}, \Delta y_{r-m}, \Delta z_{r-m}) = \sum_{i=1}^{n} [\Delta D_i - (a_i \cdot \Delta x_{r-m} + b_i \cdot \Delta y_{r-m} + c_i \cdot \Delta z_{r-m})]^2 \tag{8.17}$$

The derivation of $F_2(\Delta x_{r-m}, \Delta y_{r-m}, \Delta z_{r-m})$ can be expressed as

$$\frac{\partial F_2(\Delta x_{r-m}, \Delta y_{r-m}, \Delta z_{r-m})}{\partial \Delta x_{r-m}} = \sum_{i=1}^{n} a_i \cdot [\Delta D_i - (a_i \Delta x_{r-m} + b_i \cdot \Delta y_{r-m} + c_i \cdot \Delta z_{r-m})] \tag{8.18}$$

$$\frac{\partial F_2(\Delta x_{r-m}, \Delta y_{r-m}, \Delta z_{r-m})}{\partial \Delta x_{r-m}} = \sum_{i=1}^{n} b_i \cdot [\Delta D_i - (a_i \Delta x_{r-m} + b_i \cdot \Delta y_{r-m} + c_i \cdot \Delta z_{r-m})] \tag{8.19}$$

$$\frac{\partial F_2(\Delta x_{r-m}, \Delta y_{r-m}, \Delta z_{r-m})}{\partial \Delta x_{r-m}} = \sum_{i=1}^{n} c_i \cdot [\Delta D_i - (a_i \Delta x_{r-m} + b_i \cdot \Delta y_{r-m} + c_i \cdot \Delta z_{r-m})] \tag{8.20}$$

For obtaining the minimum value of the function $F_2(\Delta x_{r-m}, \Delta y_{r-m}, \Delta z_{r-m})$, assuming that Eqs. (8.18)–(8.20) are 0, the following equations can be listed as

$$\sum_{i=1}^{n} a_i \cdot [\Delta D_i - (a_i \Delta x_{r-m} + b_i \cdot \Delta y_{r-m} + c_i \cdot \Delta z_{r-m})] = 0 \tag{8.21}$$

$$\sum_{i=1}^{n} b_i \cdot [\Delta D_i - (a_i \Delta x_{r-m} + b_i \cdot \Delta y_{r-m} + c_i \cdot \Delta z_{r-m})] = 0 \tag{8.22}$$

$$\sum_{i=1}^{n} c_i \cdot [\Delta D_i - (a_i \Delta x_{r-m} + b_i \cdot \Delta y_{r-m} + c_i \cdot \Delta z_{r-m})] = 0 \tag{8.23}$$

Then, we obtain

$$\begin{cases} \sum_{i=1}^{n} a_i \cdot \Delta D_i = \Delta x_{r-m} \cdot \sum_{i=1}^{n} a_i^2 + \Delta y_{r-m} \cdot \sum_{i=1}^{n} a_i b_i + \Delta z_{r-m} \cdot \sum_{i=1}^{n} a_i c_i \\ \sum_{i=1}^{n} b_i \cdot \Delta D_i = \Delta x_{r-m} \cdot \sum_{i=1}^{n} a_i b_i + \Delta y_{r-m} \cdot \sum_{i=1}^{n} b_i^2 + \Delta z_{r-m} \cdot \sum_{i=1}^{n} b_i c_i \\ \sum_{i=1}^{n} c_i \cdot \Delta D_i = \Delta x_{r-m} \cdot \sum_{i=1}^{n} a_i c_i + \Delta y_{r-m} \cdot \sum_{i=1}^{n} b_i c_i + \Delta z_{r-m} \cdot \sum_{i=1}^{n} c_i^2 \end{cases} \tag{8.24}$$

The linear matrix can be derived as

$$\begin{bmatrix} \sum_{i=1}^{n} a_i \cdot \Delta D_i \\ \sum_{i=1}^{n} b_i \cdot \Delta D_i \\ \sum_{i=1}^{n} c_i \cdot \Delta D_i \end{bmatrix} = \begin{bmatrix} \sum_{i=1}^{n} a_i^2 & \sum_{i=1}^{n} a_i b_i & \sum_{i=1}^{n} a_i c_i \\ \sum_{i=1}^{n} a_i b_i & \sum_{i=1}^{n} b_i^2 & \sum_{i=1}^{n} b_i c_i \\ \sum_{i=1}^{n} a_i b_i & \sum_{i=1}^{n} b_i c_i & \sum_{i=1}^{n} c_i^2 \end{bmatrix} \cdot \begin{bmatrix} \Delta x_{r-m} \\ \Delta y_{r-m} \\ \Delta z_{r-m} \end{bmatrix} \tag{8.25}$$

Hence, the offset $(\Delta x_{r-m}, \Delta y_{r-m}, \Delta z_{r-m})$ can be expressed as

$$\begin{bmatrix} \Delta x_{r-m} \\ \Delta y_{r-m} \\ \Delta z_{r-m} \end{bmatrix} = \left(\begin{bmatrix} \sum_{i=1}^{n} a_i^2 & \sum_{i=1}^{n} a_i b_i & \sum_{i=1}^{n} a_i c_i \\ \sum_{i=1}^{n} a_i b_i & \sum_{i=1}^{n} b_i^2 & \sum_{i=1}^{n} b_i c_i \\ \sum_{i=1}^{n} a_i b_i & \sum_{i=1}^{n} b_i c_i & \sum_{i=1}^{n} c_i^2 \end{bmatrix}^T \begin{bmatrix} \sum_{i=1}^{n} a_i^2 & \sum_{i=1}^{n} \alpha_i b_i & \sum_{i=1}^{n} a_i c_i \\ \sum_{i=1}^{n} a_i b_i & \sum_{i=1}^{n} b_i^2 & \sum_{i=1}^{n} b_i c_i \\ \sum_{i=1}^{n} a_i c_i & \sum_{i=1}^{n} b_i c_i & \sum_{i=1}^{n} c_i^2 \end{bmatrix} \right)^{-1}$$
$$\cdot \begin{bmatrix} \sum_{i=1}^{n} a_i^2 & \sum_{i=1}^{n} a_i b_i & \sum_{i=1}^{n} a_i c_i \\ \sum_{i=1}^{n} a_i b_i & \sum_{i=1}^{n} b_i^2 & \sum_{i=1}^{n} b_i c_i \\ \sum_{i=1}^{n} a_i c_i & \sum_{i=1}^{n} b_i c_i & \sum_{i=1}^{n} c_i^2 \end{bmatrix}^T \cdot \begin{bmatrix} \sum_{i=1}^{n} a_i \cdot \Delta D_i \\ \sum_{i=1}^{n} b_i \cdot \Delta D_i \\ \sum_{i=1}^{n} c_i \cdot \Delta D_i \end{bmatrix} \tag{8.26}$$

In general, the mobile cranes are operated on the XOY plane and there is almost no displacement in the Z direction. The error in the Z direction is ignored for simplification. Thus, the error $(\Delta x_{r-m}, \Delta y_{r-m}, \Delta z_{r-m})$ between the estimated coordinates (x_m, y_m, z_m) and the exact coordinates (x_r, y_r, z_r) can be simplified as follows

$$\begin{cases} \sum_{i=1}^{n} a_i \cdot \Delta D_i = \Delta x_{r-m} \cdot \sum_{i=1}^{n} a_i^2 + \Delta y_{r-m} \cdot \sum_{i=1}^{n} a_i b_i \\ \sum_{i=1}^{n} b_i \cdot \Delta D_i = \Delta x_{r-m} \cdot \sum_{i=1}^{n} a_i b_i + \Delta y_{r-m} \cdot \sum_{i=1}^{n} b_i^2 \end{cases} \tag{8.27}$$

From Eq. (8.27), one has

$$\begin{cases} \Delta x_{r-m} = \dfrac{\sum_{i=1}^{n} a_i \cdot \Delta D_i \cdot \sum_{i=1}^{n} b_i^2 - \sum_{i=1}^{n} b_i \cdot \Delta D_i \cdot \sum_{i=1}^{n} a_i b_i}{\sum_{i=1}^{n} a_i^2 \cdot \sum_{i=1}^{n} b_i^2 - (\sum_{i=1}^{n} a_i b_i)^2} \\ \Delta y_{r-m} = \dfrac{\sum_{i=1}^{n} b_i \cdot \Delta D_i \cdot \sum_{i=1}^{n} a_i^2 - \sum_{i=1}^{n} a_i \cdot \Delta D_i \cdot \sum_{i=1}^{n} a_i b_i}{\sum_{i=1}^{n} a_i^2 \cdot \sum_{i=1}^{n} b_i^2 - (\sum_{i=1}^{n} a_i b_i)^2} \end{cases} \tag{8.28}$$

Hence the exact coordinates (x_r, y_r, z_r) of the CPRMC can be generated as follows

$$\begin{cases} x_r = x_m + \dfrac{\sum_{i=1}^{n} a_i \cdot \Delta D_i \cdot \sum_{i=1}^{n} b_i^2 - \sum_{i=1}^{n} b_i \cdot \Delta D_i \cdot \sum_{i=1}^{n} a_i b_i}{\sum_{i=1}^{n} a_i^2 \cdot \sum_{i=1}^{n} b_i^2 - (\sum_{i=1}^{n} a_i b_i)^2} \\ y_r = y_m + \dfrac{\sum_{i=1}^{n} b_i \cdot \Delta D_i \cdot \sum_{i=1}^{n} a_i^2 - \sum_{i=1}^{n} a_i \cdot \Delta D_i \cdot \sum_{i=1}^{n} a_i b_i}{\sum_{i=1}^{n} a_i^2 \cdot \sum_{i=1}^{n} b_i^2 - (\sum_{i=1}^{n} a_i b_i)^2} \\ z_r = z_m \end{cases} \tag{8.29}$$

8.3.2 Results and Discussion

In order to verify the localization algorithm, the following case study is perform based on multilateration method. Repetitive measurements of a certain position (5, 30, 0) for the CPRMC were conducted through simulation. Random interfering noise is applied to simulate the repetitive measuring error and noise of the Ultrasonic transmitter and receiver. The three-dimensional grid discretization of the simulated working environment is shown in Fig. 8.10.

In Z direction of the three-dimensional grid, as shown in Fig. 8.10, the height above the ground is represented with progressive color.

The resolution of three-dimensional grid is 200 × 200 × 60. Figures 8.11 and 8.12 show the three-dimensional grid map of the simulated working environment and the XOY plane grid map of the corresponding working environment, respectively.

The point P (5, 30, 0) is chosen as the position of the CPRMC. The 6 estimated position coordinates are (0, 0, 0), (0, 25, 0), (0, 50, 0), (25, 0, 0), (25, 25, 0), (25, 50, 0), respectively. Repetitive measurements of a certain position (5, 30, 0) for the

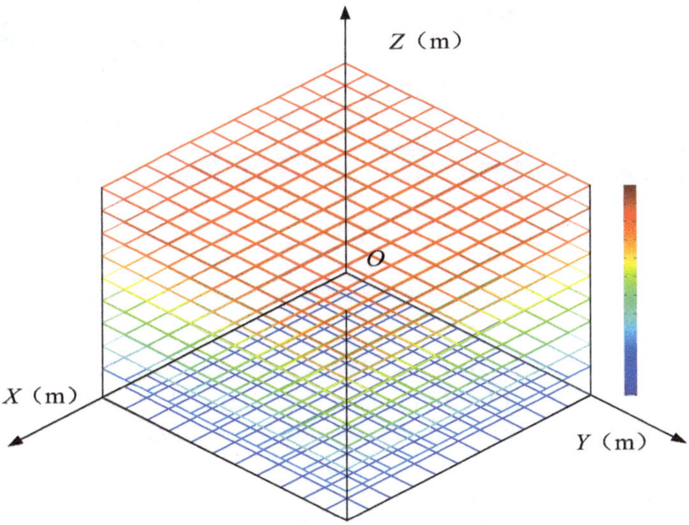

Fig. 8.10 3D grid discretization of the working environment

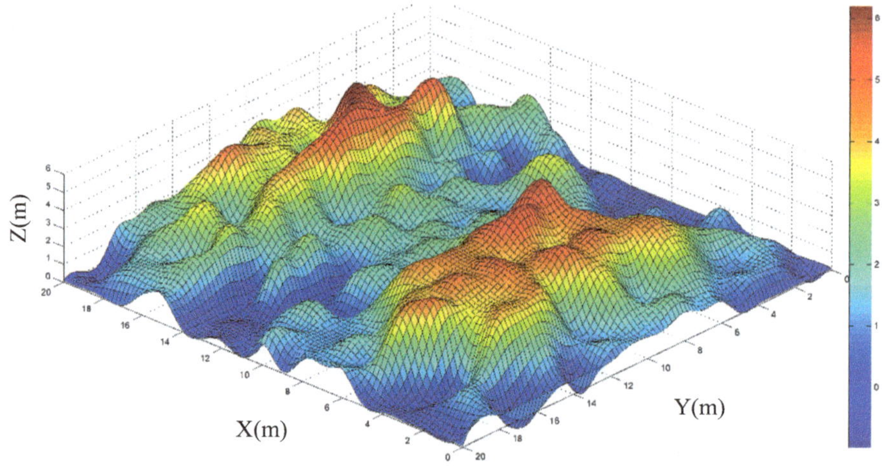

Fig. 8.11 Three-dimensional grid map of the random working environment

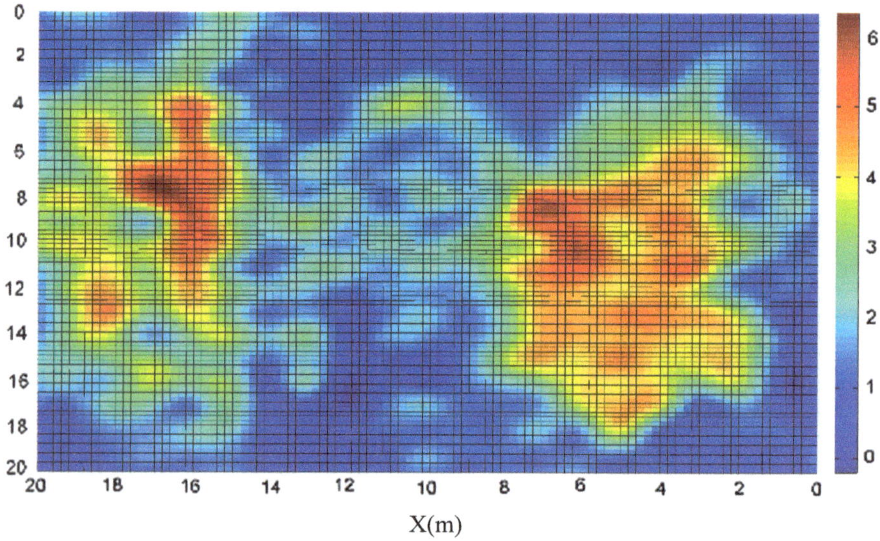

Fig. 8.12 *XOY* plane grid map of the corresponding working environment

CPRMC were conducted 120 times. The scatter distribution of the localization results is shown in Fig. 8.13. Positioning coordinates of the first 60 times are listed in Table 8.1. Then, numerical simulations are performed on the condition that the platform of the CPRMC is suspended along a desired circular trajectory. The 8 estimated position coordinates are chosen, which are (0, 25, 0), (0, 50, 0), (0, 75, 0), (0, 100, 0), (100, 25, 0), (100, 50, 0), (100, 75, 0) and (100, 100, 0), respectively.

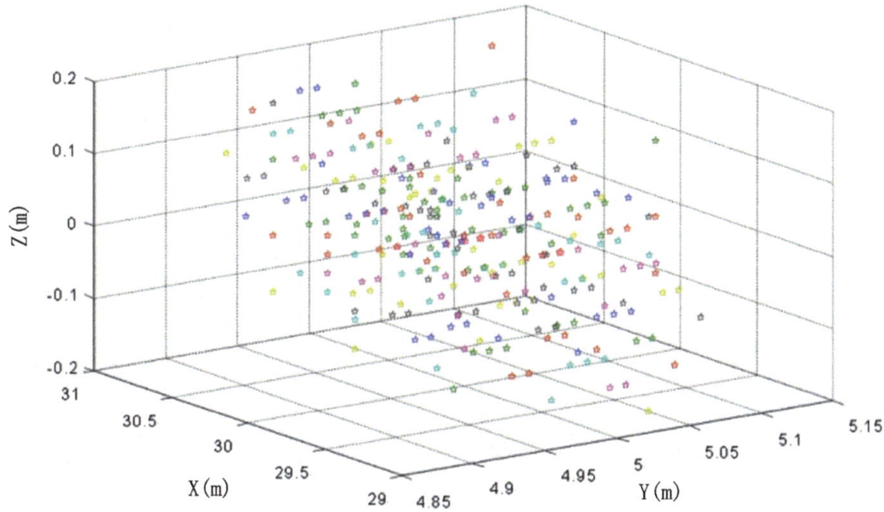

Fig. 8.13 Scatter distribution of the localization results

The localization algorithm is carried out 100 times in the process of numerical simulations.

According to Fig. 8.12 and Table 8.1, the X, Y and Z coordinates of the measured point are distributed in interval (4.84928, 5.17837), (28.89365, 31.79293) and (−0.12673, 0.10673), respectively. Compared with the true value of the position (5, 30, 0), we obtain the relative location error of the X, Y and Z coordinates are (−0.15072, 0.17837), (−1.10635, 0.97644) and (−0.12673, 0.10673), which could meet the precision requirements of the CPRMC. As shown in Figs. 8.14, 8.15 and 8.16, the localization errors between practical position and located position in X, Y and Z directions are acceptable and can meet the requirement for the path planning and obstacle avoidance. In view of the results, the improved localization algorithm is proved to be effective based on the multi-ultrasonic sensor information fusion network.

8.4 Path Planning and Obstacle Avoidance

In this section, global path planning and obstacle avoidance of the CPRMC are presented on the basis of the artificial potential field method with the grid method.

According to the artificial potential field method, there are gravitations and repulsions acting on the mobile cranes in the artificial potential field. The effect of the gravitations can guide the mobile cranes to approach the target point along the

Table 8.1 Position coordinates of the localization results

No.	X	Y	Z	No.	X	Y	Z
1	5.08449	30.99905	0.00048	2	5.14995	29.97604	−0.0225
3	5.11157	30.26086	−0.08076	4	4.88859	30.03105	−0.00374
5	5.03865	30.53172	−0.06202	6	5.01567	30.30191	−0.085
7	5.13117	29.21896	0.02716	8	4.98819	30.89592	0.00418
9	5.13825	29.64604	−0.05408	10	5.01527	29.52363	−0.07706
11	5.06533	29.67313	−0.03534	12	5.04235	29.95014	−0.05832
13	5.03785	30.8389	0.05384	14	5.11487	31.15924	0.0308
15	5.06493	30.86598	−0.0274	16	5.04195	30.43005	−0.05038
17	5.09202	30.89307	−0.00865	18	5.06904	30.70091	−0.03163
19	4.89524	31.22086	0.02356	20	4.99226	29.99105	0.00058
21	4.87232	30.49167	−0.05768	22	4.89934	30.26186	−0.08066
23	4.96941	30.76253	−0.03894	24	4.96643	30.53272	−0.06192
25	4.93852	29.55303	−0.04549	26	5.05693	29.87896	−0.08076
27	4.99656	29.18011	−0.12673	28	4.92552	29.34303	−0.045534
29	4.88269	30.07205	−0.00799	30	5.17837	29.90011	−0.02745
31	4.86652	30.97299	−0.01881	32	4.88236	30.29712	−0.05762
33	4.97228	29.60007	−0.00005	34	4.90944	29.76615	0.10673
35	5.11937	29.52716	−0.08131	36	5.01593	29.58529	0.08078
37	4.99189	30.79293	0.00788	38	5.03995	30.87513	−0.12241
39	5.01897	30.52001	−0.07336	40	5.13517	29.76397	0.02106
41	5.14605	30.3471	−0.05462	42	5.03585	29.98234	−0.10872
43	4.84928	29.97612	−0.02241	44	4.97141	30.15354	0.09382
45	4.89636	31.00032	−0.00365	46	4.95756	29.67592	0.01902
47	4.92344	30.30286	−0.08491	48	5.12497	30.89356	−0.06812
49	5.00233	29.23523	−0.09802	50	4.99632	30.68293	0.01923
51	5.07647	29.19973	0.02106	52	4.98859	29.39789	−0.08203
53	5.12981	30.97513	−0.07723	54	4.85719	29.23721	−0.21974
55	4.98332	30.72982	0.00238	56	5.02395	29.89365	−0.26723
57	5.03631	29.39892	−0.05034	58	5.07693	30.76293	0.0642
59	4.99894	29.99289	0.0289	60	5.04595	29.78192	0.03743

optimum path, while the effect of the repulsions can ensure the mobile cranes to avoid the obstacles. Under the combined action of the gravitations and repulsions, the mobile cranes are operated to move in the artificial potential field, and can avoid all the obstacles and approach the target point along the optimum path at the same time. However, the local minimum points resulted from the artificial potential field algorithm, can stop the path planning of the system. From the aforementioned analysis, the artificial potential field algorithm is combined with the grid method for the purpose of evading the local minimum points in the artificial potential field.

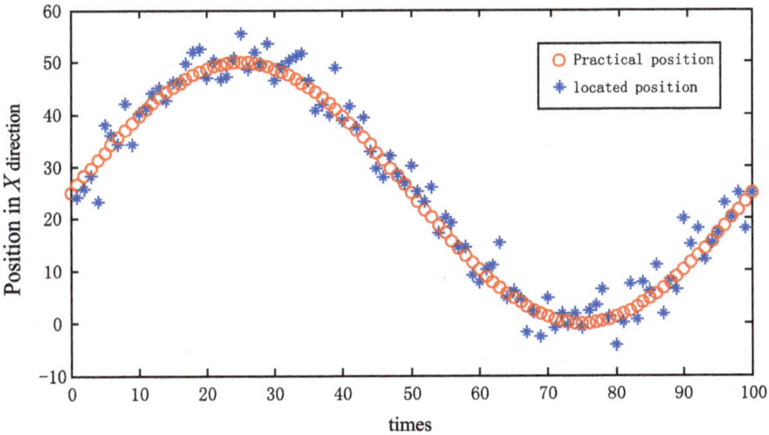

Fig. 8.14 Practical position and located position in X direction

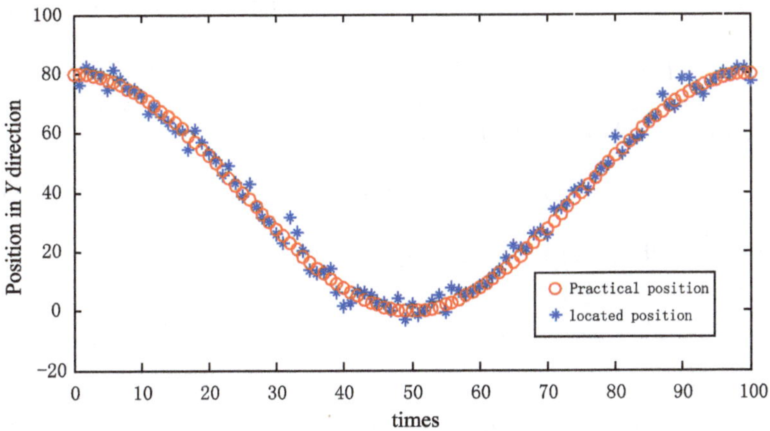

Fig. 8.15 Practical position and located position in Y direction

8.4.1 Obstacle Avoidance Planning

During the path planning process, the position coordinates of the CPRMC in the operating environment is $W(x, y, z)$. In the grid map, the obstacle space is presented by $B_{obstacle}$, and each individual grid obstacle is $B_{obstacle}(i), i = 1, 2, 3, \ldots$. Thus $B_{obstacle} = \sum B_{obstacle}(i)$. The free space presented by B_{free}, so the workspace $P = B_{free} \cup B_{obstacle}$. Assuming that the goal position of the path planning is $Goal(x, y, z)$. Since the system moves on the XOY plane and the Z-axis coordinates are usually fixed, the coordinates of the current location and the goal position

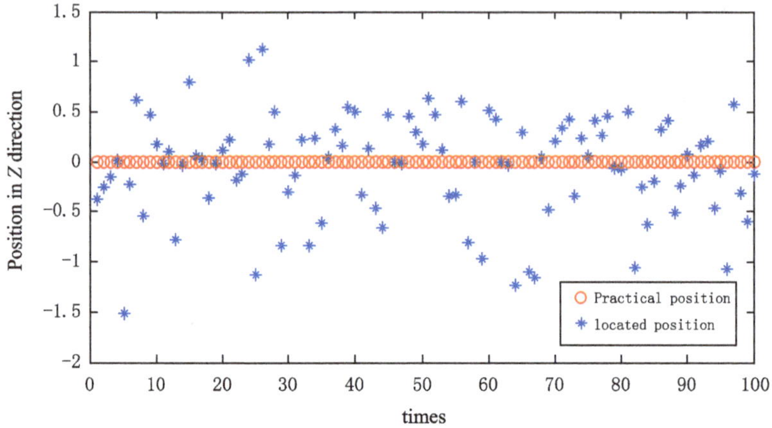

Fig. 8.16 Practical position and located position in Z direction

coordinates of the mobile robots can be expressed as $W(x, y)$ and $Goal(x_{goal}, y_{goal})$, respectively.

The virtual potential energy is defined as $U(Q_W)$, the attracted potential energy relative to $Goal(x_{goal}, y_{goal})$ is $U_{att}(Q_W)$, the exclusion potential energy relative to $B_{obstacle}(i)$ is $U_{rep}(Q_W)$. Therefore, we have

$$U(Q_W) = U_{att}(Q_W) + U_{rep}(Q_W) \tag{8.30}$$

In grip map, the gradient vector of Q_W can be expressed as

$$\nabla \boldsymbol{U}(Q_W) = \begin{bmatrix} \frac{\partial U(Q_W)}{\partial x} \\ \frac{\partial U(Q_W)}{\partial y} \end{bmatrix} \tag{8.31}$$

Combined with the gravitation F_{att}, repulsion F_{rep} and resultant force F_R, the functions can be obtained based on the definition of potential field

$$\overline{\boldsymbol{F}}_{att}(Q_W) = -grad[U_{att}(Q_W)] \tag{8.32}$$

$$\overline{\boldsymbol{F}}_{rep}(Q_W) = -grad[U_{rep}(Q_W)] \tag{8.33}$$

$$\overline{\boldsymbol{F}}_R(Q_W) = -grad[U_{att}(Q_W)] - grad[U_{rep}(Q_W)] = -\nabla U(Q_W) \tag{8.34}$$

The attracted potential energy in Q_W is

$$U_{att}(Q_W) = \frac{1}{2}\zeta[\Delta l_{W-G}]^2 = \frac{1}{2}\zeta[W(x, y) - Goal(x_{goal}, y_{goal})]^2 \tag{8.35}$$

where ζ is the positive proportional gain coefficient of gravitational potential field, which is a constant in the electrostatic field. And the distance between $W(x, y)$ and $Goal(x_{goal}, y_{goal})$ can be achieved

$$\Delta l_{W-G} = \left\| W(x, y) - Goal(x_{goal}, y_{goal}) \right\| \tag{8.36}$$

Combined with Eq. (8.32), the attractive gradient function of the attracted potential energy is

$$\overline{\boldsymbol{F}}_{att}(Q_W) = -grad[U_{att}(Q_W)] = -\zeta \Delta l_{W-G} \tag{8.37}$$

Substituting Eq. (8.36) into Eq. (8.37)

$$\overline{\boldsymbol{F}}_{att}(Q_W) = -\zeta[W(x, y) - Goal(x_{goal}, y_{goal})] = \zeta[Goal(x_{goal}, y_{goal}) - W(x, y)] \tag{8.38}$$

The electrostatic repulsion potential can be expressed as

$$U_{rep}(Q_W) = \begin{cases} \frac{1}{2}\xi(\frac{1}{\min \Delta l_{W-B}} - \frac{1}{\Delta l_k})^2, & \min \Delta l_{W-B} \le \Delta l_k \\ 0, & \min \Delta l_{W-B} > \Delta l_k \end{cases} \tag{8.39}$$

where ξ is the repulsion potential field constant in the electrostatic field.

Then we have

$$\Delta l_{W-B} = \| W(x, y) - B_{obstacle}(i) \| \tag{8.40}$$

The gradient function of the repulsive is derived as

$$\begin{aligned} \overline{\boldsymbol{F}}_{rep}(Q_W) &= -grad[U_{rep}(Q_W)] \\ &= \begin{cases} \frac{\xi}{(\min \Delta l_{W-B})^2}(\frac{1}{\min \Delta l_{W-B}} - \frac{1}{\Delta l_k}) \nabla \min \Delta l_{W-B}, & \min \Delta l_{W-B} \le \Delta l_k \\ 0, & \min \Delta l_{W-B} > \Delta l_k \end{cases} \end{aligned} \tag{8.41}$$

Thus, the potential energy of the CPRMC is

$$\begin{aligned} U(Q_W) &= U_{att}(Q_W) + \hat{U}_{rep}(Q_W) \\ &= \frac{1}{2}\zeta[W(x, y) - Goal(x_{goal}, y_{goal})]^2 + \sum \frac{1}{2}\xi(\frac{1}{\min \Delta l_{W-B}} - \frac{1}{\Delta l_k})^2 \end{aligned} \tag{8.42}$$

where $\hat{U}_{rep}(Q_W)$ denotes the resultant exclusion potential energy of all grid obstacles in potential field.

The resultant force acting on the CPRMC can be written as

$$\begin{aligned} F_R(Q_W) &= F_{att}(Q_W) + \hat{F}_{rep}(Q_W) \\ &= \zeta \Delta l_{G-W} + \sum \frac{\xi}{(\min \Delta l_{W-B})^2}\left(\frac{1}{\min \Delta l_{W-B}} - \frac{1}{\Delta l_k}\right)\nabla \min \Delta l_{W-B} \end{aligned} \quad (8.43)$$

where $\hat{F}_{rep}(Q_W)$ denotes the resultant repulsion force of all grid obstacles in potential field.

The circular window is built, and the center of the window is the local minima $W_o(x_o, y_o)$. At the local minima point $W_o(x_o, y_o)$, the resultant force $F_R(Q_{Wo}) = F_{att}(Q_{Wo}) + \hat{F}_{rep}(Q_{Wo}) = 0$, as shown in Fig. 8.17.

The distance between $W_o(x_o, y_o)$ and $Goal(x_{goal}, y_{goal})$ is

$$\Delta l_{W_o-G} = \sqrt{(x_o - x_{goal})^2 + (y_o - y_{goal})^2} \quad (8.44)$$

As shown in Fig. 8.18, the local minima point $W_o(x_o, y_o)$ and the nearest obstacle are located in the grid S_1 and S_2, respectively. To avoid stopping at the local minima point and approach the target point along the optimum path at the same time, the CPRMC are operated to move k grids along the vector $\Delta\vec{l}_R$ to the point $W_R(x_R, y_R)$ located in the grid S_k

In which

$$\Delta\vec{l}_R = \Delta\vec{l}_{S_2-S_1} + \Delta\vec{l}_{G-W_o} \quad (8.45)$$

The resultant force acting on the point $W_R(x_R, y_R)$ after the grid movement is given as

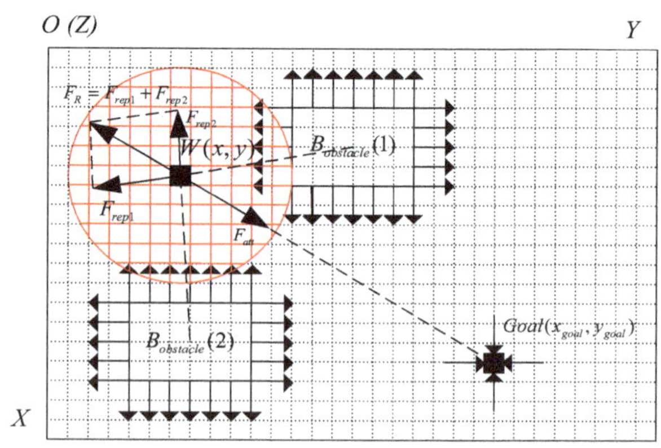

Fig. 8.17 Circular window grid

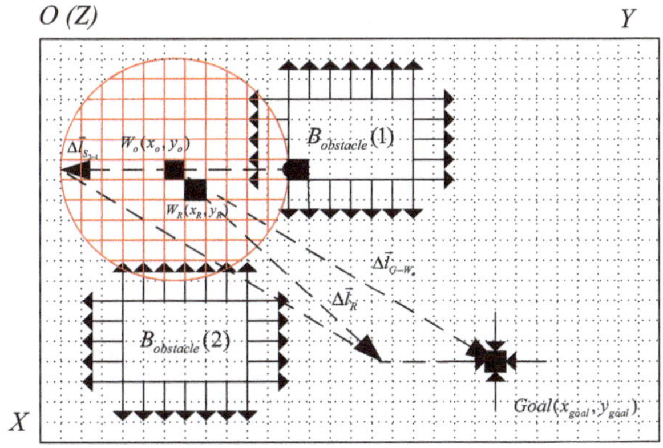

Fig. 8.18 Selecting the moving grid

$$
\begin{aligned}
F_R(Q_W) &= F_{att}(Q_W) + \hat{F}_{rep}(Q_W) \neq 0 \\
&= \zeta[Goal(x_{goal}, y_{goal}) - W(x_R, y_R)] \\
&+ \sum \frac{\xi}{(\min \Delta l_{W_R-B})^2} \left(\frac{1}{\min \Delta l_{W_R-B}} - \frac{1}{\Delta l_k}\right) \nabla \min \Delta l_{W_R-B}
\end{aligned} \tag{8.46}
$$

Therefore, after the mobile cranes smoothly go through the local minimum point in the artificial potential field, the artificial potential field method can be utilized to perform the path planning of the CPRMC based on the grid method.

Figure 8.19 shows that the CPRMC moves along the path planning results. As a whole system, the reference point of the local coordinate system of the CPRMC is the center of the platform. The platform is suspended along the trajectory generated by the artificial potential field method with the grid algorithm. As a part of the whole system, the single mobile crane adjusts its direction based on the laser sensors. The adjustment should follow the principle during the process. The single mobile crane should ensure the movement of the platform by means of the cooperative variation of the cable, the boom and the turntable. As the distance is beyond the safe threshold, the colored parts represent the process of the robots avoiding the obstacle.

Referring to Fig. 8.19, the distance between four single cranes and the obstacle can be apperceived periodically and quickly via the eight laser sensors. The distance is compared with the safe threshold to determine whether single mobile robot is in a safe position or not. Thus the moving direction of the mobile cranes is adjusted.

8.4.2 Results and Discussion

Here, simulation studies are being presented to prove the efficiency of the path planning and obstacle avoidance for the mobile cranes. The artificial potential field

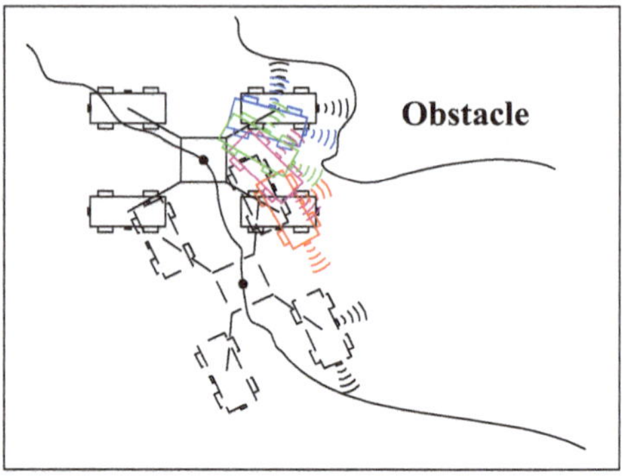

Fig. 8.19 Cooperative obstacle avoidance

algorithm and the artificial potential field algorithm with the grid method are adopted, respectively. The scanning range of the sensor is discretized into ten grids. Assuming that the CPRMC occupies nine grids, and the corresponding eight points for distance measuring are shown in Fig. 8.20. The three-dimensional grid map of the path planning is shown in Fig. 8.21. The corresponding XOY plane grid map of the working environment is shown in Fig. 8.22.

Figure 8.23 illustrates path planning under the artificial potential field method [14]. Figure 8.23 indicates that the path planning stops when using the artificial potential field method. It is because the artificial potential field method exists local minimum point problem. Figure 8.24 shows the result of the global path planning in the 3D grid map based on the artificial potential field method with the grid method. The CPRMC move along the red path from the starting point coordinate (0, 0, 0) to the target point (20, 20, 0) after avoiding several obstacles to complete the full course of the global planning successfully.

Figure 8.25 shows the global path planning under the artificial potential field method with the grid method in the XOY plane. Figure 8.26 displays the global path planning in contour line. Figure 8.27 shows the positions where the single mobile crane cannot avoid the obstacle.

As shown in Figs. 8.23, 8.24, 8.25, 8.26 and 8.27, the CPRMC succeed in avoiding all the obstacles and approaching the target point along the optimum path

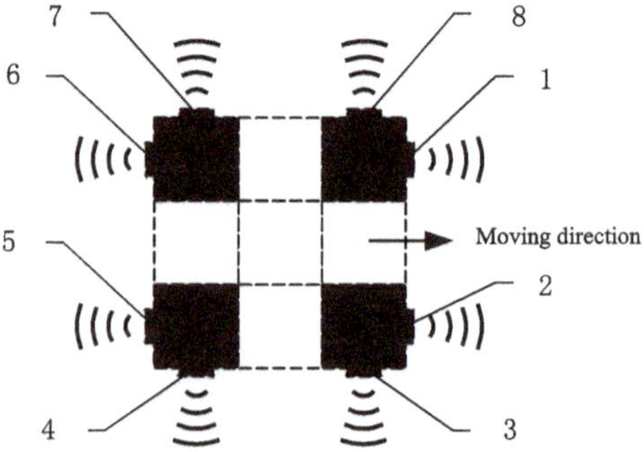

Fig. 8.20 The corresponding eight points for distance measuring

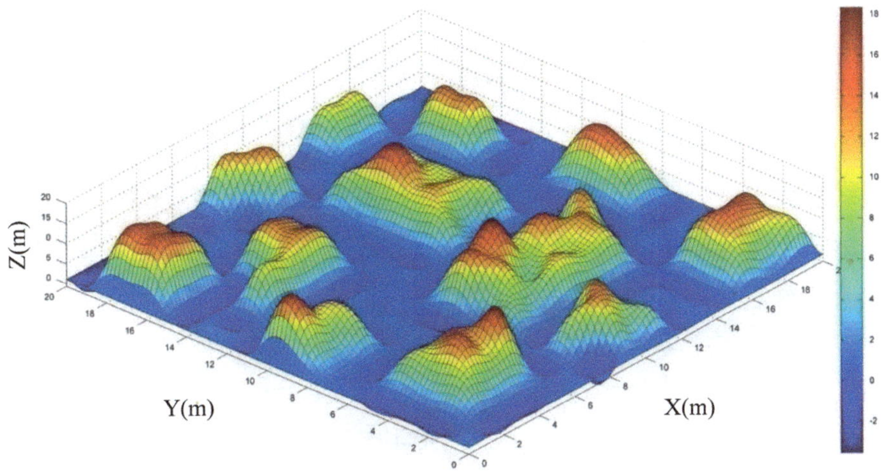

Fig. 8.21 Three-dimensional grid map of the path planning

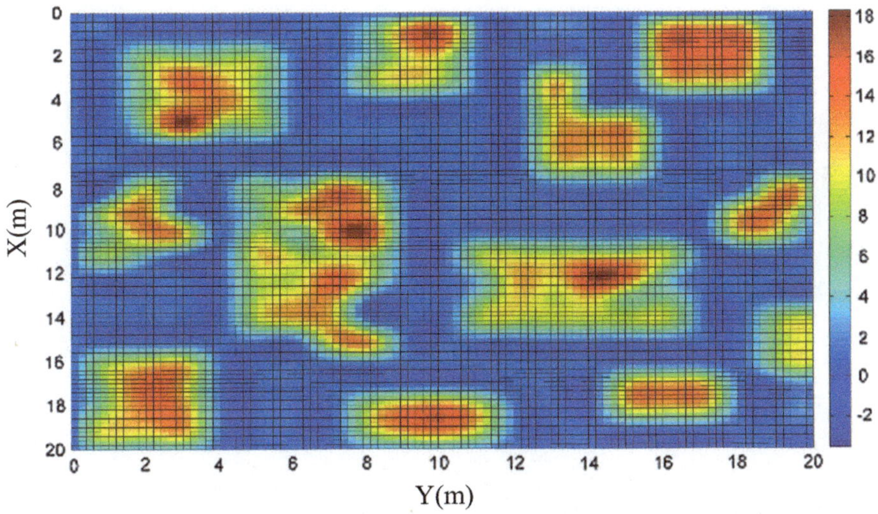

Fig. 8.22 *XOY* plane of the grid map of the working environment

at the same time. Moreover, the CPRMC evade the local minimum point in the artificial potential field and pass the three-dimensional grid map smoothly and continuously based on the artificial potential field with the grid method.

In order to display the moving process better, a co-simulation platform is carried out by using Matlab and LabVIEW. The interface of the cooperative obstacle avoidance and warning is shown in Fig. 8.28.

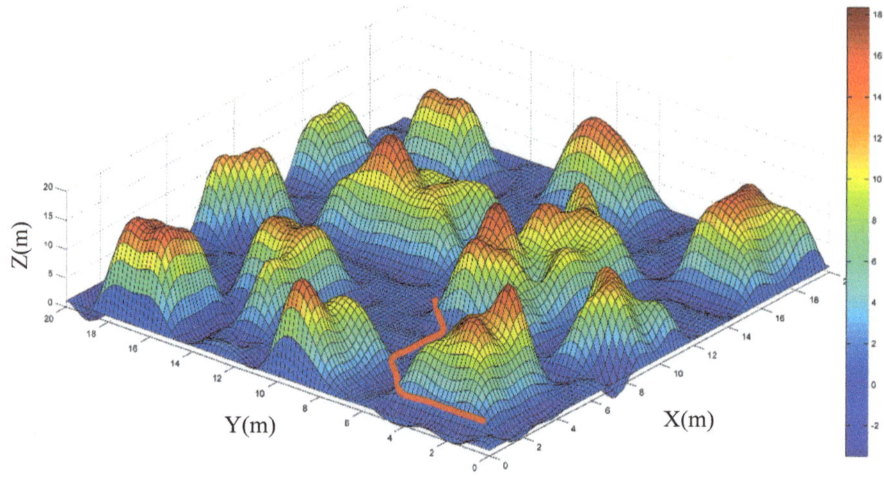

Fig. 8.23 Path planning under the artificial potential field method

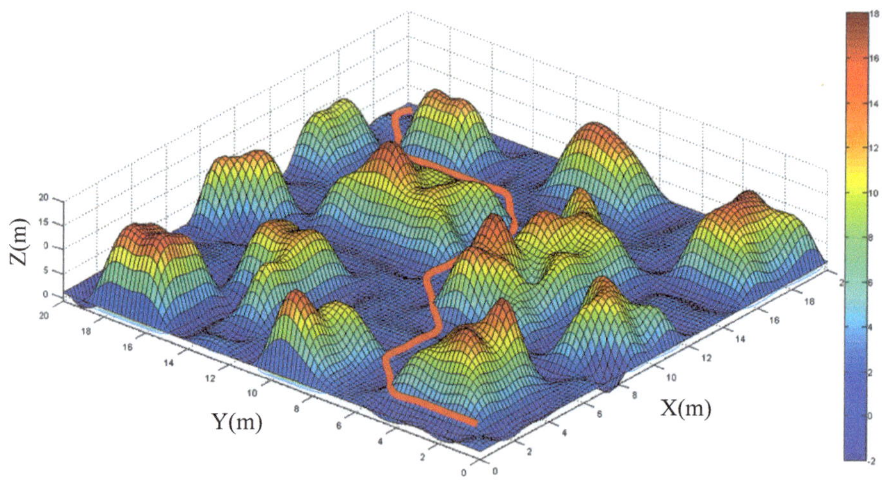

Fig. 8.24 Path planning in the 3D grid map

Referring to Fig. 8.28, the eight laser sensors corresponds to the eight points for distance measuring, and the measuring information are recorded in the eight graphics windows, where the ordinate represents the grid value of the distance. When the distance of the mobile robots and the obstacle is lower than the safety threshold, i.e. below 5 grids, the corresponding warning lamp in the eight lamps will light up. At the same time, the corresponding single mobile crane will be fine adjusted for obstacle avoidance.

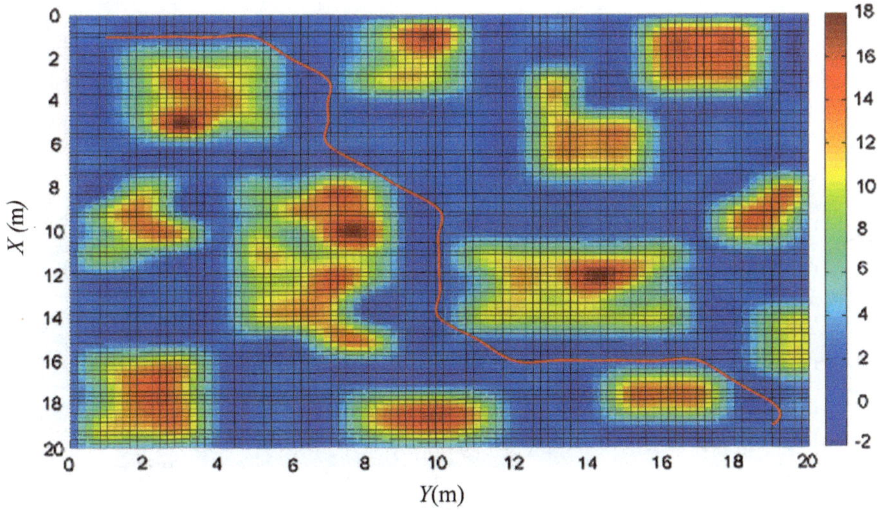

Fig. 8.25 Path planning in the *XOY* plane grid map

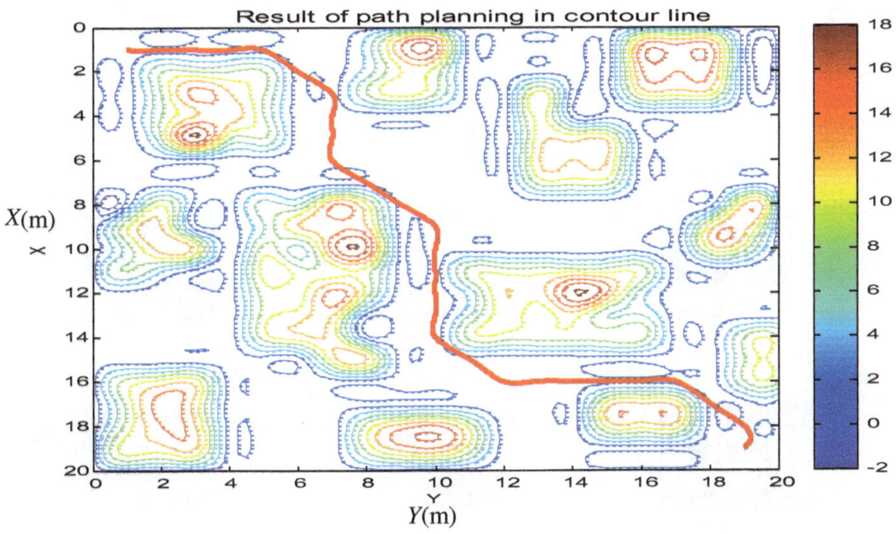

Fig. 8.26 Path planning in contour line

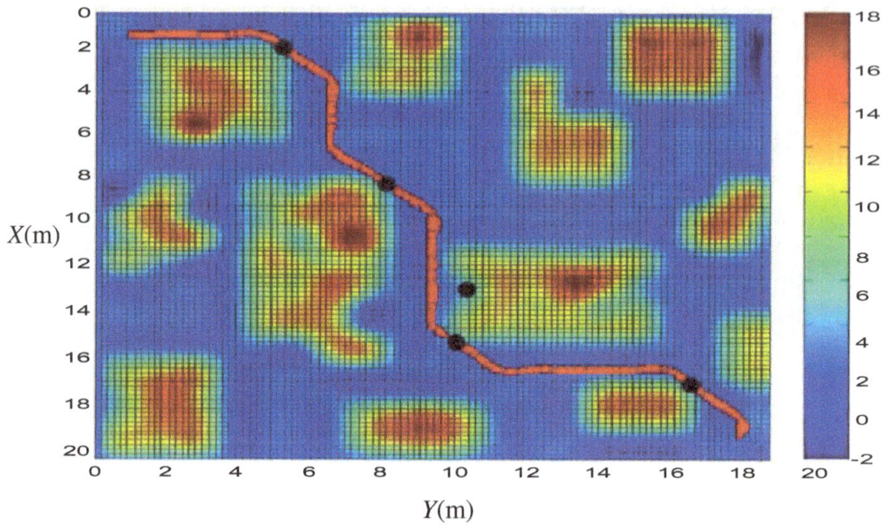

Fig. 8.27 Positions where the single mobile crane cannot avoid obstacle

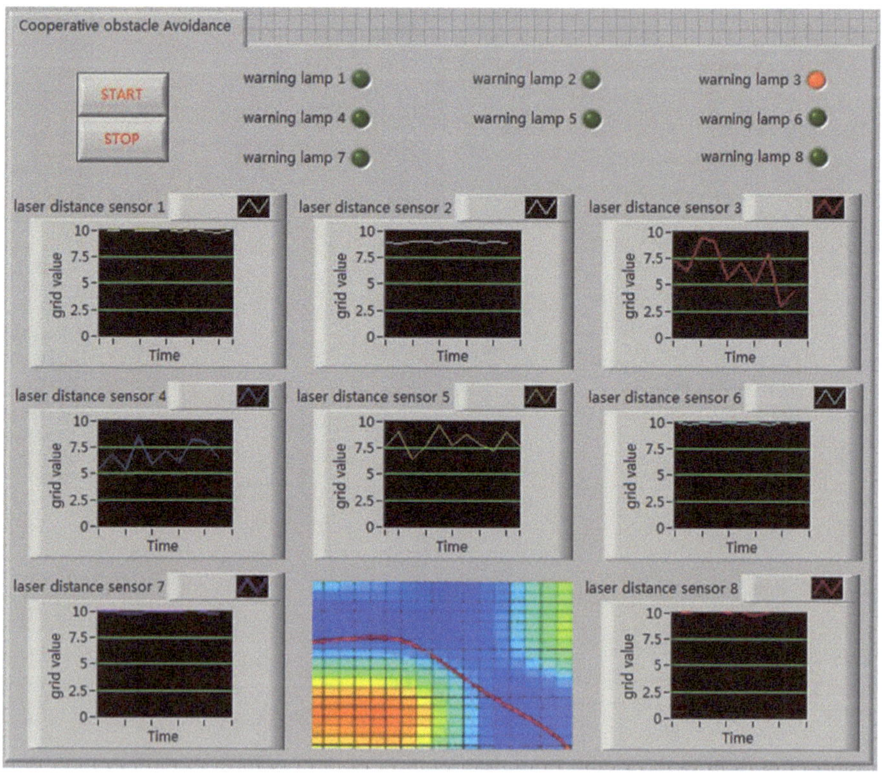

Fig. 8.28 Interface of cooperative obstacle avoidance and warning

8.5 Automatic Leveling Control

8.5.1 Leveling Control Strategy

Orientation detection, adjustment and automatic leveling control of the platform are important and common in military, mining and civil engineering. Compared with other methods, the four-point support method has better stability and anti-overturning capacity. In this research, the four-point support method is adopted to investigate automatic leveling control for the platform of the CPRMC. On the basis of the model of the CPRMC, the model of the platform under horizontal and stable condition is established (see Fig. 8.29). Under non-loaded condition the plat form of the CPRMC and the plane of the dotted rectangle are parallel to each other, and four cables are equal in length.

Under load condition, the model of the inclined platform is shown in Fig. 8.30. The points A_1, A_2, A_3, A_4 represent the hoisting points.

From Fig. 8.30, the point o is the geometric center of the cuboids $A_1A_2A_3A_4C_1C_2C_3C_4$, which is taken as the origin of the local coordinate system o′x′ y′z′ and the global coordinate system oxyz. As the platform is horizontal and stable, the coordinate system o′x′y′z′ is coinciding with the coordinate system oxyz. As the platform is inclined, the offset is generated between the local coordinate system o′x′y ′z′ and the global coordinate system oxy. In which α is the roll angle, β is the pitch angle, and γ is deflection angle. Therefore, the posture information of the platform can be expressed as $\phi = [\alpha, \beta, \gamma]^T$, and the transformation relation between the local and global coordinate system can be generated as

$$(x', y', z')^T = R(\phi) \cdot (x, y, z)^T \tag{8.47}$$

where $\boldsymbol{R}(\phi)$ is the transformational matrix of the platform.

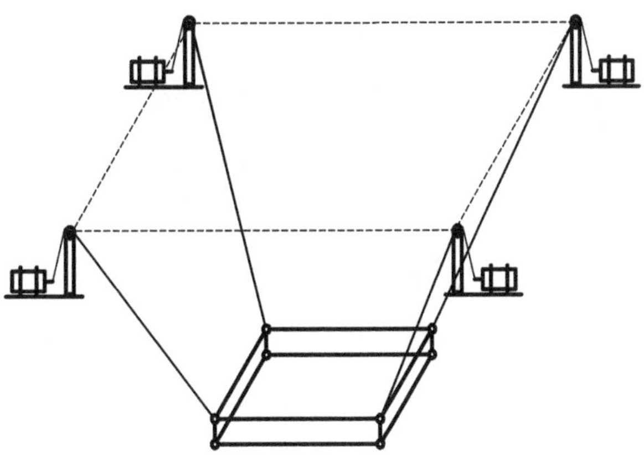

Fig. 8.29 Model of the horizontal and stable platform

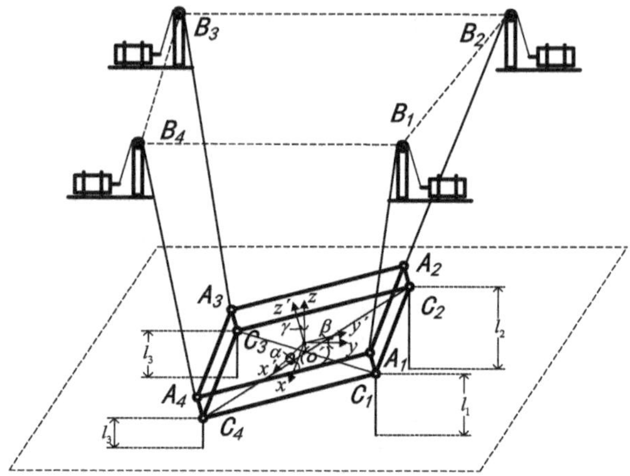

Fig. 8.30 Model of the tilted platform

$$R(\phi) = \begin{bmatrix} \cos\beta\cos\gamma & -\cos\beta\sin\gamma & \sin\beta \\ \sin\alpha\sin\beta\cos\gamma + \cos\alpha\sin\gamma & -\sin\alpha\sin\beta\sin\gamma + \cos\alpha\cos\gamma & -\sin\alpha\cos\beta \\ -\cos\alpha\sin\beta\cos\gamma + \sin\alpha\sin\gamma & \cos\alpha\sin\beta\sin\gamma + \sin\alpha\cos\gamma & \cos\alpha\cos\beta \end{bmatrix} \tag{8.48}$$

Levelness of the platform can be expressed by α and β. The change of γ will not impact on the horizontal condition of the platform. Therefore, there is no need to take γ into consideration.

Referring to Fig. 8.30, the distances between the four top points C_1, C_2, C_3, C_4 and the ground are l_1, l_2, l_3, l_4, respectively. The height of the mid-point datum for C_1, C_2, C_3, C_4 after leveling is

$$\bar{l} = \frac{l_1 + l_2 + l_3 + l_4}{4} \tag{8.49}$$

Hence, the vertical distance Δl_i between the mid-point datum and the four top points C_1, C_2, C_3, C_4 can be written as

$$\begin{aligned} C_1 &: \Delta l_1 = l_1 - \bar{l} \\ C_2 &: \Delta l_2 = l_2 - \bar{l} \\ C_3 &: \Delta l_3 = l_3 - \bar{l} \\ C_4 &: \Delta l_4 = l_4 - \bar{l} \end{aligned} \tag{8.50}$$

According to Fig. 8.30, $|A_1A_4| = l_a$, $|A_1A_2| = l_b$, $|A_1C_1| = l_c$, we have

$$\begin{aligned} C_1 : \Delta l_1 &= +l_a \sin\alpha - l_b \cos\alpha \sin\beta \\ C_2 : \Delta l_2 &= +l_a \sin\alpha - l_b \cos\alpha \sin\beta \\ C_3 : \Delta l_3 &= -l_a \sin\alpha - l_b \cos\alpha \sin\beta \\ C_4 : \Delta l_4 &= -l_a \sin\alpha - l_b \cos\alpha \sin\beta \end{aligned} \tag{8.51}$$

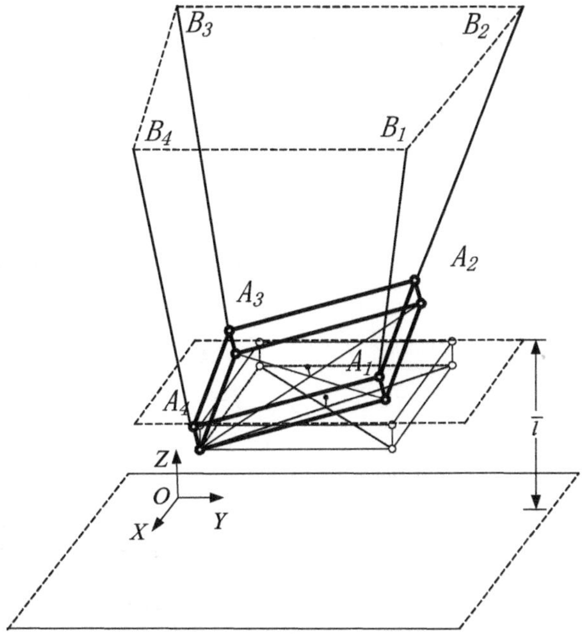

Fig. 8.31 Mid-point datum

The coordinates OXY are established, and the origin O is the projection of the hoisting point B_3 on the ground. The global coordinate frame OXYZ is depicted in Fig. 8.31.

Here

$$\begin{aligned} A_1 &: \Delta l_1 = l_1 - \bar{l} \\ A_2 &: \Delta l_2 = l_2 - \bar{l} \\ A_3 &: \Delta l_3 = l_3 - \bar{l} \\ A_4 &: \Delta l_4 = l_4 - \bar{l} \end{aligned} \tag{8.52}$$

The platform is simplified as a rectangular $A_1A_2A_3A_4$, as shown in Fig. 8.32. The first inclined condition of the platform, i.e., $\alpha \neq 0$, $\beta = 0$, is shown in Fig. 8.33. From Figs. 8.31, 8.32 and 8.33, the rectangular $A_1A_2A_3A_4$ represents the horizontal condition after leveling, assuming that the distance between the points B_1, B_2, B_3, B_4 and the ground is h. In addition, $|A_1B_1| = d_1, |A_2B_2| = d_2$, $|A_3B_3| = d_3$, $|A_4B_4| = d_4$. The projection of $|A_1B_1|$, $|A_2B_2|$, $|A_3B_3|$, $|A_4B_4|$ and $|A_4B_4|$ on the ground can be derived as follows

$$d_{\tau 1} = \sqrt{d_1^2 - [h - (\Delta l_1 + \bar{l} + l_c)]^2} \tag{8.53}$$

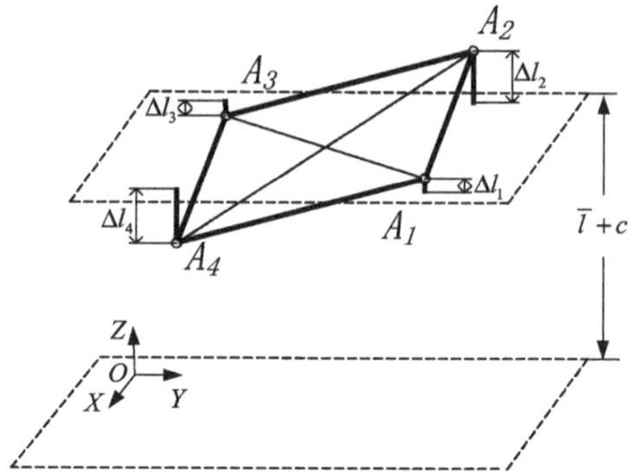

Fig. 8.32 Simplified mid-point datum

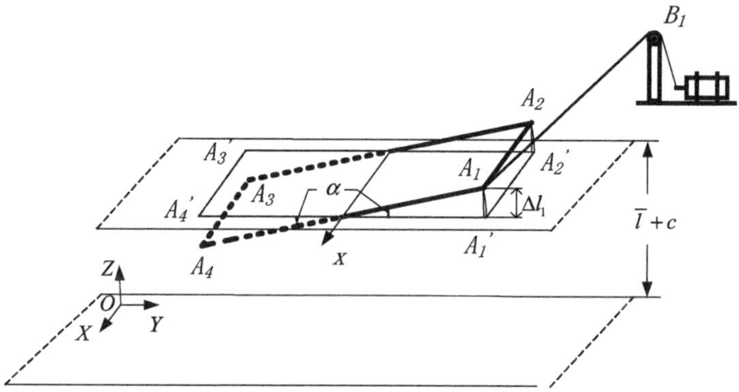

Fig. 8.33 Platform under the inclined condition ($\alpha \neq 0$, $\beta = 0$)

$$d_{\tau 2} = \sqrt{d_2^2 - [h - (\Delta l_2 + \bar{l} + l_c)]^2} \tag{8.54}$$

$$d_{\tau 3} = \sqrt{d_3^2 - [h - (\Delta l_3 + \bar{l} + l_c)]^2} \tag{8.55}$$

$$d_{\tau 4} = \sqrt{d_4^2 - [h - (\Delta l_4 + \bar{l} + l_c)]^2} \tag{8.56}$$

Since the four points A_1, A_2, A_3, A_4 are identical, the first point A_1 is chosen for analysis. The distance between the points A_1', A_2', A_3', A_4' and the points B_1, B_2, B_3, B_4 is

$$d_1' = \sqrt{d_1^2 + 2h\Delta l_1 - \Delta l_1^2 - 2\Delta l_1 \bar{l} + 2\Delta l_1 l_c} \tag{8.57}$$

Hence, the change of the cable length is given as

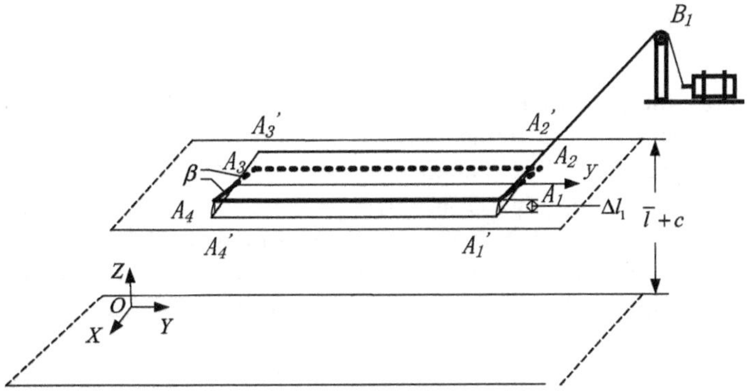

Fig. 8.34 Platform under the inclined condition ($\alpha = 0,\ \beta \neq 0$)

$$\Delta d_{1(\alpha)} = d_1 - d_1' = d_1 - \sqrt{d_1^2 + 2h\Delta l_1 - \Delta l_1^2 - 2\Delta l_1 \bar{l} + 2\Delta l_1 l_c} \quad (8.58)$$

The second inclined condition of the platform, i.e., $\alpha = 0$, $\beta \neq 0$, is presented in Fig. 8.34.

Similarly, the change of the cable length is given by the following expression

$$\Delta d_{1(\beta)} = d_1 - d_1' = d_1 - \sqrt{d_1^2 + 2h\Delta l_1 - \Delta l_1^2 - 2\Delta l_1 \bar{l} + 2\Delta l_1 l_c} \quad (8.59)$$

For simplicity, the third inclined condition of the platform, i.e. $\alpha \neq 0$, $\beta \neq 0$, can be regarded as the composition of the above two conditions. As a result, the error Δd_i between the actual length d_i before leveling and the actual length d_i' after leveling is

$$\Delta d_i = d_i - d_i' = d_i - \sqrt{d_i^2 + 2h\Delta l_i - \Delta l_i^2 - 2\Delta l_i \bar{l} + 2\Delta l_i l_c} \quad (8.60)$$

In order to adjust the inclination of the platform, a PID controller is adopted to adjust the angular displacement $f_i(k)$ of gear motors for the cable driving devices, as shown in Fig. 8.35. The orientation-leveling controller is made up of four identical PID controllers based on auto-leveling strategy.

The PID control law is given as follows

$$f_i(k) = K_p \cdot \Delta d_i(k) + K_i \cdot T_s \cdot \sum_{n=1}^{k} \Delta d_i(n) + K_d \cdot [\Delta d_i(k) - \Delta d_i(k-1)]/T_k \quad (8.61)$$

Here, K_p, K_i, K_d is proportional gain, integral gain and differential gain, respectively. T_k is the kth sampling period.

8.5.2 Result

For evaluating the proposed collaborative leveling control scheme, some results, regarding the automatic leveling control performance of the platform for the CPRMC, are presented. The parameters were selected as follows: $l_a = 0.8$ m, $l_b = 0.5$ m, $l_c = 0.2$ m, $h = 5$ m, $l_1 = 1.15$ m, $l_2 = 1.20$ m, $l_3 = 1.10$ m, $l_4 = 1.05$ m, $d_1 = 7.13$ m, $d_2 = 7.57$ m, $d_3 = 6.88$ m, $d_4 = 6.98$ m. The parameters of the PID controller are chosen as follows: proportional gain $K_p = 0.2$, integral gain $K_i = 0.02$, differential gain $K_d = 0.0005$, and sampling time $T_k = 0.5$ s.

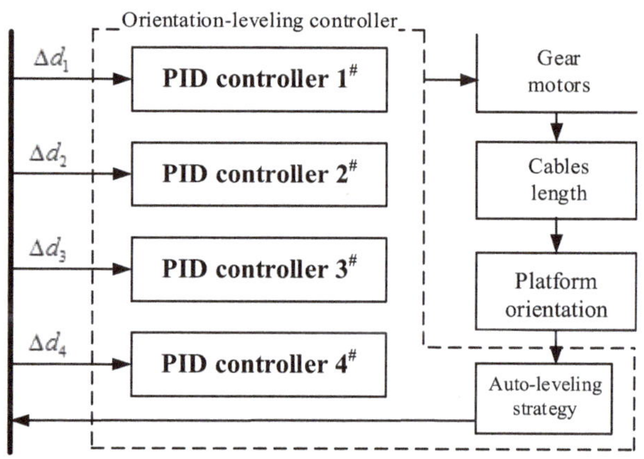

Fig. 8.35 Orientation-leveling controller

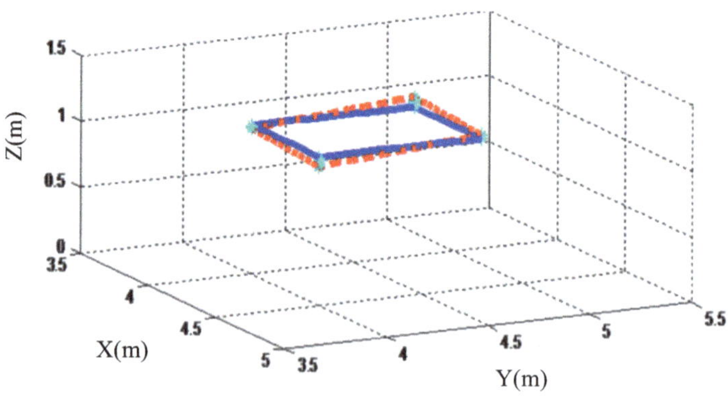

Fig. 8.36 Change in the platform model before and after leveling control

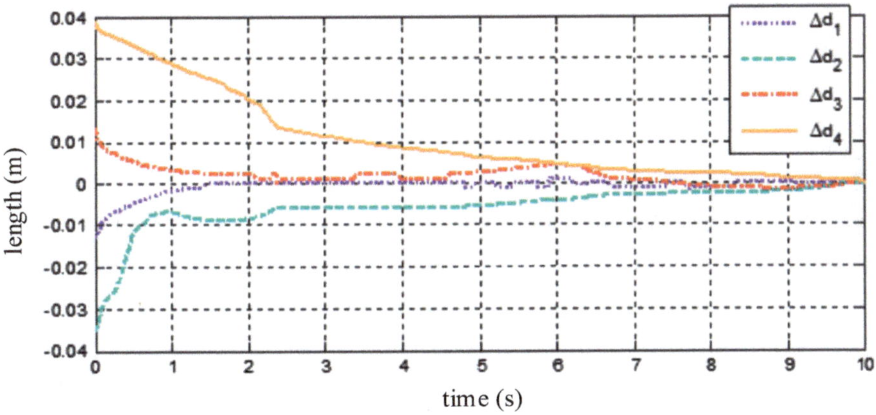

Fig. 8.37 Error changes of cable length

The change of the platform and errors of the cable length for the CPRMC are shown in Figs. 8.36 and 8.37, respectively. Figure 8.38 represents the changes of the cable length. The curves of the orientation, i.e. the changes of roll angle and pitch angle of the platform is shown in Fig. 8.39. In addition, Fig. 8.40 shows the curves of the height l_i between four vertexes of the platform and ground. Tables 8.2, 8.3, 8.4 and 8.5 compare orientation and height of the platform before and after automatic leveling control, respectively.

Referring to Figs. 8.37, 8.38, 8.39 and 8.40 and Tables 8.2, 8.3, 8.4 and 8.5, it can be generated that the platform's orientation reached a preliminary level in 8 s. As can be seen, the leveling error of roll angle α is $-0.14°$, and the leveling error of pitch angle β is $0.21°$. The relative height error between the platform vertexes and the ground is 0.81, 1.64, −0.42, −1.10%, respectively. The automatic leveling

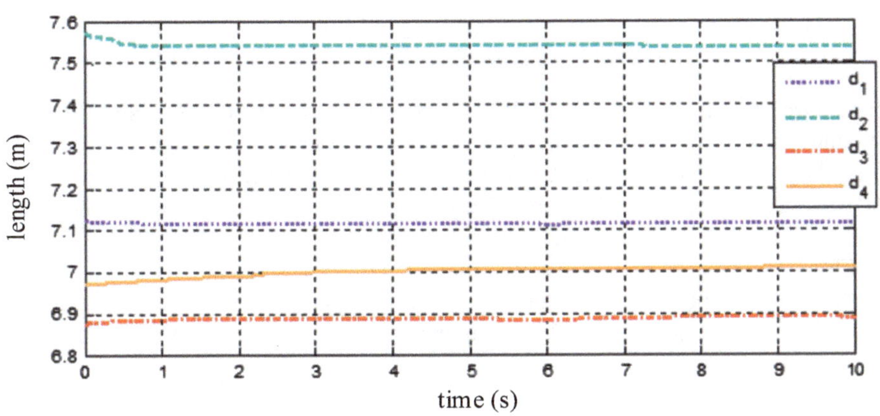

Fig. 8.38 Changes of cable length

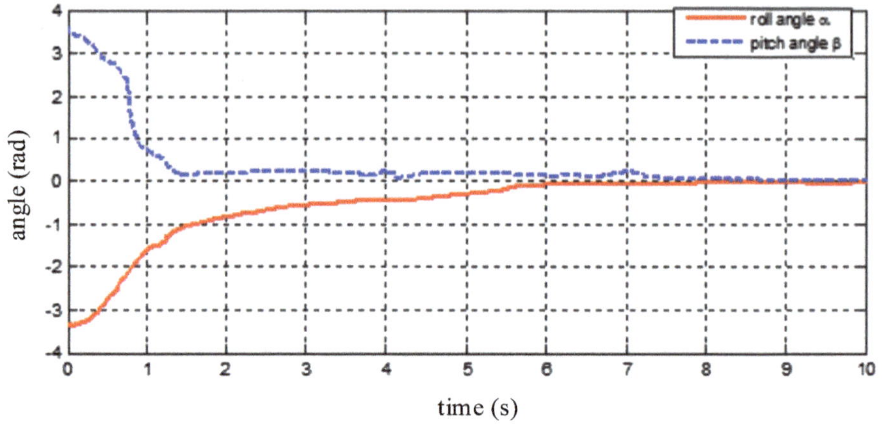

Fig. 8.39 Orientation changes of the platform

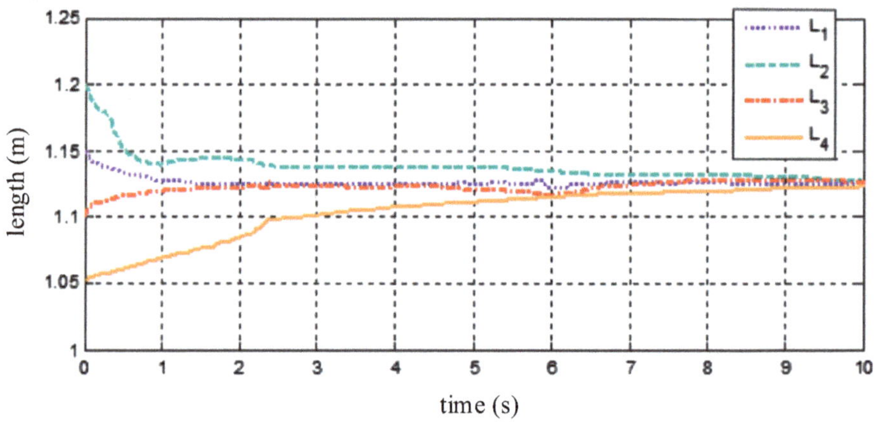

Fig. 8.40 Height changes between the platform and the ground

procedure ends in 9.72 s, when all the errors are at a lower level. Note that, the leveling error of roll angle α is 0.062°, the leveling error of pitch angle β is −0.073°, the relative error of the height between the platform vertexes and the ground is −0.08, 0.20, 0.05, −0.10%, respectively. The obtained results validate the effectiveness of the PID control strategy.

From the above results, it may be concluded that the results demonstrate that the method of cooperative static automatic leveling is qualified to decrease the errors of cables length fast, and can realize the automatic leveling of the platform accurately.

Table 8.2 Platform orientation when leveling control time is 8 s

Platform orientation	Before leveling	After leveling
Roll angle $\alpha(°)$	−2.27	−0.14
Pitch angle $\beta(°)$	3.53	0.21
Automatic leveling time (s)	8	

Table 8.3 Height between the platform and the ground

Height of the platform	Before leveling	After leveling
l_1 (m)	1.15	1.1341
l_2 (m)	1.20	1.1435
l_3 (m)	1.10	1.1203
l_4 (m)	1.05	1.1126
Automatic leveling time (s)	8	

Table 8.4 Platform orientation of the platform before and after leveling

Platform orientation	Before leveling	After leveling
Roll angle $\alpha(°)$	−2.27	0.062
Pitch angle $\beta(°)$	3.53	−0.073
Automatic leveling time (s)	9.72	

Table 8.5 Height between the platform and the ground before and after leveling

Height of the platform	Before leveling	After leveling
l_1 (m)	1.15	1.1241
l_2 (m)	1.20	1.1272
l_3 (m)	1.10	1.1256
l_4 (m)	1.05	1.1239
Automatic leveling time(s)	9.72	

8.6 Conclusions

In this study, we have introduced a cooperative cable parallel robot system for four mobile cranes, which combines the mobile crane with the cable parallel robot in a way to provide a solution for handling complex tasks that are difficult or even impossible for a single crane. Design model and cooperative localization scheme of the CPRMC are presented, and the improved localization algorithm is designed based on multilateration method. The scatter distribution of the localization results shows the improved localization algorithm is effective. The global path planning of the CPRMC is performed on the basis of the artificial potential field method with the grid method. The sensor technology is applied to the cooperative obstacle avoidance, and a co-simulation platform is built for cooperative obstacle avoidance analysis. The four-point collaborative leveling method is adopted for automatic leveling control for the platform of the CPRMC with PID controller. The leveling error of roll angle α of the platform is 0.062°, and the leveling error of pitch angle β of the platform is −0.073°. The relative height errors between the platform vertexes and the ground are −0.08, 0.20, 0.05, −0.10%, respectively. The results demonstrated that the automatic leveling control performance of the platform for the CPRMC has been remarkably improved by applying the proposed collaborative leveling method.

References

1. Zi B, Lin J, Qian S (2015) Localization, obstacle avoidance planning and control of a cooperative cable parallel robot for multiple mobile cranes. Robot Comput-Integr Manuf 34:105–123
2. Zhuang Y, Wang Z, Yu H et al (2013) A robust extended H∞ filtering approach to multi-robot cooperative localization in dynamic indoor environments. Control Eng Pract 21 (7):953–961
3. Gholami M, Cai N, Brennan RW (2013) An artificial neural network approach to the problem of wireless sensors network localization. Robot Comput-Integr Manuf 29(1):96–109
4. Sit TCH, Liu Z, Ang MH Jr et al (2007) Multi-robot mobility enhanced hop-count based localization in ad hoc networks. Robot Auton Syst 55(3):244–252
5. Rone W, Ben-Tzvi P (2013) Mapping, localization and motion planning in mobile multi-robotic systems. Robotica 31(01):1–23
6. Lee HC, Lee SH, Choi MH et al (2012) Probabilistic map merging for multi-robot RBPF-SLAM with unknown initial poses. Robotica 30(02):205–220
7. Cimino M, Pagilla PR (2011) Optimal location of mouse sensors on mobile robots for position sensing. Automatica 47(10):2267–2272
8. Sharma B, Vanualailai J, Singh S (2012) Lyapunov-based nonlinear controllers for obstacle avoidance with a planar n-link doubly nonholonomic manipulator. Robot Auton Syst 60 (12):1484–1497
9. Chyan GS, Ponnambalam SG (2012) Obstacle avoidance control of redundant robots using variants of particle swarm optimization. Robot Comput-Integr Manuf 28(2):147–153
10. Trianni V, Nolfi S, Dorigo M (2006) Cooperative hole avoidance in a swarm-bot. Robot Auton Syst 54(2):97–103
11. Moon J, Prasad JVR (2011) Minimum-time approach to obstacle avoidance constrained by envelope protection for autonomous UAVs. Mechatronics 21(5):861–875
12. Liu Y, Fang S, Otsubo H et al (2013) Simulation and research on the automatic leveling of a precision stage. Comput Aided Des 45(3):717–722
13. Qinghua X (2012) Research on Localization Algorithm in Wireless Sensor Networks. Nat Sci J Harbin Normal Univ 2:015
14. Ge SS, Cui YJ (2000) New potential functions for mobile robot path planning. IEEE Trans Robot Autom 16(5):615–620

Chapter 9
Development, Coordinated Control and Experiments of the Cooperative Cable Parallel Manipulators for Multiple Mobile Cranes

9.1 Preamble

This chapter addresses the dynamics and trajectory tracking control of cooperative multiple mobile cranes. Compared with a single mobile crane, cooperative cable parallel manipulators for multiple mobile cranes (CPRMC) are more complex in configuration, which have the characters of both series and parallel manipulators. Therefore, for the CPRMC, the forward as well as the inverse kinematics and dynamics include the difficulties of both series and parallel manipulators. However, the closed kinematic chain brings about potential benefits, including sufficient accuracy, higher cost performance, better lifting capacity and security et al. First, the forward and inverse kinematics of the CPRMC with point-mass are derived with elimination method, and the complete dynamic model of the CPRMC is established based on Lagrange equation and the complete kinematics. Second, consider the repetitive tasks and high security and precision requirement, a robust iterative learning controller is designed for trajectory tracking on the basis of the linearization of the dynamics. Third, taking the engineering practice into consideration, two case studies are simulated with the same expected trajectory but different weights of the loads. Finally, the designed controller is compared with traditional PD control algorithm via numerical simulation. The results demonstrate the feasibility and superiority of the CPRMC and designed controller, and provide a theoretical basis for the cooperative of multiple mobile cranes.

Different types of cranes are widely used for material transportation in many industrial fields, such as steel enterprises, electric power construction, container terminals and many manufacturing segments. A single crane with a cable suspended load has been widely studied in the literature. Owing to the under-actuated nature, the crane dynamics are not fully input-to-output linearizable [1]. The severely nonlinear properties together with extraneous disturbances cause undesired swings, especially at launch and break, which badly reduce the efficiency and

B. Zi and S. Qian, *Design, Analysis and Control of Cable-Suspended Parallel Robots and Its Applications*, DOI 10.1007/978-981-10-1753-7_9

security [2]. Thus, fast and accurate suspending of loads with minimum swing is the objective of various crane controllers.

The last decades have seen the rapid development of world economy. Due to the ever-increasing quantity of cargo, hazardous terrain and weather, some complex tasks are impossible to handle with a single traditional crane, inevitably [3]. Hence, a new generation of cranes, cable parallel crane-type robots are proposed. The closed kinematic chain in the cable parallel crane-type robots makes the configuration more complex, however, brings about potential benefits including better load control, sufficient accuracy, larger capacity and workspace [4–7]. Obviously, the cable parallel crane-type robots have many advantages compared with traditional cranes. However, it should be noted that, the cable parallel crane-type robots are still not widely used in engineering practice. One of the reasons is most of the cable parallel crane-type robots are redesigned for specific tasks instead of general ones. Meanwhile, the cooperative multiple cranes attracts more attention recently for the redesigning cost is saved [8]. Thus, as more reconfigurable and cost-effective cable parallel crane-type robots, the cooperative multiple cranes have broad application prospects, as shown in Fig. 9.1.

It is well known that the inverse kinematics of parallel manipulators is simple in comparison with series ones. Inversely, the forward kinematics of parallel manipulators is more difficulty. In the CPRMC, each single mobile crane can be considered as a series manipulator with three degrees of freedom (DOFs). Moreover, the cables together with the load can be treated as a cable parallel robot, of which the vertexes are moved by the corresponding mobile crane. Hence, for the CPRMC, the forward as well as the inverse kinematics includes the difficulties of both series and parallel manipulators [9]. As a version of cable parallel crane-type robot, cooperative multiple mobile cranes has also been studied in previous researches, in particular in [10, 11]. A cost-effective, safe, and compliant solution for the operation of the CPRMC is the urgent need to reach potential advantages of both cranes and the cable parallel robot [10]. And there are still some important unresolved problems, especially

Fig. 9.1 CPRMC in engineering applications

complete kinematics and dynamics. Specifically, quasi-static models of the CPRMC is studied in [10, 11], with the assumption that the load together with the mobile cranes have motions that give rise to negligible inertial forces. In addition, the dynamics of the platform in the CPRMC was analyzed based on D'Alembert principle, regardless of the dynamics of cranes [8]. Indeed it is impossible to make exact assertions about the results without explicitly modeling and analyzing the complete dynamics of the system. Different approaches can be used to solve the dynamics of manipulators, such as Kane equation [12], Newton-Euler equations [13], principle of virtual work [14] and Lagrange equation [4], etc.

In addition, control of cooperative multiple mobile cranes is another key problem to be solved. Many recent researches have discussed the control of multiple robots. For instance, a multiple unmanned aerial vehicles control scheme was developed in [15], considering the full nonlinear position/orientation model of a multi-quadrotors system. Zi et al. [16] designed an adaptive iterative learning tracking controller for a hybrid driven cable parallel manipulator. An inverse kinematic control strategy for cooperative dual cranes was presented in [17] regardless of the ship motion on which the cranes are attached. Li et al. [18] proposed a new version of adaptive robust control for multiple mobile manipulators carrying a common object in a cooperative manner.

The major contribution of this chapter lies in the solution of the complete kinematics and dynamics of the cooperative cable parallel manipulators for multiple mobile cranes, which is the foundation for the further research of this new promising and significant engineering equipment. Moreover, an effective and suitable controller is designed considering the engineering practice. The rest of the paper is structured as follows. In Sect. 9.2, the kinematics of the cooperative three mobile cranes with point-mass is established with elimination method, including both forward and inverse kinematics. Based on Lagrange equation and the kinematics, the complete dynamic model of the CPRMC is derived in Sect. 9.3. In Sect. 9.4, a robust iterative learning controller (ILC) of the CPRMC is designed on the basis of the linearization of the dynamics. Numerical simulation is conducted in Sect. 9.5, in order to investigate the dynamics and the designed robust ILC of the CPRMC. Finally Sect. 9.6 provides concluding remarks and thoughts on future work.

9.2 Kinematics

9.2.1 Description of Mechanism

Figure 9.2 shows the general three-dimensional CAD model of the CPRMC, which is established based on the engineering applications in Fig. 9.1. The simple schematic sketches of the whole CPRMC and single mobile crane are provided in

Fig. 9.2 CAD model of the CPRMC

Figs. 9.3 and 9.4, respectively. As shown in Fig. 9.3, the global frame OXYZ is attached to the ground at the slewing center O_1 of the first mobile crane in order to define the position of the payload, as well as the orientation of three cranes. The three cables of cranes are connected together at the hoisting point in the load. The CPRMC can be considered as a nine-input and three-output system, in which the three cables are capable of controlling the three Cartesian DOFs of the load but not including the rotation. However, taking the engineering application into consideration, we are only interested in the Cartesian DOFs in this study and the mentioned configuration can meet most engineering tasks. Obviously, the CPRMC is a redundant actuated manipulator since the DOFs of the load are less than the inputs of the system. It is important to note that although the load is fully constrained, its maximum acceleration in the vertical direction will be limited by the acceleration due to the gravity.

Referring to Fig. 9.4, the local frame of each single mobile crane $o_ix_iy_i$ are attached to the corresponding turrets at the slewing center O_i (i = 1, 2, 3). Links A_iB_i and C_iD_i denote the telescopic booms and the luffing hydraulic cylinders, respectively. The length of boom is L_i (i = 1, 2, 3). The lower pivot point B_i of the boom as well as the lower pivot point C_i of luffing hydraulic cylinder are fixed in the crane turret of respective mobile crane. O_iE_i represent the slewing axis of crane turret. In addition, the horizontal distance between down pivot-point of boom B_i and the slewing axis is f. The length of links B_iD_i are l_{si} (i = 1, 2, 3). The horizontal and vertical distance between down pivot-point of cylinder C_i and the slewing center O_i are s and h, respectively.

9.2.2 *Forward Kinematics of Cable Parallel Manipulator with Movable Vertexes*

A single mobile crane can be treated as a 3-DOF rigid serial manipulator with 3 inputs and 3 outputs, of which the end effecter can reach any point in designed workspace. Hence, the parameters of nine inputs of driving devices are chosen as the generalized global coordinates of the whole system for kinematics and dynamics analysis, which are the slewing angles ϕ_i (i = 1, 2, 3) of three groups

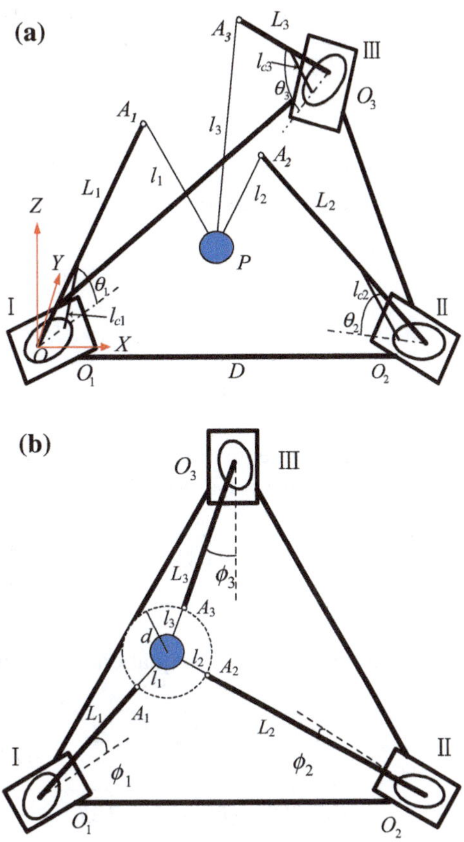

Fig. 9.3 Schematic sketch (**a**) and a top view (**b**) of the CPRMC

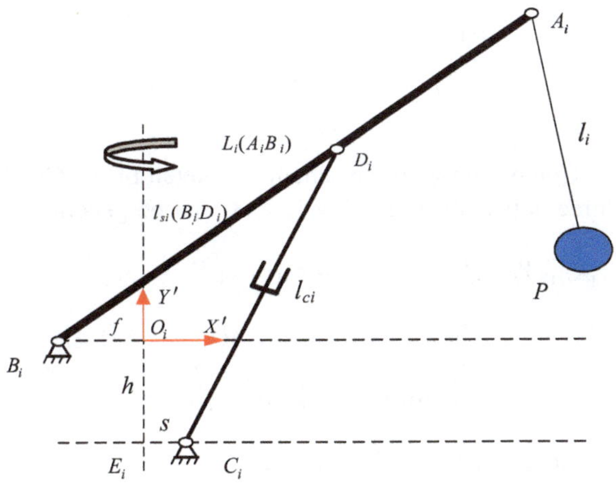

Fig. 9.4 Schematic sketch of a mobile crane

of turrets, lengths of three groups of cables l_i (i = 1, 2, 3) and hydraulic cylinders l_{ci} (i = 1, 2, 3).

$$\boldsymbol{q} = [\phi_1, \phi_2, \phi_3, l_1, l_2, l_3, l_{c1}, l_{c2}, l_{c3}]. \tag{9.1}$$

According to the structure of the CPRMC in Fig. 9.3, the global coordinates of each mobile crane are obtained as

$$[X_{O1}, Y_{O1}, Z_{O1}] = [0 \quad 0 \quad 0]^T \tag{9.2}$$

$$[X_{O2}, Y_{O2}, Z_{O2}] = [D \quad 0 \quad 0]^T \tag{9.3}$$

$$[X_{O3}, Y_{O3}, Z_{O3}] = [D \sin(\pi/6) \quad D \cos(\pi/6) \quad 0]^T. \tag{9.4}$$

The constraint equations of the pivot point D_i (x_{Di}, y_{Di}) (i = 1, 2, 3) on the plane $A_iB_iC_i$ consisted by the boom and hydraulic cylinder can be expressed as

$$\begin{cases} (x_{Di}l_i + s)^2 + y_{Di}l_1^2 = l_s^2 \\ (x_{Di}l_i + s - f)^2 + (y_{Di}l_i + h)^2 - lc_i^2. \end{cases} \tag{9.5}$$

By solving the above equation, one can obtain

$$\begin{aligned} x_{Di} = & \frac{-2sf + f^2 + h^2 - l_{ci}^2 + l_s^2}{2f} - \frac{1}{2f(f^2 + h^2)} h\left(f^2 h + h^3 - hl_{ci}^2 + hl_s^2\right. \\ & \left. - f\sqrt{\left(f^2 + h^2 - l_{ci}^2 + 2l_{ci}l_s - l_s^2\right)\left(-f^2 - h^2 + l_{ci}^2 + 2l_{c1}l_s + l_s^2\right)}\right) \end{aligned} \tag{9.6}$$

$$\begin{aligned} y_{Di} = & \frac{1}{2(f^2 + h^2)} \left(-f^2 h - h^3 + hl_{ci}^2 - hl_s^2\right. \\ & \left. + f\sqrt{\left(f^2 + h^2 - l_{ci}^2 + 2l_{ci}l_s - l_s^2\right)\left(-f^2 - h^2 + l_{ci}^2 + 2l_{ci}l_s + l_s^2\right)}\right). \end{aligned} \tag{9.7}$$

Combining with the schematic sketch of the CPRMC, the global coordinates of three upper pivots D_i (i = 1, 2, 3) can be generated as

$$[X_{D1}, Y_{D1}, Z_{D1}] = [X_{O1} + x_{D1} \cos(\phi_1 + \pi/6), Y_{O1} + x_{D1} \sin(\phi_1 + \pi/6), y_{D1}] \tag{9.8}$$

$$[X_{D2}, Y_{D2}, Z_{D2}] = [X_{O2} + x_{D2} \cos(\pi/6 - \phi_2), Y_{O2} + x_{D2} \sin(\pi/6 - \phi_2), y_{D2}] \tag{9.9}$$

$$[X_{D3}, Y_{D3}, Z_{D3}] = [X_{O3} + x_{D3} \cos \phi_3, Y_{O3} + x_{D3} \sin \phi_3, y_{D3}]. \tag{9.10}$$

Hence, the global coordinates of three top points of booms both on the plane $A_iB_iC_i$ and in space can be derived as

$$\begin{cases} x_{Ai} = \frac{L(x_{Di}+s)}{l_s} - s \\ y_{Ai} = \frac{L y_{Di}}{l_s} \end{cases} \quad (9.11)$$

$$[X_{A1}, Y_{A1}, Z_{A1}] = [X_{O1} + x_{A1}\cos(\phi_1 + \pi/6), Y_{O1} + x_{A1}\sin(\phi_1 + \pi/6), y_{A1}] \quad (9.12)$$

$$[X_{A2}, Y_{A2}, Z_{A2}] = [X_{O2} + x_{A2}\cos(\pi/6 - \phi_2), Y_{O2} + x_{A2}\sin(\pi/6 - \phi_2), y_{A2}] \quad (9.13)$$

$$[X_{A3}, Y_{A3}, Z_{A3}] = [X_{O3} + x_{A3}\sin\phi_2, Y_{O3} + x_{A3}\cos\phi_2, y_{A3}]. \quad (9.14)$$

It is well known that the inverse kinematics of parallel manipulators is simple compared with series ones. However, the forward kinematics of parallel manipulators is more difficulty relatively. Therefore, as mentioned in Sect. 8.1, the vertexes of cable towers in traditional cable parallel manipulators are mainly fixed related to the ground. In addition, some assumptions are made for simplification of modelling, such as the equal of the height of the cable towers. Meanwhile, the workspace and flexibility of the CPR is reduced.

Let [X Y Z] be the coordinate of the load. Then, the constraint equation of the load and three cables in the CPR can be generated as follows based on the forward kinematics of single mobile crane in Eqs. (9.12)–(9.14)

$$l_i^2 = (X - X_{Ai})^2 + (Y - Y_{Ai})^2 + (Z - Z_{Ai})^2 \quad (9.15)$$

where the cable length l_i is measured from the hoisting point P to the pulley center A_i.

Elimination method is used to solve the equations with respect to the global coordinates of three top points of booms, iterative substitution is defined as

$$[\Delta X, \Delta Y, \Delta Z]^T = [X - X_{A1}, Y - Y_{A1}, Z - Z_{A1}]^T \quad (9.16)$$

$$[\Delta X_1, \Delta Y_1, \Delta Z_1]^T = [X_{A2} - X_{A1}, Y_{A2} - Y_{A1}, Z_{A2} - Z_{A1}]^T \quad (9.17)$$

$$[\Delta X_2, \Delta Y_2, \Delta Z_2]^T = [X_{A3} - X_{A1}, Y_{A3} - Y_{A1}, Z_{A3} - Z_{A1}]^T. \quad (9.18)$$

Then, the length of the cables can be expressed as

$$\begin{bmatrix} l_1^2 \\ l_2^2 \\ l_3^2 \end{bmatrix} = \begin{bmatrix} \Delta X^2 + \Delta Y^2 + \Delta Z^2 \\ (\Delta X - \Delta X_1)^2 + (\Delta Y - \Delta Y_1)^2 + (\Delta Z - \Delta Z_1)^2 \\ (\Delta X - \Delta X_2)^2 + (\Delta Y - \Delta Y_2)^2 + (\Delta Z - \Delta Z_2)^2 \end{bmatrix}. \quad (9.19)$$

Via expansion and subtraction, the following equation can be derived as

$$l_1^2 - l_2^2 + \Delta X_1^2 + \Delta Y_1^2 + \Delta Z_1^2 = 2(\Delta X_1 \Delta X + \Delta Y_1 \Delta Y + \Delta Z_1 \Delta Z) \quad (9.20)$$

$$l_1^2 - l_3^2 + \Delta X_2^2 + \Delta Y_2^2 + \Delta Z_2^2 = 2(\Delta X_2 \Delta X + \Delta Y_2 \Delta Y + \Delta Z_2 \Delta Z). \tag{9.21}$$

Equations (9.20) and (9.21) can be rewritten through substitution of the right part as follows:

$$\Delta X = \frac{u\Delta Y_2 - v\Delta Y_1 + \Delta Z(\Delta Z_2 \Delta Y_1 - \Delta Z_1 \Delta Y_2)}{\Delta X_1 \Delta Y_2 - \Delta X_1 \Delta Y_2} = u_1 + u_2 \Delta Z \tag{9.22}$$

$$\Delta Y = \frac{u\Delta X_2 - v\Delta X_1 + \Delta Z(\Delta Z_2 \Delta X_1 - \Delta Z_1 \Delta X_2)}{\Delta X_1 \Delta Y_2 - \Delta X_1 \Delta Y_2} = u_3 + u_4 \Delta Z \tag{9.23}$$

where

$$u = \frac{1}{2}\left(l_1^2 - l_2^2 + \Delta X_1^2 + \Delta Y_1^2 + \Delta Z_1^2\right) \tag{9.24}$$

$$v = \frac{1}{2}\left(l_1^2 - l_3^2 + \Delta X_2^2 + \Delta Y_2^2 + \Delta Z_2^2\right) \tag{9.25}$$

$$u_1 = \frac{u\Delta Y_2 - v\Delta Y_1}{\Delta X_1 \Delta Y_2 - \Delta X_1 \Delta Y_2} \tag{9.26}$$

$$u_2 = \frac{\Delta Z_2 \Delta Y_1 - \Delta Z_1 \Delta Y_2}{\Delta X_1 \Delta Y_2 - \Delta X_1 \Delta Y_2} \tag{9.27}$$

$$u_3 = \frac{u\Delta X_2 - v\Delta X_1}{\Delta X_1 \Delta Y_2 - \Delta X_1 \Delta Y_2} \tag{9.28}$$

$$u_4 = \frac{\Delta Z_2 \Delta X_1 - \Delta Z_1 \Delta X_2}{\Delta X_1 \Delta Y_2 - \Delta X_1 \Delta Y_2}. \tag{9.29}$$

Thus, substitution variable ΔZ can be generated with respect to the global coordinates of the vertexes and the lengths of three cables

$$l_1^2 = (u_1 + u_2 \Delta Z)^2 + (u_3 + u_4 \Delta Z)^2 + \Delta Z^2. \tag{9.30}$$

Through solving Eq. (9.30) and ignoring the unreasonable one of the two solutions, one can obtain substitution variable ΔZ, and then the global coordinates of the load [X, Y, Z] can be derived by combining with Eqs. (9.19)–(9.23).

9.2.3 *Inverse Kinematics of Cable Parallel Manipulator with Movable Vertexes*

As mentioned, the CPRMC is a redundant actuated system. Furthermore, there are infinite groups of inputs that can meet the given movement of the load. Generally, limited solutions can be obtained through adding certain constraints, such as shortest stroke, minimum power and force, etc. Taking the engineering application into consideration, two certain constraints are proposed, as depicted in Fig. 9.3b. In detail, in order to avoid collision during the cooperation of mobile cranes, suppose the horizontal distance between the top points of the boom A_i (i = 1, 2, 3) and hoisting point of load P is d. Besides, the boom and the corresponding cable are constrained in the same direction from the top view of the CPRMC, in order to minimize the lateral forces in cables and avoid side tumbling of cranes.

Hence, the coordinates of the top points of the booms A_i can be obtained by

$$\begin{bmatrix} X_{A_i} \\ Y_{A_i} \\ Z_{A_i} \end{bmatrix} = \begin{bmatrix} X + d(X_{Oi} - X)/\sqrt{(X - X_{Oi})^2 + (Y - Y_{Oi})^2} \\ Y + d(Y_{Oi} - Y)/\sqrt{(X - X_{Oi})^2 + (Y - Y_{Oi})^2} \\ \sqrt{L^2 - \left(\sqrt{(X - X_{Oi})^2 + (Y - Y_{Oi})^2} - d + s\right)^2} \end{bmatrix}. \tag{9.31}$$

On the basis of the schematic sketch of a mobile crane in Fig. 9.4, the coordinates of the pivot points D_i relative to the global frame O can be achieved as

$$\begin{bmatrix} X_{D_i} \\ Y_{D_i} \\ Z_{D_i} \end{bmatrix} = \begin{bmatrix} X_{Ai} + \frac{(L-l_s)(X_{Oi}-X_{Ai})}{L-Ls/\left(\sqrt{(X-X_{Oi})^2+(Y-Y_{Oi})^2}+s-d\right)} \\ Y_{Ai} + \frac{(L_i-l_s)(Y_{Oi}-Y_{Ai})}{L-Ls/\left(\sqrt{(X-X_{Oi})^2+(Y-Y_{Oi})^2}+s-d\right)} \\ \frac{Z_{Ai}l_s}{L} \end{bmatrix}. \tag{9.32}$$

Thus the following equation is concluded as

$$l_{ci} = \sqrt{\left(\sqrt{(X_{Di} - X_{Di})^2 + (Y_{Di} - Y_{Oi})^2} - f\right)^2 + (Z_{Di} + h)^2} \tag{9.33}$$

$$l_i = \sqrt{(X - X_{Ai})^2 + (Y - Y_{Ai})^2 + (Z - Z_{Ai})^2} \tag{9.34}$$

$$\begin{bmatrix} \phi_1 \\ \phi_2 \\ \phi_3 \end{bmatrix} = \begin{bmatrix} \arctan\left(\frac{Y_{A1}/X_{A1}-\tan(\pi/6)}{1+(Y_{A1}/X_{A1})\tan(\pi/6)}\right) \\ \arctan\left(\frac{(Y_{A2}-Y_{O2})/(X_{A2}-X_{O2})+\tan(\pi/6)}{1-((Y_{A2}-Y_{O2})/(X_{A2}-X_{O2}))\tan(\pi/6)}\right) \\ \arctan\frac{(X_{A3}-X_{O3})}{(Y_{O3}-Y_{A3})} \end{bmatrix}. \tag{9.35}$$

9.3 Dynamics

There are different approaches for solving the dynamics of manipulators, including Newton-Euler equation, Kane equation, Lagrange equation, etc. Based on D'Alembert principle, Lagrange equation takes the whole mechanical system as the object, which is simple in form. Moreover, the dynamic model established based on Lagrange equation is standard in form, which can be directly used for further study of the mechanism, such as control and load distribution, etc. In this article, Lagrange equation is adopted to solve the dynamics of the CPRMC. The three main parts of the CPRMC are analyzed, including the load, the booms and the turrets.

Based on the forward kinematics of cable parallel manipulator with activities vertexes, the velocity of the load can be written in differential matrix form as

$$\begin{bmatrix} v_X \\ v_Y \\ v_Z \end{bmatrix} = \begin{bmatrix} \sum_{i=1}^{3} \left(\frac{\partial X}{\partial l_i} l'_i + \frac{\partial X}{\partial l_{ci}} l'_{ci} + \frac{\partial X}{\partial \phi_i} \phi'_i \right) \\ \sum_{i=1}^{3} \left(\frac{\partial Y}{\partial l_i} l'_i + \frac{\partial Y}{\partial l_{ci}} l'_{ci} + \frac{\partial Y}{\partial \phi_i} \phi'_i \right) \\ \sum_{i=1}^{3} \left(\frac{\partial Z}{\partial l_i} l'_i + \frac{\partial Z}{\partial l_{ci}} l'_{ci} + \frac{\partial Z}{\partial \phi_i} \phi'_i \right) \end{bmatrix}. \tag{9.36}$$

Hence, the kinetic energy of the load mass m can be expressed as

$$K_m = \frac{1}{2} m \left(v_X^2 + v_Y^2 + v_Z^2 \right) = \sum_{i=1}^{9} \frac{q'_i S_i}{2} \tag{9.37}$$

where

$$S_i = m \left(\frac{\partial X}{\partial q_i} v_X + \frac{\partial Y}{\partial q_i} v_Y + \frac{\partial Z}{\partial q_i} v_Z \right), \quad i = 1, 2 \ldots 9. \tag{9.38}$$

The gravitational potential energy of the load is as follows

$$P_m = mgZ. \tag{9.39}$$

The Lagrange equation of the load is established as

$$\frac{\partial}{\partial t} \left(\frac{\partial K_m}{\partial q'} \right) - \frac{\partial K_m}{\partial q} = \boldsymbol{\tau}_m + \frac{\partial P_m}{\partial q} \tag{9.40}$$

where $\boldsymbol{\tau}_m$ stands for the general force on the nine actuators of the CPRMC with respect to the mass of the payload m,

$$\boldsymbol{\tau}_m = \left[\tau_{m1}, \tau_{m2}, \tau_{m3}, \tau_{m4}, \tau_{m5}, \tau_{m6}, \tau_{m7}, \tau_{m8}, \tau_{m9}\right]^T. \tag{9.41}$$

Referring to Eq. (9.36) to Eq. (9.39), we obtain

$$\frac{\partial}{\partial t}\left(\frac{\partial K_m}{\partial \dot{q}_i}\right) = \frac{\partial S_i}{\partial t} = \sum_{i=1}^{9}\left(\frac{\partial S_i}{\partial \dot{q}_i}\ddot{q}_i + \frac{\partial S_i}{\partial q_i}\dot{q}_i\right). \tag{9.42}$$

By substituting Eq. (9.42) into Eq. (9.40), the left part of the Lagrange equation of the load can be expressed in the standard form as

$$\frac{\partial}{\partial t}\left(\frac{\partial K_m}{\partial \boldsymbol{q}'}\right) - \frac{\partial K_m}{\partial \boldsymbol{q}} = \boldsymbol{M}_m\ddot{\boldsymbol{q}} + \boldsymbol{C}_m\dot{\boldsymbol{q}} \tag{9.43}$$

where

$$\boldsymbol{M}_{mij} = \frac{\partial S_i}{\partial q_j'}, \quad i = 1, 2\ldots 9; \quad j = 1, 2\ldots 9 \tag{9.44}$$

$$\boldsymbol{C}_{mij} = \frac{\partial S_i}{\partial q_j} + \frac{\partial S_j}{2\partial q_i}, \quad i = 1, 2\ldots 9; \quad j = 1, 2\ldots 9. \tag{9.45}$$

In addition, the angles of the booms around the down pivots are

$$\theta_i = \arcsin\frac{Z_{Ai}}{L}, \quad i = 1, 2, 3. \tag{9.46}$$

One should notice that, the movement of each boom includes two parts, which are rotation around the lower pivot with the angle θ_i (i = 1, 2, 3) and rotation around the slewing center of the turret with the angle ϕ_i (i = 1, 2, 3). Therefore, the kinetic energy of the boom contains two parts, which can be represented as

$$K_L = \frac{1}{2}\sum_{i=1}^{3}\left(\bar{J}_{Li}\dot{\theta}_i^2 + \tilde{J}_{Li}\dot{\phi}_i^2\right), \quad i = 1, 2, 3 \tag{9.47}$$

$$\dot{\theta}_i = \frac{1}{\sqrt{L^2 - Z_{Ai}^2}}\sum_{j=1}^{9}\left(\frac{\partial Z_{Ai}}{\partial q_j}q_j'\right), \quad i = 1, 2, 3 \tag{9.48}$$

$$\bar{J}_{Li} = \frac{1}{3}m_L L^2, \quad i = 1, 2, 3 \tag{9.49}$$

where $\bar{J}_{Li}$ is the rotary inertia of the booms around the corresponding lower pivots. $\dot{\theta}_i$ represents the angle velocity of the boom around the lower pivot.

During the rotation of booms around the slewing center of the turrets, the rotary inertia of the booms around the slewing center are variable with the angle θ_i. Therefore, the rotary inertia around the slewing center can be derived as

$$\begin{aligned}\tilde{J}_{Li} &= \frac{1}{3} m_L L^2 (1 - \sin^2 \theta_i) \\ &= \frac{1}{3} m_L (L^2 - Z_{Ai}^2).\end{aligned} \tag{9.50}$$

From Fig. 9.4, the coordinates of the lower boom pivots can be represented as

$$[X_{B1}, Y_{B1}, Z_{B1}] = [X_{O1} - s\cos(\phi_1 + \pi/6), Y_{O1} - s\sin(\phi_1 + \pi/6), 0] \tag{9.51}$$

$$[X_{B2}, Y_{B2}, Z_{B2}] = [X_{O2} - s\cos(\pi/6 - \phi_2), Y_{O2} - s\sin(\pi/6 - \phi_2), 0] \tag{9.52}$$

$$[X_{B3}, Y_{B3}, Z_{B3}] = [X_{O3} - s\sin\phi_3, Y_{O3} - s\cos\phi_3, 0]. \tag{9.53}$$

The assumption is made that the telescopic booms of mobile cranes are homogeneous, thus the center of gravity of each boom in global coordinate system can be expressed as

$$[X_{Li}, Y_{Li}, Z_{Li}] = \left[\frac{(X_{Ai} + X_{Bi})}{2}, \frac{(Y_{Ai} + Y_{Bi})}{2}, \frac{(Z_{Ai} + Z_{Bi})}{2}\right], \quad i = 1, 2, 3. \tag{9.54}$$

Thus, the gravitational potential energy of the booms are

$$P_L = \sum_{i=1}^{3} m_L g Z_{Li}, \quad i = 1, 2, 3. \tag{9.55}$$

Therefore, the Lagrange equation of the booms can be established in the standard form as

$$\tau_L = \frac{\partial}{\partial t}\left(\frac{\partial K_L}{\partial \boldsymbol{q}'}\right) - \frac{\partial K_L}{\partial \boldsymbol{q}} + \frac{\partial P_L}{\partial \boldsymbol{q}} = \boldsymbol{M}_L \ddot{\boldsymbol{q}} + \boldsymbol{C}_L \dot{\boldsymbol{q}} + \boldsymbol{G}_L \tag{9.56}$$

where

$$M_{Lij} = \frac{\partial}{\partial q_j'}\left(\frac{\partial K_L}{\partial q_i'}\right), \quad i = 1, 2, 3 \cdots 9; \quad j = 1, 2, 3 \tag{9.57}$$

$$C_{Lij} = \frac{\partial}{\partial q_j}\left(\frac{\partial K_L}{\partial q_i'}\right) + \frac{\partial}{2\partial q_i}\eta_j + \delta_{ij}, \quad i = 1, 2, 3 \cdots 9; \quad j = 1, 2, 3 \tag{9.58}$$

$$\eta_i = \sum_{n=1}^{3} \frac{\bar{J}_L}{L^2 - Z_{An}^2} \frac{\partial Z_{An}}{\partial q_i} \dot{\theta}_n, \quad i = 1, 2, 3 \tag{9.59}$$

$\boldsymbol{\delta}$ is a 9 × 9 matrix generated in the process of deriving the Coriolis matrix of the 9-input and 3-output system. For simplicity in form, $\boldsymbol{\delta}$ is written in partitioned matrix in Eq. (9.60) including two block matrixes $\bar{\boldsymbol{\delta}}$ and $\tilde{\boldsymbol{\delta}}$, which is represented by Eq. (9.61) and Eq. (9.62), respectively.

$$\boldsymbol{\delta} = \left[\bar{\boldsymbol{\delta}} \vdots \tilde{\boldsymbol{\delta}}\right] \tag{9.60}$$

$$\bar{\boldsymbol{\delta}} = 0_{9,6} \tag{9.61}$$

$$\tilde{\boldsymbol{\delta}}_{ij} = \frac{m_L}{6} \left[\frac{\partial \left(L^2 - Z_{aj}^2 \right)}{\partial q_i} \dot{\phi}_j \right], \quad i = 7, 8, 9; \quad j = 1, 2, 3. \tag{9.62}$$

Referring to Figs. 9.3 and 9.4, the kinetic energy of the turrets are K_r and P_r, respectively.

$$K_r = \frac{1}{2} \sum_{i=1}^{3} J_r \dot{\phi}_i^2, \quad i = 1, 2, 3 \tag{9.63}$$

$$P_r = 0 \tag{9.64}$$

where J_r is the rotary inertia of the turrets around the corresponding slewing center.

Therefore, the Lagrange equation of the turrets can be established in the standard form as

$$\boldsymbol{\tau}_r = \frac{\partial}{\partial t} \left(\frac{\partial K_r}{\partial \boldsymbol{q}'} \right) - \frac{\partial K_r}{\partial \boldsymbol{q}} + \frac{\partial P_r}{\partial \boldsymbol{q}} = \boldsymbol{M}_r \ddot{\boldsymbol{q}} + \boldsymbol{C}_r \dot{\boldsymbol{q}} + \boldsymbol{G}_r \tag{9.65}$$

where

$$\boldsymbol{M}_r = \boldsymbol{diag}(0, 0, 0, 0, 0, 0, J_r, J_r, J_r) \tag{9.66}$$

$$\boldsymbol{C}_r = 0. \tag{9.67}$$

Hence, the Lagrange equation of the CPRMC is established in the standard form as

$$\boldsymbol{\tau} = \frac{\partial}{\partial t} \left(\frac{\partial K}{\partial \boldsymbol{q}'} \right) - \frac{\partial K}{\partial \boldsymbol{q}} + \frac{\partial P}{\partial \boldsymbol{q}} = \boldsymbol{M} \ddot{q} + \boldsymbol{C} \dot{q} + \boldsymbol{G} \tag{9.68}$$

where

$$\tau = [\tau_1, \tau_2, \tau_3, \tau_4, \tau_5, \tau_6, \tau_7, \tau_8, \tau_9]^T \tag{9.69}$$

$$M = M_m + M_L + M_r \tag{9.70}$$

$$C = C_m + C_L + C_r \tag{9.71}$$

$$G = G_m + G_L + G_r. \tag{9.72}$$

where τ is the generalized force of the CPRMC, **M** is the inertia matrix of the system, which is a positive definite matrix, **C** is the vector of Coriolis and centripetal terms, **G** is the vector of gravity terms.

9.4 Robust ILC Design

ILC is based on the notion that the performance of a system that executes repetitive tasks can be improved by learning from previous iteration. It is promising control scheme for rehabilitation robots due to the repetitiveness of the therapy mode. ILC has been successfully applied to industrial robots control [16]. As is known, mobile cranes are widely used for loading, mounting, carrying large heavy-duty loads in engineering practice. It should be noted that most of the tasks are repetitive and require high security and precision. In this section, An robust ILC method is designed for trajectory tracking control of the CPRMC. Take the non-repetitive interference and uncertain dynamic terms into consideration, the dynamic equation of the CPRMC can be written as

$$M(q^j(t))\ddot{q}^j(t) + C(q^j(t), \dot{q}^j(t))\dot{q}^j(t) + G(q^j(t), \dot{q}^j(t)) + d^j(t) = \tau^j(t) \tag{9.73}$$

$$d^j(t) = \Delta M(q^j(t))\ddot{q}^j(t) + \Delta C(q^j(t), \dot{q}^j(t))\dot{q}^j(t) + \Delta G(q^j(t), \dot{q}^j(t)) + \tau_a^j(t) \tag{9.74}$$

where $t \in [0, T]$ is the time, $j \in N$ denotes the iteration number. $\tau_a^j(t)$ denotes the unknown disturbance, and $\tau^j(t)$ is the input torque. The dynamics of the CPRMC is of the following two characteristics, specifically, the inertia matrix M a symmetric, bounded, and positive definite matrix and the matrix $\dot{M} - 2C$ is a skew symmetric. Hence, along the desired trajectory $(qd(t), \dot{q}d(t), \ddot{q}d(t))$, Eq. (9.73) can be linearized with Taylor formula as

$$M(t)\ddot{e} + (C - C_1)\dot{e} + Fe + n(\ddot{e}, \dot{e}, e, t) = H - (M\ddot{q} + C\dot{q} + G) \tag{9.75}$$

where e stands for the positional tracking error matrix of the actuators in the CPRMC, including the cable length, slewing angle and hydraulic cylinder length. For the j-th iteration, $e^j(t) = q_d(t) - q^j(t)$,

$$n(\ddot{e},\dot{e},e,t) = -\left.\frac{\partial M}{\partial q}\right|_{q_d}\ddot{e}e - \left.\frac{\partial C}{\partial q}\right|_{q_d,\dot{q}_d}\square ee - \left.\frac{\partial G}{\partial q}\right|_{q_d,\dot{q}_d}\dot{e}\dot{e} + O_M(\cdot)\ddot{q} + O_C(\cdot)\dot{q} - O_G(\cdot). \tag{9.76}$$

Substituting Eq. (9.73) into Eq. (9.75), one can obtain

$$M(t)\ddot{e}^j(t) + (C(t) + C_1(t))\dot{e}^j(t) + F(t)e^j(t) - d_1^j(t) = H(t) - T^j(t) \tag{9.77}$$

where

$$d_1^j(t) = -n(\ddot{e}^j, \dot{e}^j, e^j, t) + d^j(t) \tag{9.78}$$

$$F(t) = \left.\frac{\partial M}{\partial q}\right|_{q_d(t)}\ddot{q}_d(t) - \left.\frac{\partial C}{\partial q}\right|_{q_d(t),\dot{q}_d(t)}\dot{q}_d(t) - \left.\frac{\partial G}{\partial q}\right|_{q_d(t)} \tag{9.79}$$

$$H(t) = M(\dot{q}_d(t))\ddot{q}_d(t) + C(q_d(t), \dot{q}_d(t))\dot{q}_d(t) + G(q_d(t)). \tag{9.80}$$

Hence

$$M(t)\ddot{e}^{j+1}(t) + (C(t) + C_1(t))\dot{e}^{j+1}(t) + F(t)e^{j+1}(t) - d_1^{j+1}(t) = H(t) - T^{j+1}(t). \tag{9.81}$$

The assumptions are made that the desired trajectory is of the third-order continuity with respect to time. In addition, the same initial conditions are input for each iteration. In the iterative operational domain, all information from the current and previous operations is utilized as feedforward. For system described by (9.73), the control law in the j-th iterative operation is defined as follows:

$$T^j(t) = K_p^j e^j(t) + K_d^j \dot{e}^j(t) + T^{j-1}(t) + E\mathrm{sgn}(\varepsilon y^{j-1}), \quad j = 0, 1, \cdots, N \tag{9.82}$$

$$E \geq \left\|\varepsilon d_1^j(t)\right\| = \left\|d_1^{j+1}(t) - d_1^j(t)\right\| \tag{9.83}$$

$$y^j = \dot{e}^j + \Lambda e^j \tag{9.84}$$

where $K_p^j = \beta(j)K_p^0$, $K_d^j = \beta(j)K_d^0$, $\beta(j+1) > \beta(j) > 1$, $\beta(j)$ is the regulatory factor of the control gains which acts in the j-th iteration, K_p^j and K_d^j are the proportional and the differential gains of the j-th iteration, respectively. K_p^0 and K_d^0 are the initial PD control gains obtained by traditional experience tuning. The control gain matrixes are adjusted from iteration to iteration. $E\,\mathrm{sgn}(\varepsilon y^{j-1})$ is the robust term of the (j − 1)-th iteration.

The block diagram of robust ILC for the CPRMC is shown in Fig. 9.5. The prototype of the CPRMC is shown in Fig. 9.6.

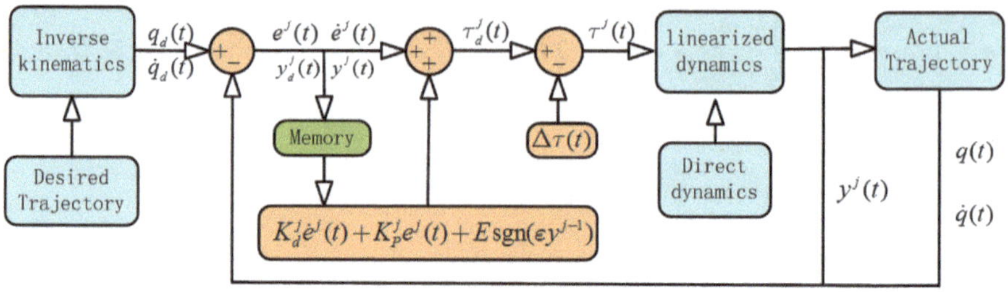

Fig. 9.5 Robust ILC structure of the CPRMC

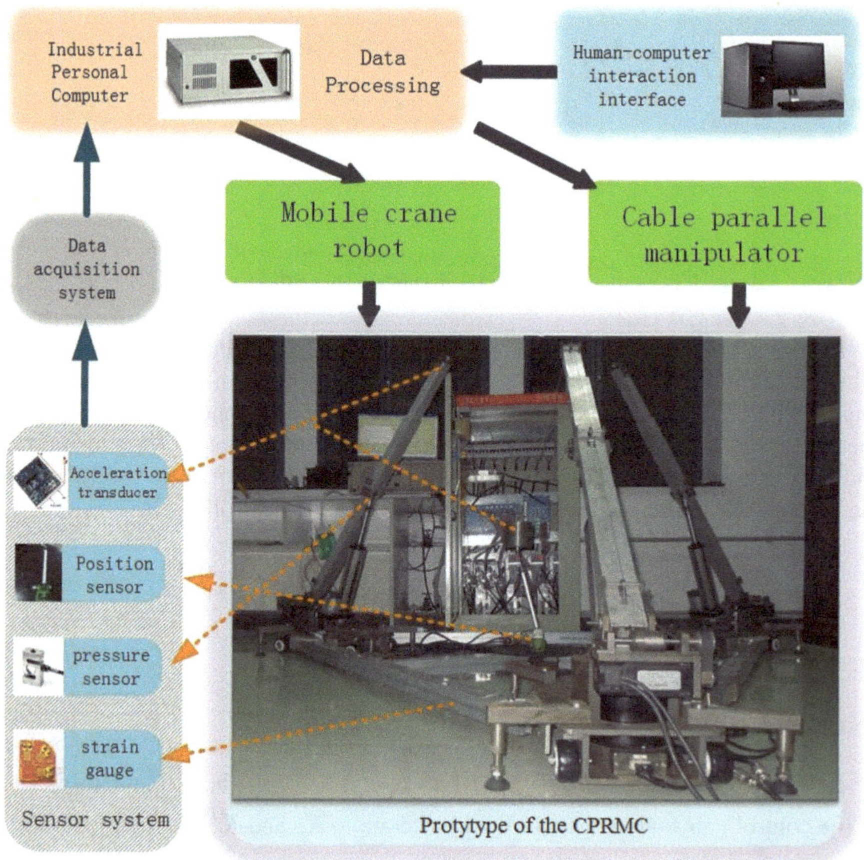

Fig. 9.6 Prototype of the CPRMC

9.5 Results and Discussion

In order to investigate the dynamics and the designed robust ILC controller of the CPRMC, numerical simulation is conducted with Matlab software.

First of all, by referring to the engineering practice, a certain case study is simulated. Specifically, all of the luffing angle in the multi-crane system are constrained (i.e., $\theta_i = [\theta]$), thus the CPRMC is simplified as a 6-input and 3-output system, so that the computational cost and simulation time can be reduced obviously. Then the designed robust ILC is implemented on the basis of the CPRMC dynamics from iteration to iteration until the total error along the trajectory and smoothness are acceptable according to the requirements of specific engineering tasks. The main parameters of the CPRMC are listed in Table 9.1.

During the simulation, the load is suspended in the workspace generated in the preceding part along the given spatial trajectory formulated as

$$\begin{cases} x = \sin\frac{\pi t}{2} + 10 \\ y = \cos\frac{\pi t}{2} + \frac{5\sin(\pi/3)}{2} \\ z = \frac{t}{2} + 5 \end{cases}. \tag{9.85}$$

The related trajectories are obtained online through simulation of the specific engineering case represented in Eq. (9.85). The actual process of the operation is similar to that of the widely used industrial robots. The first step is online, aimed to obtain the torque trajectories of all actuators according to the concrete target and task, specifically the mass and desired trajectory of the payload for the CPRMC. The designed robust ILC is implemented on the basis of the CPRMC dynamics from iteration to iteration until the torque trajectories meet the tracking performance

Table 9.1 Parameters of the CPRMC

System parameters	Value
Mass of the payload $\mathbf{m}$	3×10^3 kg
Acceleration due to gravity $\mathbf{g}$	9.81 m/s^2
Mass of boom $\mathbf{m_l}$	2×10^3 kg
the rotary inertia of the turrets $\mathbf{J_r}$	4×10^3 kg m^2
Distance between each mobile crane and the reference point $\mathbf{D}$	20 m
Length of boom $\mathbf{L}$	20 m
Vertical distance between the lower hinge joints of boom and cylinder $\mathbf{h}$	1.5 m
Horizontal distance between the lower hinge joints of boom and cylinder $\mathbf{f}$	2.45 m
Horizontal distance between the lower hinge joint of boom and the slewing axle $\mathbf{s}$	0.25 m
Distance between the lower hinge joints and the center of gravity of boom $\mathbf{l_g}$	10 m
Distance between the lower hinge joint of boom and the upper hinge joint of cylinder $\mathbf{l_s}$	7 m

requirement. The second step is offline. The nine actuators of the CPRMC are actuated according to the obtained torque trajectories in the online step.

The repetitive disturbances on the 6 joints are expressed as

$$\Delta\boldsymbol{\tau}(t) = \boldsymbol{diag}(\Delta\tau_1, \Delta\tau_1, \Delta\tau_1, \Delta\tau_2, \Delta\tau_2, \Delta\tau_2) \tag{9.86}$$

where $\Delta\tau_1 = 200\sin(5t)(1 - e^{-t})$, $\Delta\tau_2 = 100\sin(20t)$.

The initial proportional and the differential gain of the designed robust ILC are obtained by traditional experience tuning

$$\boldsymbol{K}_p^0 = 2\boldsymbol{K}_d^0 = \boldsymbol{diag}(3000, 3000, 3000, 1500, 1500, 1500). \tag{9.87}$$

The gain of the robust term is the updated according to the linearized dynamics of the CPRMC

$$\boldsymbol{E} = \left\| \boldsymbol{d}_1^{j+1}(t) - \boldsymbol{d}_1^{j}(t) \right\|. \tag{9.88}$$

Figures 9.7, 9.8, 9.9, 9.10, 9.11 and 9.12 show the tracking performance of the 6 active joints, including the three cable length l_1, l_2, l_3 and the slewing angle ϕ_1, ϕ_2, ϕ_3. The smooth curves indicate that the designed controller can avoid impact stress effectively, which is important for the stability of the system, especially prolonging service life the hydraulic driving system. Figures 9.13 and 9.14 represent the maximum tracking errors of cable length and slewing angle in different iterations with iteration times on the horizontal axis and maximum tracking errors on the vertical axis. As shown, the tracking performance was considerably improved with the increase of the iteration. Only after the fifth iteration, the Maximum tracking errors of the cable lengths l_1, l_2, l_3 are reduced to 0.027, 0.025 and 0.024 (m), respectively. And the Maximum tracking errors of slewing angle ϕ_1, ϕ_2, ϕ_3 are reduced to 0.0046, 0.0038 and 0.0015 (rad), respectively.

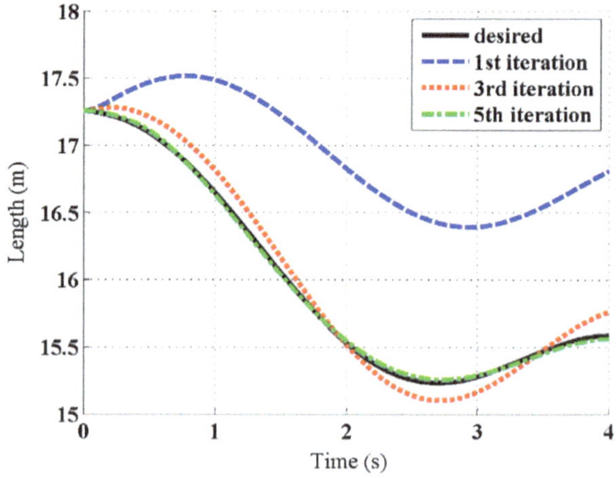

Fig. 9.7 Tracking performance of cable length l_1

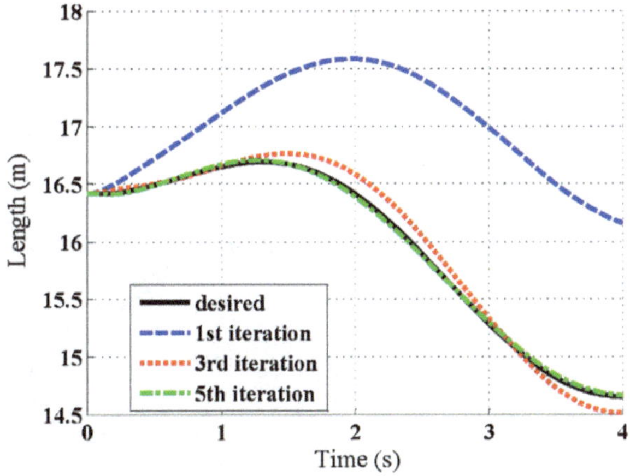

Fig. 9.8 Tracking performance of cable length l_2

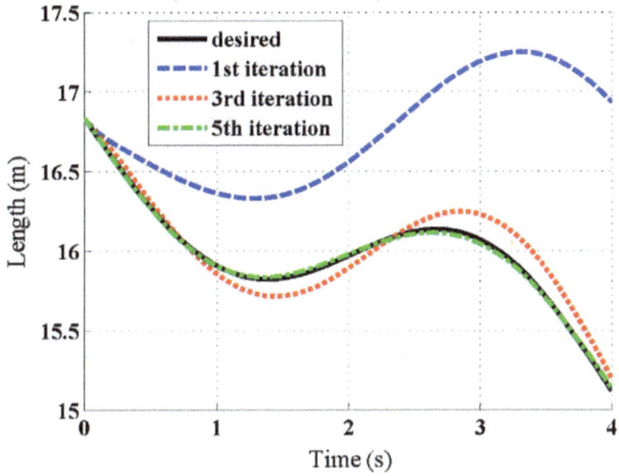

Fig. 9.9 Tracking performance of cable length l_3

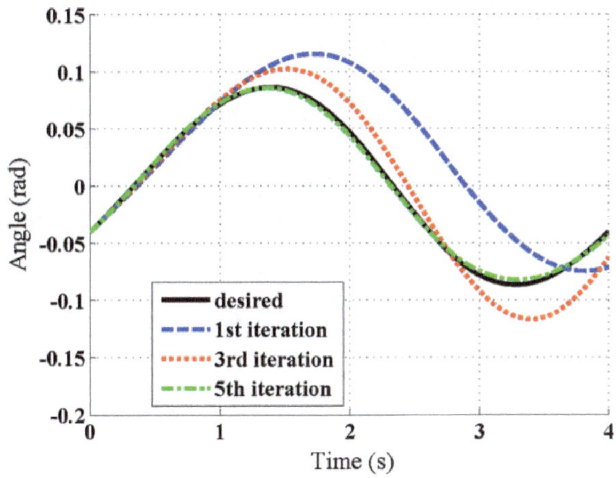

Fig. 9.10 Tracking performance of slewing angle φ_1

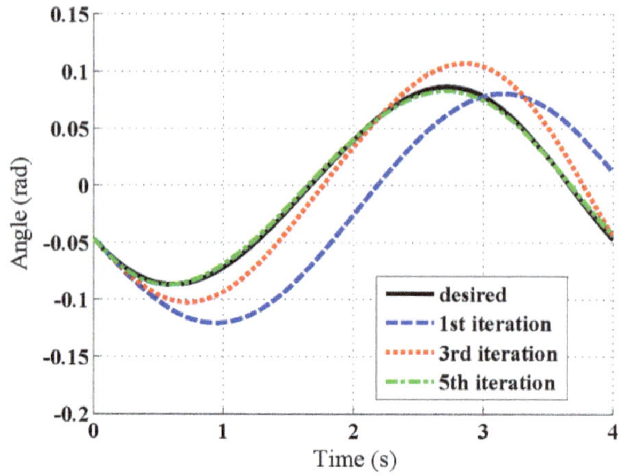

Fig. 9.11 Tracking performance of slewing angle φ_2

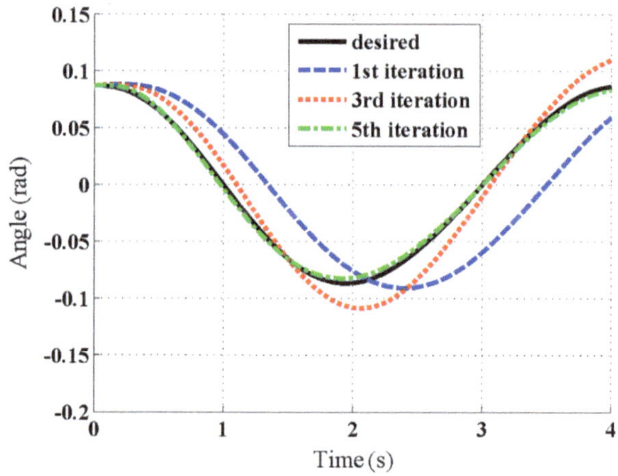

Fig. 9.12 Tracking performance of slewing angle φ_3

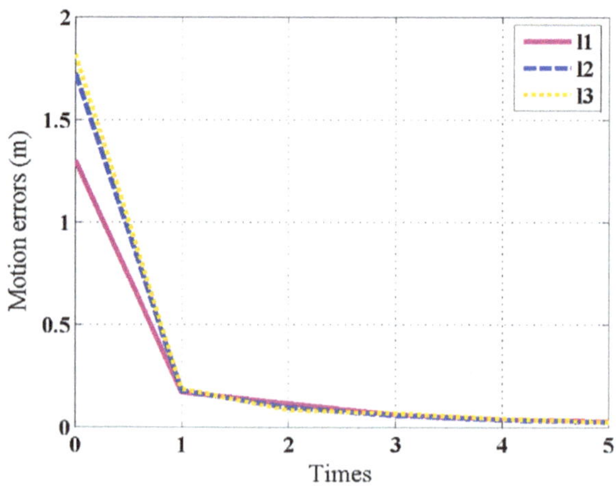

Fig. 9.13 Maximum tracking errors of cable length in different iterations

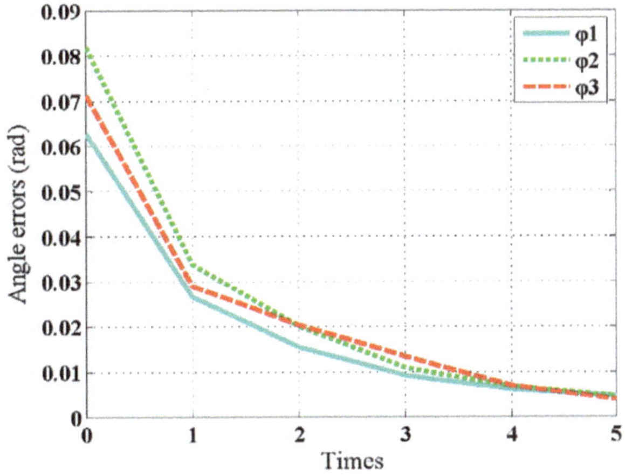

Fig. 9.14 Maximum tracking errors of slewing angle in different iterations

The velocity tracking performance of the 6 active joints are shown in Figs. 9.15, 9.16, 9.17, 9.18, 9.19 and 9.20. Figures 9.21 and 9.22 show the maximum velocity tracking errors of the cable length and slewing angle in different iterations with iteration times on the horizontal axis and maximum tracking errors on the vertical axis, respectively. As is shown in Figs. 9.15, 9.16, 9.17, 9.18, 9.19 and 9.20, the velocity of the 6 active joints change smoothly in different iteration. Figures 9.21 and 9.22 indicate that the Maximum velocity tracking errors of cable lengths are reduced from 0.9534, 0.8023, 0.7885 to 0.08912, 0.1233, 0.08507 (m/s), respectively. The similar decreasing trend of the velocity tracking errors can be found for the slewing angles. The maximum velocity tracking errors of the slewing angles are reduced from 0.0926, 0.1039, 0.0883 to 0.0092, 0.0131, 0.0122 (rad/s), respectively.

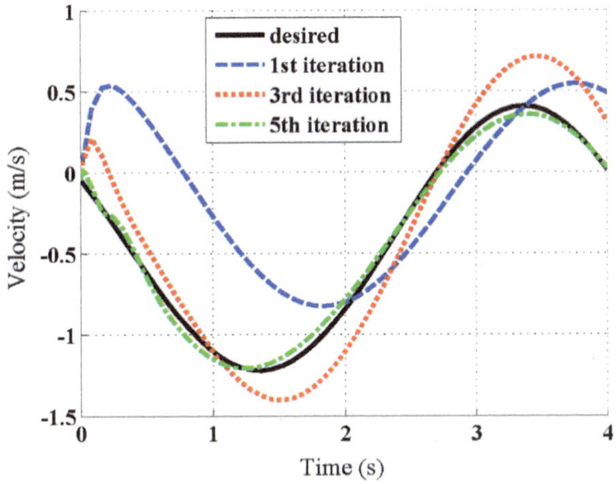

Fig. 9.15 Velocity tracking performance of cable length l_1

Fig. 9.16 Velocity tracking performance of cable length l_2

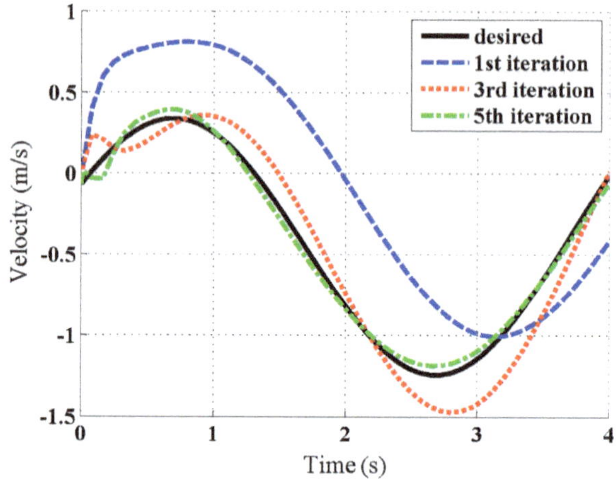

Fig. 9.17 Velocity tracking performance of cable length l_3

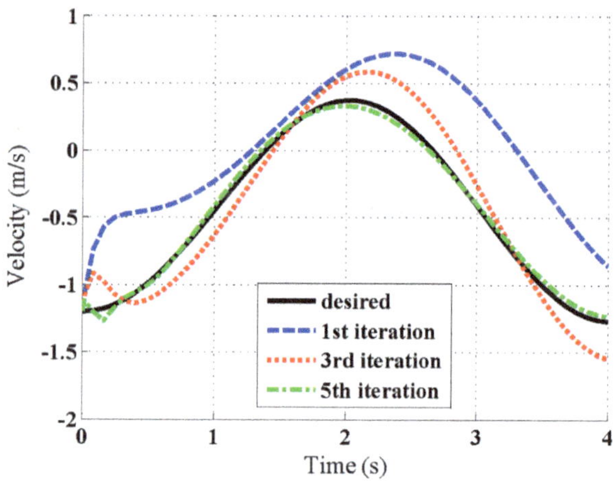

Fig. 9.18 Velocity tracking performance of slewing angle φ_1

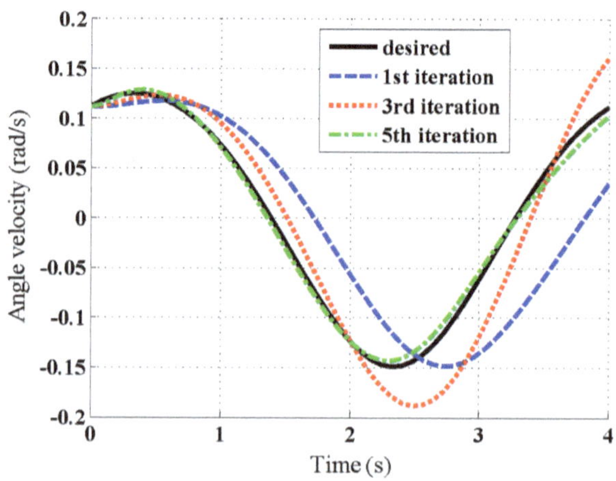

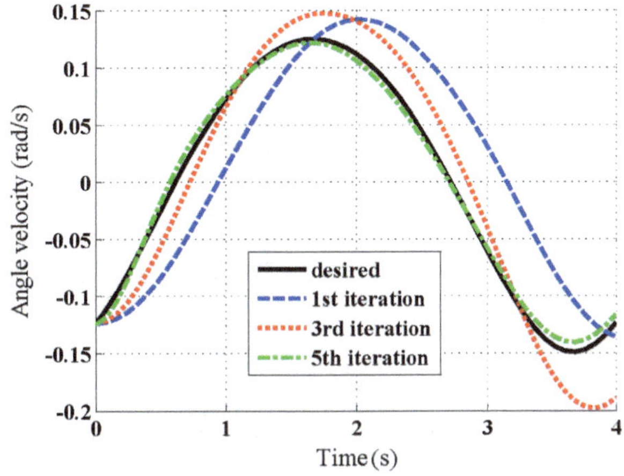

Fig. 9.19 Velocity tracking performance of slewing angle φ_2

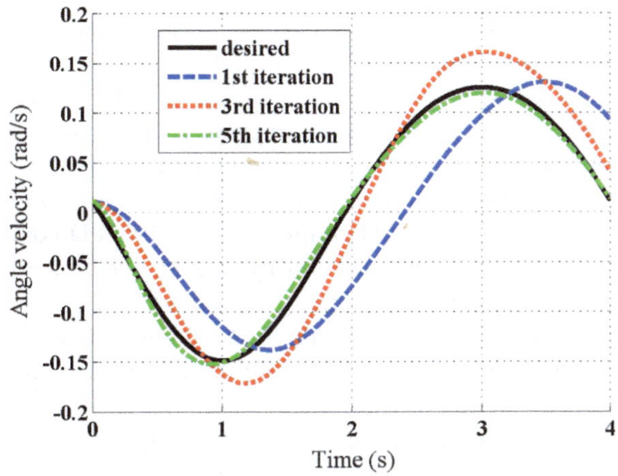

Fig. 9.20 Velocity tracking performance of slewing angle φ_3

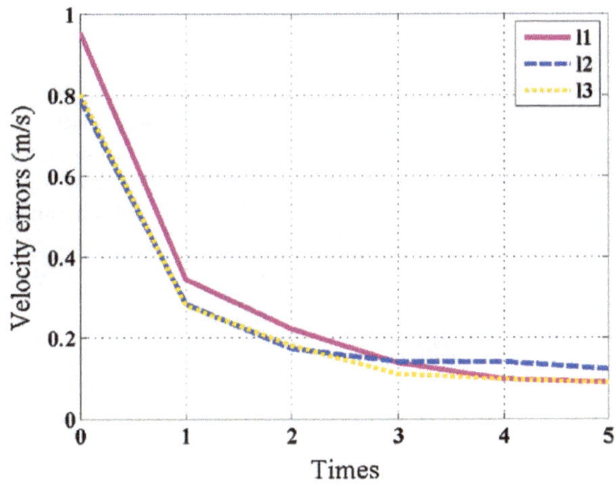

Fig. 9.21 Maximum velocity tracking errors of cable length in different iterations

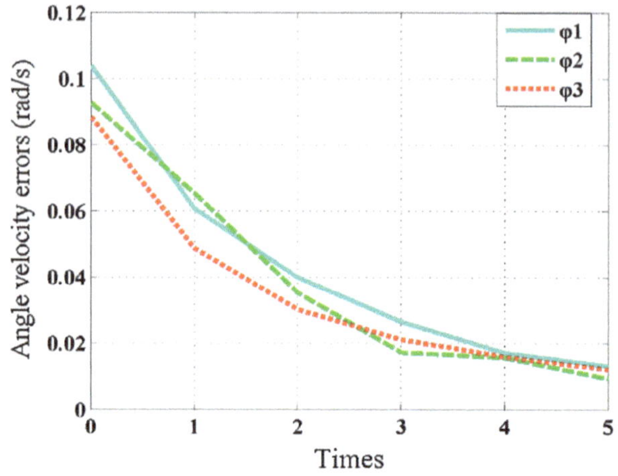

Fig. 9.22 Maximum velocity tracking errors of slewing angle in different iterations

Figure 9.23 shows the desired trajectory and tracking trajectories of the load in different iterations. The tracking trajectory closes to the desired trajectory with iteration. After the fifth iteration, the actual trajectory tracks the desired trajectory with high accuracy and the tracking error of the load converge close to zero. The results indicate that tracking performance of the robust ILC is acceptable for cooperation of multiple mobile cranes in engineering practice.

The traditional PID controller as well as PD controller are widely accepted for automated machines in industry and engineering for the reliable and robust performance under poor working conditions [19]. In order to investigate the effect, the robust ILC is compared with the traditional PD controller. Specifically, robust ILC and traditional PD controller are implemented in the CPRMC, respectively, in the light-duty case (i.e., the mass of the load is 3 tons) under the same initial proportional gain $\boldsymbol{K}_p^0$ and differential gain $\boldsymbol{K}_d^0$. The resultant trajectories of the fifth iteration of the robust ILC and traditional PD controller are shown in Fig. 9.24. After the

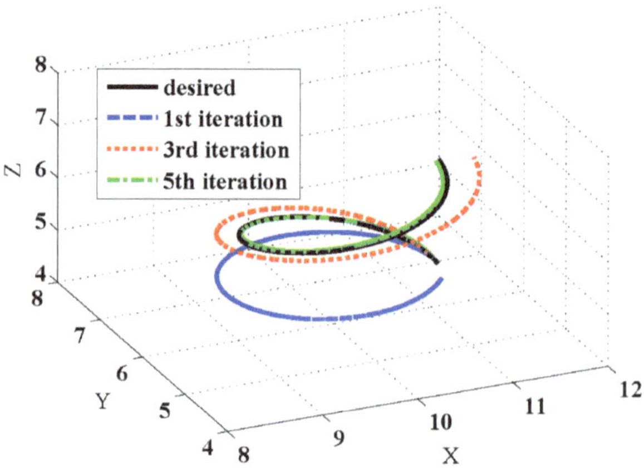

Fig. 9.23 Tracking performance of the load under robust ILC

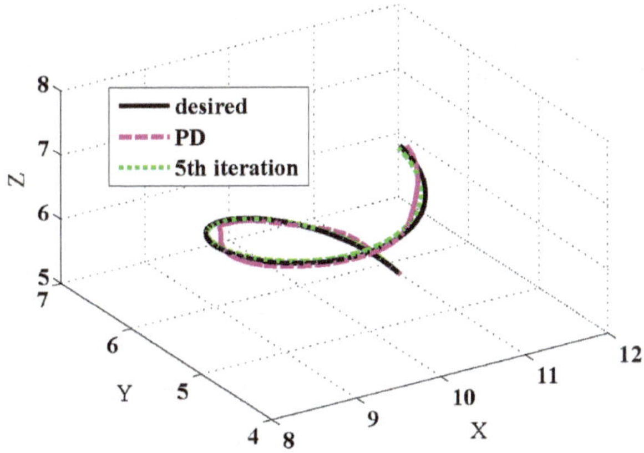

Fig. 9.24 Tracking performance of the load under robust ILC and PD controller

fifth iteration, the actual trajectory tracks the desired trajectory with much higher accuracy than adopting traditional PD controller. Figures 9.25 and 9.26 display the lateral forces on the cranes with robust ILC and traditional PD controller for the 5th iteration. As shown, the lateral forces on the three cranes with traditional PD controller are 250.39, 348.76 and 307.84 N. However, the corresponding lateral forces reduced to 78.08, 47.56 and 43.27 N at the fifth iteration with robust ILC, which is significant for the stability of the system, especially prolonging service life of the supporting devices in cranes. The results shown in Figs. 9.24, 9.25 and 9.26 indicate that the designed control is significantly useful compared with traditional PD control, and acceptable for multi-cranes.

In addition, two case studies are simulated with the same expected trajectory but different load weights of 10 and 3 tons, which are heavy-duty and light-duty conditions, respectively. The simulation results at the fifth iteration with the robust

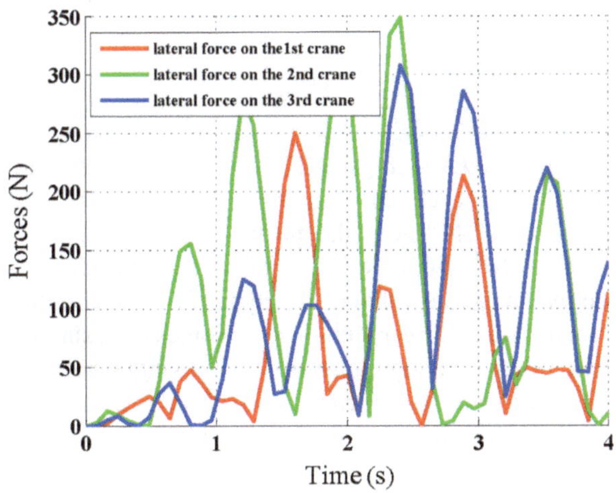

Fig. 9.25 The lateral forces on the cranes under light-duty conditions with traditional PD control

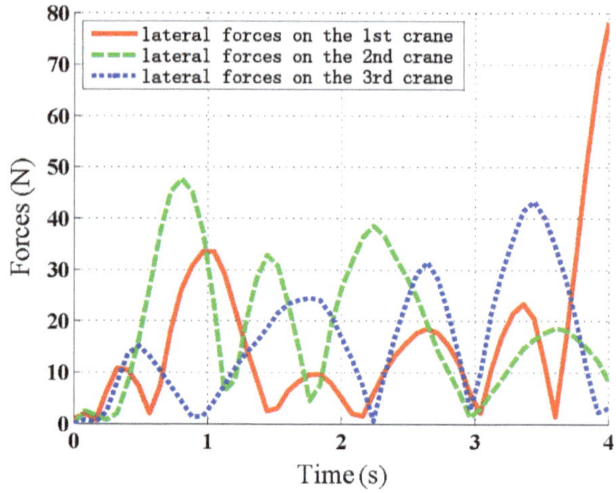

Fig. 9.26 The lateral forces on the cranes under light-duty condition with robust ILC for the 6th iteration

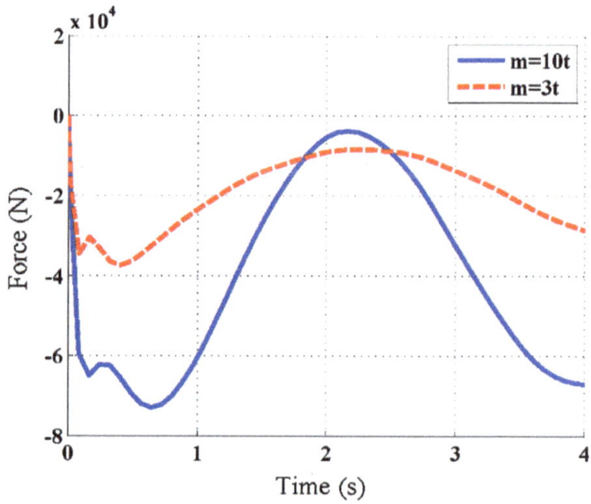

Fig. 9.27 Required force in the 1st cable under light-duty and heavy-duty conditions for the fifth iteration

ILC are shown in Figs. 9.27, 9.28, 9.29, 9.30, 9.31 and 9.32. The max absolute value of the forces on the three cables are 7.301×10^4, 7.451×10^4, 7.058×10^4 N under the heavy-duty condition, while reduced to 3.744×10^4, 3.171×10^4, 3.056×10^4 N under the light-duty condition. One can obtain that, the max absolute value of the dynamic forces on the three cables are much smaller than even the corresponding static forces on a single crane. Obviously, the CPRMC have better carrying capability than a single crane, which are more safe and stable during engineering practice. The similar decreasing trend can be found for the max absolute value of the torques on the slewing motors. In detail, under the heavy-duty

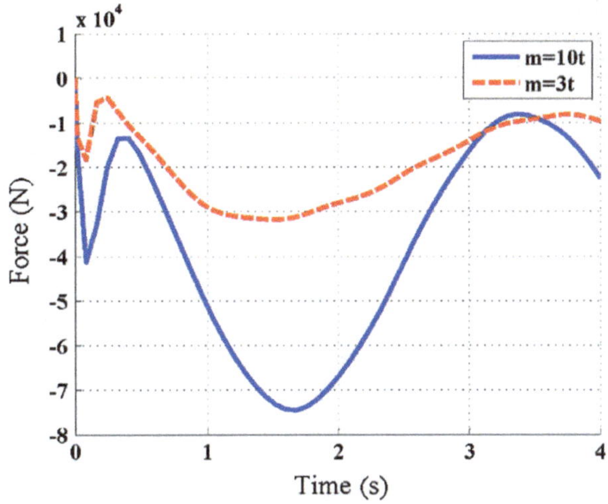

Fig. 9.28 Required force in the 2nd cable under light-duty and heavy-duty conditions for the fifth iteration

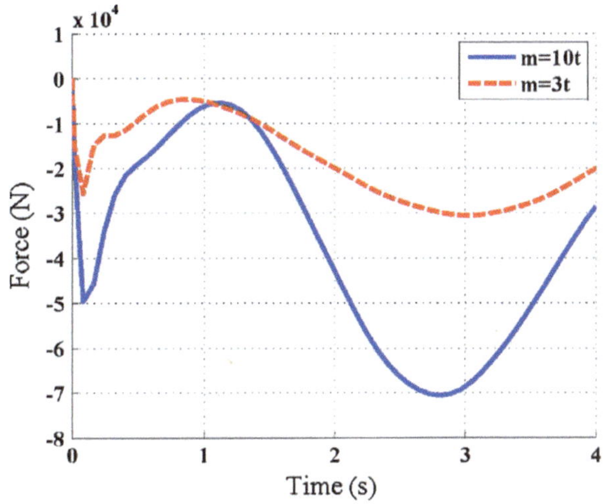

Fig. 9.29 Required force in the 3rd cable under light-duty and heavy-duty conditions for the fifth iteration

condition, the max absolute value of the torques on the slewing motors are 3536, 2991 and 3272 N·M,while measured 2411, 2909 and 3220 N·M under the light-duty condition. However, the reduced proportion is much more less than under the heavy-duty condition.

Figures 9.33 and 9.34 illustrate the corresponding driving power of the 5th iteration of the robust ILC and traditional PD controller, respectively, including the power of the three winches and three slewing motors. The corresponding max absolute values are listed in Table 9.2. Compared with the max absolute values with the traditional PD control method, the values of each the driving power is smaller when the designed robust ILC is implemented. The results indicate that the three cable winches undertake the most of power for the CPRMC system.

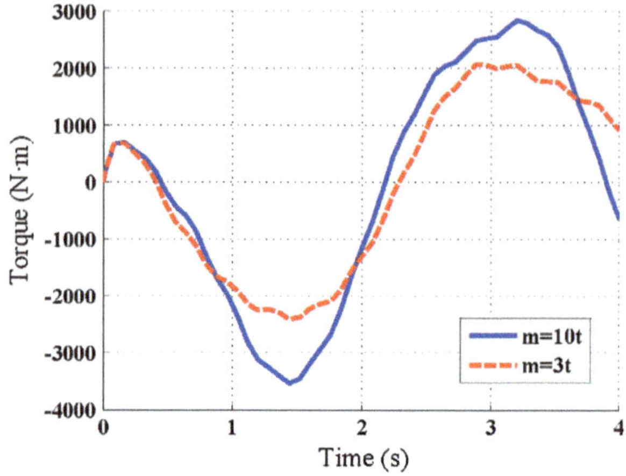

Fig. 9.30 Required force on the 1st slewing motor under light-duty and heavy-duty conditions for the fifth iteration

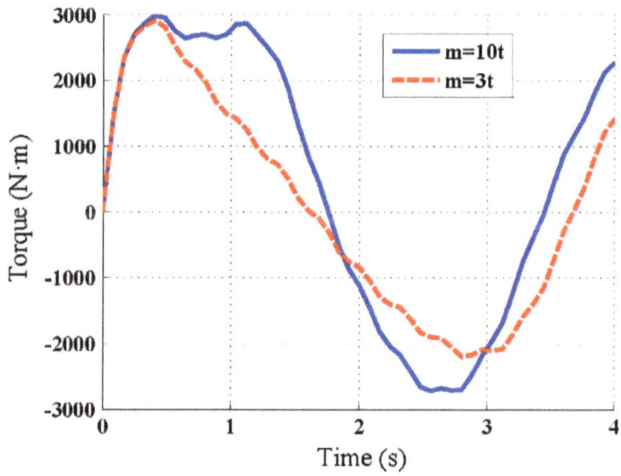

Fig. 9.31 Required force on the 2nd slewing motor under light-duty and heavy-duty conditions for the fifth iteration

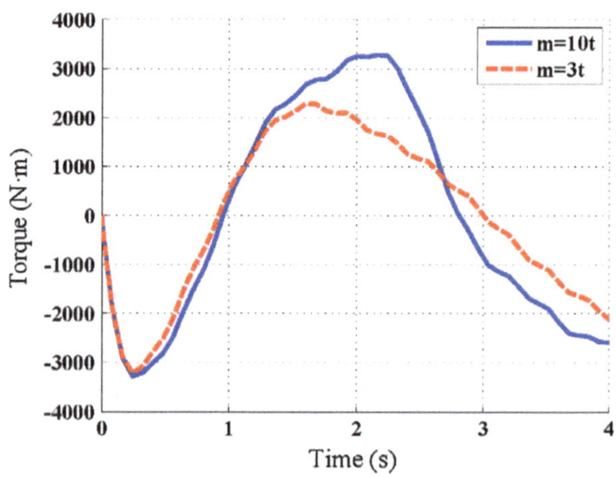

Fig. 9.32 Required force on the 3rd slewing motor under light-duty and heavy-duty conditions for the fifth iteration

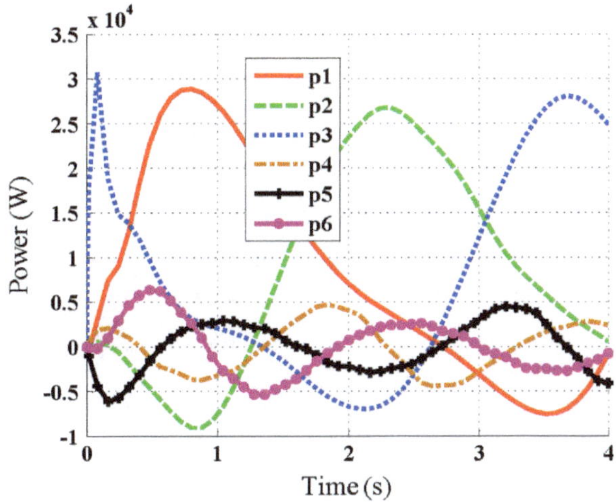

Fig. 9.33 The driving power of the fifth iteration of the robust ILC

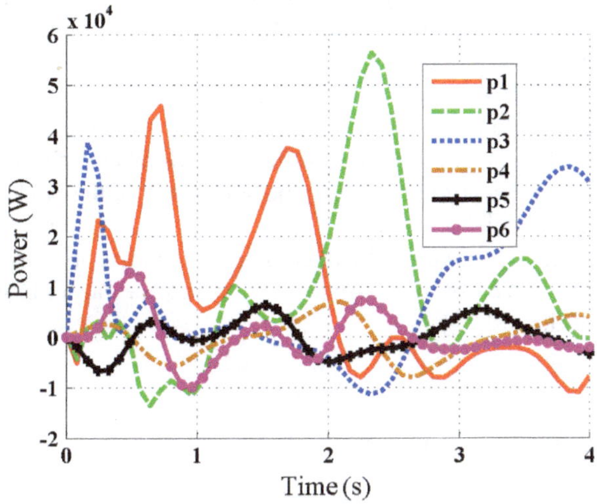

Fig. 9.34 The driving power of the traditional PD controller

Table 9.2 The driving power of the fifth iteration of the robust ILC and traditional PD controller

Power	The 1st winch	The 2nd winch	The 3rd winch	The 1st slewing motor	The 2nd slewing motor	The 3rd slewing motor	Unit
Robust ILC	28.88	26.74	30.76	4.71	6.10	6.39	kW
PD	45.85	56.39	38.71	7.99	6.65	12.81	kW

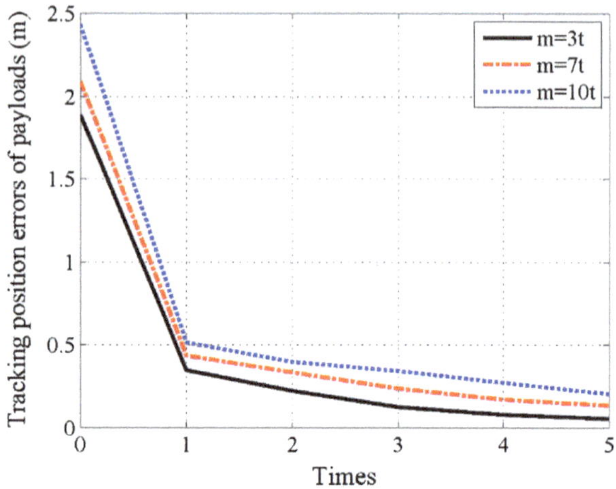

Fig. 9.35 The maximum tracking position error of different payloads

Effect of varying mass should be considered since the CPRMC is designed to lift loads whose masses are different, such as containers in the terminals and wind power equipment.

The simulations of the engineering case presented in Eq. (9.85) with different masses of payloads are carried out with the proposed ILC controller. The maximum tracking position error of different payloads are shown in Fig. 9.35. One can obtain that the maximum tracking position error of the payload increases with the payload mass, which are 0.0550, 0.1336 and 0.2024 m when the payload mass are 3, 7 and 10 t, respectively. The tracking position error in heavy-duty working condition can be reduced by increasing the iterative times according to the accuracy requirement of the task.

9.6 Conclusions

The kinematics of the CPRMC is derived via iterative substitution. The complete dynamic model of the nine-input and three-output system is established based on Lagrange equation and the kinematics. On the basis of the linearization of the dynamics, a robust iterative learning control strategy is presented for trajectory tracking control of cooperative multiple mobile cranes. A case study is analyzed that the CPRMC is actuated by slewing motors and cable winches with the luffing angles constrained. Numerical simulations of tracking the same desired path are carried out with different loads and control schemes. The results indicate the better lifting capacity of the CPRMC, compared with an under-actuated single crane. The robust ILC shows a better position and velocity tracking performance than the traditional PD control scheme under the same proportional and differential gains. In addition, the lateral forces on the cranes and corresponding driving power are reduced considerably, which validate the stability and energy-efficiency of the proposed controller. In our future work, we will extend the proposed approach to more common engineering cases so as to take the algorithm into practice. In addition, the experimental prototype is being conducted for further research, including the optimal path planning and dynamic load carrying capacity of cooperative multiple mobile cranes.

References

1. Sun N, Fang Y (2014) Nonlinear tracking control of underactuated cranes with load transferring and lowering: theory and experimentation. Automatica 50(9):2350–2357
2. Qian S, Zi B, Ding H (2016) Dynamics and trajectory tracking control of cooperative multiple mobile cranes. Nonlinear Dyn 83(1–2):89–108
3. Zi B, Qian S, Ding H, et al (2012) Design and analysis of cooperative cable parallel manipulators for multiple mobile cranes. Int J Adv Robot Syst 50(9):2350–2357
4. Zi B, Duan BY, Du JL et al (2008) Dynamic modeling and active control of a cable-suspended parallel robot. Mechatronics 18(1):1–12
5. Korayem MH, Bamdad M, Tourajizadeh H et al (2013) Development of ICASBOT: a cable-suspended robot's with Six DOF. Arab J Sci Eng 38(5):1131–1149
6. Khosravi MA, Taghirad HD (2014) Robust PID control of fully-constrained cable driven parallel robots. Mechatronics 24(2):87–97
7. Tang X, Chai X, Tang L et al (2014) Accuracy synthesis of a multi-level hybrid positioning mechanism for the feed support system in FAST. Robot Comput-Integr Manuf 30(5):565–575
8. Eberharter JK, Schneider K (2010) Control for synchronizing multi-crane lifts. In: 2010 IEEE international conference on control applications. IEEE, pp 1301–1306
9. Hu B (2014) Complete kinematics of a serial–parallel manipulator formed by two Tricept parallel manipulators connected in serials. Nonlinear Dyn 78(4):2685–2698
10. Zi B, Lin J, Qian S (2015) Localization, obstacle avoidance planning and control of a cooperative cable parallel robot for multiple mobile cranes. Robot Comput-Integr Manuf 34:105–123
11. Qian S, Zi B, Zhang D et al (2014) Kinematics and error analysis of cooperative cable parallel manipulators for multiple mobile cranes. Int J Mech Mater Des 10(4):395–409
12. Liu J, Li Y, Zhang Y et al (2014) Dynamics and control of a parallel mechanism for active vibration isolation in space station. Nonlinear Dyn 76(3):1737–1751
13. Wu J, Chen X, Wang L et al (2014) Dynamic load-carrying capacity of a novel redundantly actuated parallel conveyor. Nonlinear Dyn 78(1):241–250
14. Staicu S, Liu XJ, Li J (2009) Explicit dynamics equations of the constrained robotic systems. Nonlinear Dyn 58(1–2):217–235
15. Parra-Vega V, Sanchez A, Izaguirre C et al (2013) Toward aerial grasping and manipulation with multiple UAVs. J Intell Rob Syst 70(1–4):575–593
16. Zi B, Cao J, Qian S (2014) Iterative learning tracking control of a hybrid-driven based three-cable parallel manipulator. Adv Mech Eng 6:248385
17. Leban FA, Diaz-Gonzalez J, Parker GG et al (2015) Inverse kinematic control of a dual crane system experiencing base motion. IEEE Trans Control Syst Technol 23(1):331–339
18. Li Z, Li J, Kang Y (2010) Adaptive robust coordinated control of multiple mobile manipulators interacting with rigid environments. Automatica 46(12):2028–2034
19. Li HZ, Li L, He L et al (2012) PID plus fuzzy logic method for torque control in traction control system. Int J Automot Technol 13(3):441–450

Kolibri 360 GmbH
Ferdinand-Jühlke-Straße 7
99095 Erfurt, Deutschland
produktsicherheit@kolibri360.de